Australians in the Air

Australians in the Air

GREG COPLEY

RIGBY

To Marjorie Lindsay Copley,
who has loved the air and its wings.

National Library of Australia
Cataloguing-in-Publication entry

Copley, Gregory R.
Australians in the air;
Index.
Bibliography.
ISBN 0 7270 0139 6.
1. Aeronautics—Australia—History.
I. Title.

629.130994

RIGBY LIMITED • ADELAIDE • SYDNEY
MELBOURNE • BRISBANE • PERTH
First published in 1976

Wholly designed and set up in Australia
Set in Monophoto Bembo by Modgraphic Pty Ltd,
Adelaide, South Australia
Printed by the Griffin Press, Adelaide, South Australia

Contents

List of Colour Illustrations

NOTE: All black-and-white illustrations are listed in the Index. Please see page 283

Acknowledgments

The author and publishers render most grateful thanks to Sir Norman Brearley, founder of Australia's first regular airline and airmail service and one of the great names in Australian civil and military aviation. Sir Norman showed a keen and inspiring interest in the book from its earliest stages. He read the greater part of the work in manuscript form, gave much useful help and advice out of his store of aviation lore and knowledge accumulated over more than sixty years, loaned original photographs and other material, and kindly consented to quotation from his book *Australian Aviator*.

Three more great men of Australian aviation, Sir Reginald Ansett, Sir Lawrence Wackett, and the late Sir Hudson Fysh, gave freely of their time and recollections when the author interviewed them at various times during the past several years.

Group Captain Keith Isaacs, RAAF (retd.), now of Canberra and a well-known air historian, gave invaluable help in the selection of paintings and photographs from the Australian War Memorial collections and helped in other ways. Personnel of the AWM, especially the Director, Mr W. R. Lancaster (now retired), and Mr W. Sweeting, Acting Director, dealt most patiently with queries, arranged for photography and photographic processing, and kindly granted permission for quotation from the Australian War Memorial publications *RAAF Saga* and *These Eagles*.

Air Commodore E. W. Tonkin, Officer Commanding RAAF Base Point Cook; Wing Commander W. E. Richardson; and the personnel of the RAAF Museum at Point Cook were all extremely generous in aiding research, arranging for photography, and providing photographs from the Museum files.

Mr K. G. Knight, Federal Secretary of the Royal Flying Doctor Service of Australia, not only arranged for the loan of photographs illustrating RFDS activities but also checked the chapter on the RFDS and made numerous helpful comments.

The late Robin Miller (Mrs Harold Dicks) graciously consented to quotation from her book *Flying Nurse*, and her father H. C. "Horrie" Miller, co-founder of MacRobertson Miller Airways and one of the great pioneers of Australian civil aviation, gave similar permission regarding his book *Early Birds* and lent some of his photographs. Mr R. D. Cohen, Public Relations Executive of MacRobertson Miller Airways, provided photographs and information, and Mr K. C. Holyman supplied material relating to Australian National Airways.

Mr H. K. Millicer, Principal Lecturer in Aeronautics in the Royal Melbourne Institute of Technology, supplied valuable data and original photographs relating to the Victa Airtourer story, and Mr Kenneth O. Blight, Sales Promotions Manager of Victa Limited, supplied additional material on this wholly Australian aviation venture.

Government departments and instrumentalities and Australian-based aircraft manufacturers approached for information all proved most helpful. Special mention must be made of Hawker Aircraft Limited and Hawker Siddeley for the provision of historical photographs; the Commonwealth Aircraft Corporation Pty Ltd, the Government Aircraft Factory, and the Department of Supply for substantial help with information and illustrations relating to Australian aircraft design and production including the Nomad project; the Department of Air, RAAF Public Relations, Defence Public Relations, and the Australian Government Publishing Service for photographs, information, and permission to use *Mission Vietnam*, by George Odgers, as a reference work for Chapter 18.

Mr John Godwin, whose meticulous research into the story of Ray Parer and John McIntosh resulted in his book *Battling Parer*, kindly helped to obtain relevant photographs. His book was invaluable reference material.

Mr W. C. Clarke, Senior Photographer of Qantas Airways Limited, provided recent and historic photographs relating to Qantas operations. The Secretaries of Royal Aero Clubs in each State provided valuable information and the management of East–West Airlines supplied historical data on that organisation. Mr Kym Bonython kindly gave permission for reproduction of a portion of a letter to the Adelaide *Advertiser* and supplied additional relevant material. Mr Terry Gwynne-Jones, the air historian, provided information about some early flyers.

Mention must be made of the Principal Librarians and staff of those two great institutions, the Mitchell Library in Sydney and the National Library in Canberra, who dealt efficiently and expeditiously with requests for illustrations. The Editors and Pictorial Editors of Australian newspapers and journals including *Aircraft*, the Melbourne *Herald*, the *West Australian*, and the Adelaide *Advertiser* gave valuable assistance.

The author and publishers also wish to thank Mr David McGrath, one of Australia's most significant aviation historians, for access to his notes and for discussions with the author over a number of years. His guidance in many areas was invaluable. Miss Diana Thompson, of Sydney, also gave freely of her time to aid in research, turning up many hitherto unpublished notes, as well as published sources which otherwise would have gone unnoticed. The books of press clippings kept by Mrs Marjorie Copley during the late 1920s and early 1930s proved extremely valuable, as did her research, on the author's behalf, into Western Australian aviation history.

Grateful thanks should also be rendered to the very many aviation figures, both prominent and otherwise, who over the years gave freely of their time and ideas to the author in Australia and abroad.

Photographs were obtained from a number of sources and in most cases these are accredited in the captions. In a few instances the publishers experienced difficulty in locating the original sources or present copyright holders of textual or pictorial material. Full accreditation has been given wherever possible but in case of any errors or omissions the actual holder of copyright is requested to contact the publishers.

Preface

It is a cool, damp Californian night. Wind comes in frozen gasps from the hills and chops across San Francisco Bay. A Qantas Boeing 747 rises into the dark, its red tail fin illuminated to show a winged kangaroo, nosing away from the airfield and on to a famous trail. Most of those aboard will be unconscious of the past which still hangs heavy in the air beneath.

For along this Southern Cross route came the first men to fly right across the Pacific. And they were Australians: Kingsford Smith and Ulm. They and men like Bert Hinkler led Australia to the zenith of its aviation spirit. I feel their presence now, around this bay and across the miles of ocean, as freshly as one still scents new-baked bread an hour after the ovens have been shut for the day.

Every country aerodrome in Australia has a special kind of edge or excitement. There is a contentment and optimism borne in the sight and sound of a hangar door rumbling open at dawn in springtime.

Australia, to me, is a country which is still "up in the air", without being ethereal. Many of the original pioneers of Aerial Australia still live. Newer entries in the field unconsciously emulate their verve and dash, and so the swagger of the Aussie flyer persists. It is thus in no other land.

Too many of these men have gone. Smithy is dead, and so is Ulm. Harry Hawker died in an aircraft crash. And Bert Hinkler's last landing was into an Italian mountainside. They went in full flower, and would have wished brevity upon the mournings. Still, they are gone, and younger men, such as I, would have given much to talk with them.

Now I can but stare at the petrified, encased "Southern Cross" Fokker tri-motor monoplane at Brisbane Airport and repeat to myself that this was the aircraft which flew the Pacific first. And the Tasman, and landed near to its final repose to the roar not just of its engines, but of the cheering crowd.

But it is foolish to regret the lateness of my birth, or the early death of these men. I should be, and largely am, content in having met and talked with so many of these early birds, and gleaned some second-hand feeling for those feats.

And, in a time when the constructions of national boundaries mean less and less, I feel inordinately proud of the Australian aviation heritage. For there are still many Australians in the air who would do great things for the world, without seeking first the reasons why, or seeking later the reward.

G.R.C.
San Francisco, 1976.

Foreword

by Sir Norman Brearley,
C.B.E., D.S.O., M.C.,
A.F.C., F.R.Ae.S.

At last an accurate record of Australians in the air!

All too often authors seem to repeat incorrect reports previously published of the development of Australian aviation and, unfortunately, these tend to become history. But now we have a young Australian author, Gregory Copley, who has flown with many of the world's greatest airlines in his search for information. He has attended the famous air displays at Farnborough in England, Paris, and many other places.

He has discussed all aspects of flying with some of the world's leading scientists and aircraft design teams, and these experiences have developed in him a sound knowledge of his subject.

I have been honoured by his invitation for me to write this foreword to his book. Although my pioneering activities are given much prominence, which causes me some embarrassment, I know that in most cases the facts recorded in this book can be confirmed by official reports. Because of the author's reliable portrayal of these incidents I have pleasure in writing this foreword.

It is difficult to overcome the false legends that, if unchallenged, tend to become history. In 1970 I hoped that suitable recognition could be given to the fiftieth anniversary of the founding of Australia's first regular airmail service in 1921, by the issue of special postage stamps etc. in 1971. It was therefore extraordinary to find that the Postmaster-General, who lived in Queensland, had already issued special stamps in 1970 to celebrate what he incorrectly claimed was the fiftieth anniversary of Qantas, Australia's second regular air service. Qantas was then only in its forty-eighth year of service. Thus, false legends are reinforced and tend to mislead those who are unaware of the facts.

Both Federal and State Governments played important parts in the early days when civil flying commenced in a serious way in Australia. Some of these actions were most helpful in the pioneering days, while others were just the reverse.

The book recalls both the triumphs and the tragedies of some of those who played important parts in the development of this new form of transport. Many names come to mind of those who deserve recognition, but it would need additional volumes to include them. The jet age is now accepted by millions of people as a normal way of life. Flying at 600 miles an hour with each plane carrying hundreds of passengers is so commonplace that we may forget how it all started. It commenced during the lifetime of Australians who can still remember the first flights by the Wright brothers in 1903, just seventy-one years ago.

Eighteen years later the first regular airline in Australia commenced operating in Western Australia, signalling the birth of the vast network of airlines that now cover our continent. Those were the days when the last of the Cobb & Co's horse-drawn coaches were regularly employed carrying the Royal Mail to some outback Australian communities.

This historically-correct book covers the period when the first airline in Australia used aeroplanes capable of carrying two passengers at a speed of seventy-five miles an hour. It faced the hazards of unprepared landing grounds and of operating into tropical areas where cyclonic storms of over 100 miles an hour were frequently recorded in the summer months.

In its present form the book portrays the history of aviation from its very early days, but it concentrates on the Australian scene when the pioneers faced the future in aircraft of comparatively poor performance. These aircraft demanded special pilot skills to ensure safety in their operations. It concludes by referring to the sacrifices made by some of those who made this progress possible.

I regard this as a great Australian book.

Sir Norman Brearley

1 First Flaps of the Wings

Australians can claim a place among the "Magnificent Men," although they sometimes found it very difficult to get their Flying Machines off the ground

Australia may claim a link with aviation almost two centuries old, because a man who helped establish the colony had some interest in flight. He was Sir Joseph Banks, the botanist and scientist who sailed on Captain James Cook's first voyage to Australia. Botany Bay and the *Banksia* shrub were both named after him. So when looking at either the Bay or the blossom, Australians may think of Banks himself as an early enthusiast of flight.

When Benjamin Franklin was in Paris as a representative of the rebelling American colonies, he corresponded with Banks on the subject of flight. He wrote on 27 July 1783 telling Banks of the world's first unmanned balloon flights, made by Montgolfier, and said:

"Doctor Blagden will acquaint you with the experiment of a vast Globe sent up into the Air, much talked of here, and which, if prosecuted, may furnish new means of knowledge." Ben Franklin, as George Washington's chief of espionage, may also have seen the balloon as a step into technological warfare.

Someone once asked Franklin what was the use of his electrical experiments. He replied: "What is the use of a new-born infant?" and the same might have been asked about balloons.

On 19 September, some months after his letter to Banks, a sheep, cock, and duck lurched from earth in the first "staffed" Montgolfier balloon, at Versailles. They were being used for the same purpose as the monkeys sent into orbit in the nose-cone satellites of American missiles and the dogs in Soviet sputniks: to discover whether a creature of earth could survive in an unknown element.

Man followed the animals upon the paths of earth. So he followed them into the firmament.

On 21 November 1783 the Marquis d'Arlandes and Monsieur Pilâtre de Rozier made the first manned flight: five miles across country in a Montgolfier balloon raised by heated air. This was only five years before the First Fleet brought the first convicts and settlers to Australia, and when the nineteenth century dawned some Austalians already were dreaming of aviation. They were remarkably advanced dreams.

Francis Forbes, of Skillatur, at Muswellbrook, New South Wales, proposed in the early 1800s a heavier-than-air machine—not merely a balloon such as Montgolfier had devised—and a powered one at that. "The screw-fans or vanes must turn on an axis lying in the plane of the kite and they must be placed behind the kite, or at the side, to propel it forward" he wrote. The first heavier-than-air powered machines followed similar rules almost a century afterwards. Ron J. Gibson, the former Qantas Airways' historian, said in a private historical study that Francis Forbes was the elder son of Sir Francis Forbes, the first Chief Justice of New South Wales.

Newspapers traditionally receive many crank letters, and the editors of the *Sydney Morning Herald* may well have thought that the man who signed his letter "Aeronaut", in 1843, was one such writer. But his letter was published in the edition of 1 June 1843, saying that he had invented a powered balloon some twenty years earlier—in the 1820s! He felt that the ordinary balloon, which drifted aimlessly, did not "accomplish the great end of periodical or exact transit". So, as early as the 1820s, a man could see the eventual use of aviation as a point-to-point transit system.

The "new-type" balloon would have reverted to the old principle of rarefied air, created by "portable gas and argand burners". In order to make the balloon transport more exact he thought "we might ascend by mechanical contrivance by which means instead of being at the mercy of atmospheric currents, we might make them subservient to our wishes". But nowhere is there evidence that "Aeronaut" built his dream craft.

Thirteen years lapsed in which the colony was untroubled by would-be flying fools. The peace was profoundly broken on 15 December 1856 by one Pierre Maigré, who brought a hot-air balloon to the Sydney Domain. Crowds amassed to watch the take-off; specially-built grandstands were packed; and the show turned into a riot when the 10,000 spectators saw the balloon refuse to leave the ground.

The rowdy Sydneysiders were unsympathetic when Maigré discovered a rent in his balloon during the gas-filling operation. He repaired it and continued filling. The crowd became noisier and more sceptical, and hustled closely around the balloon which by that time was ready for take-off. However, a score of onlookers clung to the guy-ropes, preventing it from lifting. The ignorant crowd ignored Maigré's pleas for them to release the craft. They continued to berate him; he became afraid for his safety, and jumped from the basket to dash towards near-by Government House and there take refuge in the South Lodge. The crowd turned from him for a moment, fell back to the open Domain to burn the balloon and all the grandstands and railings which had been set up for the event, and then again sought Maigré. They hurled rocks through the windows of Government House in an attempt to bring him from its precincts. In the process they knocked one of the fence posts on to eleven-year-old Tommy Downs who had been swept along in the excitement. He died of a fractured skull.

The first man to gain real significance in Australian aviation was Dr William Bland, a Sydney surgeon, inventor, and author. He arrived in Australia as a convict in 1814 but was a free man when, in March 1851, he drew up detailed plans of his "Atmotic Ship".

Bland's Atmotic (or Aermotic) Ship was to be powered by two air-screws driven by a steam engine, and he said it would reduce the tedious months of voyage from England to Australia down to four or five days! A steam-powered airship? Nothing new, you say? Well, the first such powered airship did not take to the air until eighteen months after that, on 24 September 1852, and then not in Australia.

The Atmotic Ship looked, more than anything else, like the parent of the Zeppelin, which was to be developed and flown by Count Zeppelin some fifty years later, in 1900. Bland's ship never reached the construction stage.

His plans and models were on display in the Crystal Palace, London, by 1854, and he worked to convince Australia's colonial authorities that they should take an interest in the craft for military and commercial purposes. He had no record of success. Those pre-occupied powers, if they had any interest in flight at all, could see no further than free-flight ballooning.

Bland was no crank, or if he was then he was no ordinary crank. He was a member of the New South Wales Legislative Council when he designed his ninety-foot dirigible which theoretically could have lifted one-and-a-half tons as well as the three-and-a-half-ton load of the car which held the engine.

Ron Gibson wrote: "Rumour has it that the French Emperor was restrained by his Finance Minister from buying Bland's model at the Paris Universal Exhibition. There have also been suggestions that Zeppelin borrowed much from Bland in the construction of his dirigibles, but I have not been able to find evidence of this."

Bland died in 1868, ten years after Australia's first successful lighter-than-air flight, and ten years before young Lawrence Hargrave was to become fascinated by heavier-than-air flight.

His plans had been rejected by Emperor Napoleon III of France, Queen Victoria, and President Abraham Lincoln. Bland would have been delighted had Lincoln used his Atmotic Ship in the American Civil War.

The first Australian flight in an unpowered, lighter-than-air vehicle was that of William Dean, an Englishman, in a balloon made for George Coppin, by H. T. Coxwell, of Tottenham, England. The solo flight was to have been made with a companion, C. H. Brown.

The aerostat, named the *Australasian,* had a capacity of 31,000 cubic feet of coal-gas, and was partly filled at the Melbourne Gas Works and then taken to Mr Coppin's home for the topping up. It stood forty feet across and sixty feet high, including the "car", when it was ready in Cremorne Gardens on 1 February 1858.

How nervous must Dean and Brown have been when they climbed into the basket and waited for something to happen. The craft faltered unsteadily for a second or two before it became obvious that it would not lift the pair of them. Brown jumped out before the stays were released, allowing the *Australasian* to lift. Dean floated seven miles across Melbourne.

Still, Brown was not cheated of his flight. He took the *Australasian* off a fortnight later on 15 February, again from Cremorne Gardens, floating for forty-four minutes between Collingwood Stockade and Brunswick. He claims to have risen more than 10,000 feet, but at times he must have floated very low, for history has penned that at one stage he was "attacked by onlookers".

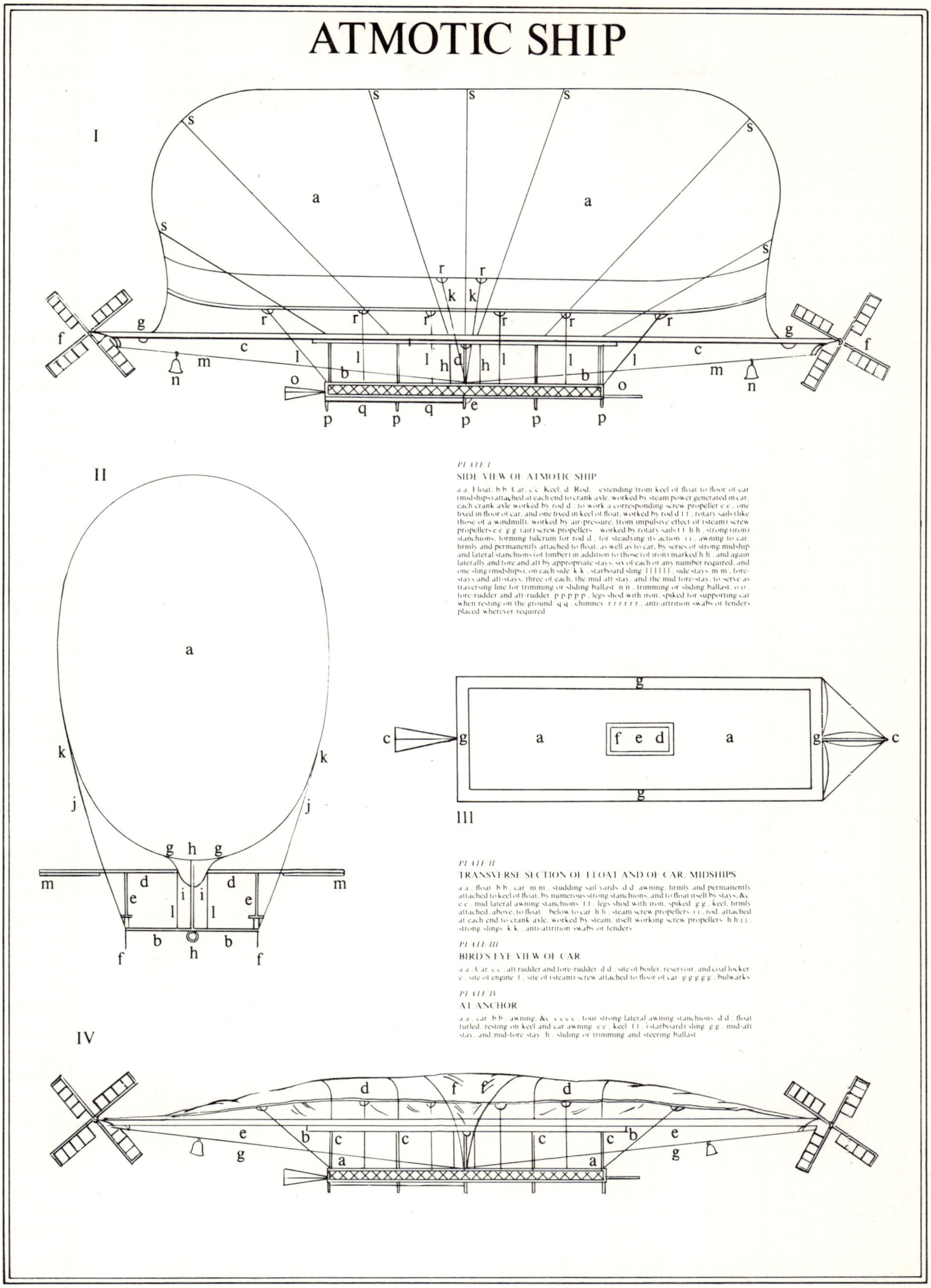

Bland's Atmotic Ship, redrawn from original material in the National Library, Canberra

ABOVE: Tom Gale takes off on South Australia's first balloon flight, 1871 (*Illustrated Sydney News*)

TOP: Tom Gale's balloon ascent in Sydney, 1870 (*Illustrated Sydney News*)

BOTTOM: Australia's first air wreck: L'Estrange's balloon explodes near Sydney (*Australasian Sketcher*)

Dean and Brown were able to make a dual-passenger flight when they ascended in the cool, moist—and therefore more dense—night air of 25 March 1858. They took off at 10 p.m. and landed at Emerald Hill an hour later.

They flew the *Australasian* again in Melbourne, but there seems no clear record of any flight by its owner, Mr Coppin. Still, Queen Isabella of Spain might have paid for the ships of Christopher Columbus but she didn't voyage to America in them.

Dean and Brown went to Sydney after the Melbourne success, and made that city's first flight. They were waved off from the Domain by Governor Denison on 13 December 1858 and floated above Sydney Harbour.

Strangely, despite the obviously brilliant minds of men such as Bland and Hargrave, there were not many more flights between then and the end of the century.

Tom Gale and John Allen made balloon flights in Sydney in 1870 and 1871 before Gale went to South Australia to make that State's first flight.

According to Group Captain Keith Isaacs, the aviation historian, another balloonist attracted particular attention on 14 April 1879. Harry L'Estrange was demonstrating his balloon *Aurora* near Victoria Barracks in Melbourne when the fabric of the aerostat burst open and the vehicle plunged from 9,000 feet toward the earth. L'Estrange parachuted and escaped unhurt. The *Aurora* already had a chequered career, and reputedly had been used to carry documents "from the beleaguered capital to the provinces" during the Franco-Prussian War.

The first parachute jump in Australia was thus reluctantly made by Harry L'Estrange, and the first deliberate jump was not made until nine years later, on 8 December 1888, when J. T. Williams parachuted 6,000 feet to the Ashfield Recreation Ground in Sydney's western suburbs, from Captain Hendon's balloon. This was almost 100 years after the first successful use of a parachute, in Europe on 22 October 1797.

E. W. Cole, a well-known Melbourne bookshop proprietor, took an ardent interest in flying in 1882, and it probably did his store's public image considerable good. On 31 October he announced that he would pay a prize of £1,000 to the person who, within two years, invented a machine which could fly 100 miles. The machine could be propelled electrically, chemically, or mechanically, and "shall come and stop in front of the Book Arcade, Bourke Street, Melbourne, as easily and safely as a carriage stops there now".

Mr Cole doubtless whiled away his lunch-hours imagining the crowd around his bookstore on the day that the flying machine landed outside. He was in little danger of having to pay out the £1,000.

During this era there were dedicated men in various parts of the world to whom the balloon was a mere clumsy compromise. They dreamed of flying like the birds which flirt with all the subtle variations of nature's airways. Some killed themselves in this endeavour, many were mocked, many more were unknown. The names of a very few have survived as being the true pioneers of flight.

One of these survivors was Lawrence Hargrave, who may be claimed as an Australian despite his birth in England in 1850. His father, a barrister, emigrated to Australia in 1856, leaving Lawrence in England to be educated at a boarding school. By 1865 Hargrave senior was a Supreme Court judge, and in 1866 Lawrence joined him in Australia.

Young Lawrence became an engineering draughtsman, but in 1872 he began five adventurous years prospecting and exploring in New Guinea and Queensland. In 1877 he returned to Sydney, was elected a member of the Royal Society of New South Wales, married, and took a job as assistant astronomical observer at Sydney Observatory.

He was becoming obsessed by the problem of flight—real, manned, heavier-than-air, navigable flight. In 1883 he retired at an early age to devote his life to research, which he began by studying air-currents and sea-waves and the actions of birds and various insects. In 1884 he read a paper to the Royal Society on "The Trochoided Plane", defining his research so far, and in the same year began to experiment with model aircraft.

Built from wood and paper, these were propelled by twisted rubber bands like a modern child's toy or by tiny engines powered by steam, clockwork, or compressed air. Some of the models had flapping wings; others had air-screws.

One of the models powered by compressed air is now in the National Museum in Washington, DC, USA. Hargrave's flapping-wing device was the first to work successfully since Giovanni Alfonso Borelli, the Italian physicist and astronomer, developed the principle of human flight by that method in 1680.

The internal combustion engine was then in its earliest infancy, so Hargrave was denied this vital key to powered flight. But he did build a rotary engine, powered by compressed air, seventeen years before the invention of the Gnome rotary aircraft engine in France.

In 1892 he proved one of the great principles of flight: that a curved wing surface gives greater "lift" than a flat one because of the flow of air round the curves. He confirmed this by experimenting with kites, and on 12 November 1894, on the beach at Stanwell Park in New South Wales, he actually lifted himself sixteen feet from the ground by using four kites employing curved aerofoils. The principle he proved followed the old theorem by Bernoulli that "where speed is greatest, pressure is least"—the basic theory of flight.

But Hargrave was still denied his dream: that of controlled heavier-than-air flight. He continued his experiments and found that a vertical tail fin increased aerial stability, and in 1899 he went to England to address the Aeronautical Society of Great Britain on his discoveries.

A true scientist, he never attempted to patent any of his discoveries or to keep them to himself. He allowed anyone to make use of them and there can be little doubt that they played a substantial part in the later successful construction of aircraft.

Hargrave's box-kite design was snapped up by other enthusiasts around the world. One of them was the Brazilian Alberto Santos-Dumont, whose box-kite-design aeroplane made Europe's first powered heavier-than-air flight in 1906. Voisin, Farman, and "Colonel" S. F. Cody used Hargrave's theories in aircraft which made them famous. Cody's box-kite gliders were demonstrated to the Royal Navy at Spithead in 1903, and to the British Army's Balloon School in 1904. There Cody made a dashing figure with his broad hat, as he and his favourite white horse towed the box-kite aloft. Suspended under the kite was a basket holding an observer. Other men, like the American Octave Chanute, Samuel Langley, and of course, the Wright brothers, used the theories proven by Hargrave on the slopes of Stanwell Park, even though they may have arrived at them by different routes.

Hargrave died on 6 July 1915. He had lived long enough to see powered flight become a reality and, worse than that, to be used for war. Probably he had watched the two British box-kite aircraft which toured Australia in 1911, and compared them with the kites which had raised him into the air. The principles were much the same—but the aircraft had petrol engines.

If Hargrave had been encouraged and helped by the Australian Government, then an Australian and not an American might have been the first to fly a heavier-

Top: Lawrence Hargrave preparing for one of his experiments with a man-lifting kite at Stanwell Park

Bottom: Three of Lawrence Hargrave's experimental aircraft models. That on the right shows his development of the curved wing (*Board of Trustees, Museum of Applied Arts and Sciences, Sydney*)

than-air vehicle. But even such meagre recognition as a memorial had to wait until twenty-five years after his death. It stands now on the summit of the slopes at Stanwell Park, where the chill winds still thrust up the hillsides waiting again for someone to exploit their "lift". Australia has been notably ungrateful to its pioneers, so perhaps it is not surprising that the Government and military were blind to Hargrave's potential.

Among the few who did recognise Hargrave's genius was Percy Sinclair Pilcher, a Royal Navy officer who was one of the first men to complete a true glider flight. He heard Hargrave lecture in England in 1899, and as a result of this designed a powered triplane. But for a quirk of fate, England would have been the mother of the first powered flight—in a Pilcher-Hargrave triplane.

Pilcher had his engine aboard by the summer of 1899 and planned his first flight test at Market Harborough in Leicestershire for 30 September. But that day was foggy, and the flight was postponed. Instead Pilcher decided to give an exhibition in his *Hawk* glider. He was, after all, a product of Her Majesty's Navy, and had no wish to disappoint the spectators who had gathered to see the powered flight.

He took off well, and the *Hawk* rose to thirty feet. An ominous crack signified that a vital strut had snapped, and the *Hawk* smashed hard on to the ground. Two days later Pilcher died from injuries. His Hargrave-inspired powered aeroplane was never to claim the record for man's first self-propelled heavier-than-air flight.

That came about near Kill Devil Hill, Kitty Hawk, North Carolina, in the United States, on 17 December 1903, with the flight of another man who had benefited from Hargrave's research. Orville Wright took his flyer off the ground for twelve seconds, for a distance of 120 feet and at 31 m.p.h. The aircraft which he and his brother Wilbur had built at their Dayton, Ohio, bicycle shop was powered by a modified 13-hp Pope-Toledo engine.

The Wright brothers had been gliding over the earth at Kitty Hawk for some years before they joined an engine to a wing. It seems certain that their own experiments owed something to Hargrave's researches.

While men obsessed with the idea of heavier-than-air flight had been trying to get off the ground, ballooning had gone on apace in Australia. There were, though, isolated pockets of enthusiasm for gliders.

Sydney seemed to be the centre for ballooning in those growing years of the colonies. It was not until 24 March 1891 that the first ascent was made in Perth, by "Professor Price". The balloon lifted with awesome presence to about 150 feet. Price was hanging—on a trapeze—below the basket. The craft slumped earthward and the basket and balloon became snared in a tree, wrecking the whole thing. Price was undeterred. He arranged further exploits with the intrepid Miss Millie Viola, in her balloon, and on 21 July 1891 they took off from the grounds of Perth's Occidental Hotel. Millie was to have jumped, with parachute, over Perth, but a gentle breeze nudged the balloon out across the broad expanse of the Swan River on to some mud-flats in South Perth.

"Why didn't you jump?" reporters asked.

"I can't swim," she replied. But she was not about to give up and nor was Price. They got the balloon off the mud and prepared for blast-off on 3 August. A proud beast was the balloon *Westralia*. The crowd was excited—especially the young lad on the guy-ropes. He pulled too hard, and the craft tipped on to a furnace and was destroyed within three seconds.

Millie was distraught and returned home to Albany. But the urge to fly returned and she obtained yet another craft and actually made an ascent, and then jumped with parachute over picturesque Albany on the south-west coast. An updraught of air caught the canopy and carried her over woodlands and she became entangled in a tree.

Meanwhile, in Melbourne, a university professor named W. C. Kernot was making a valuable, reasoned study of aeronautics. He built two wind tunnels to help illustrate and discover the secrets of the atmosphere. They were the first wind tunnels built in the country, though in ensuing years such devices became one of the prime sources of research in aeronautics.

Crowd-thrilling aviators still plied their precarious trade. "Professor Fernandez" made a number of balloon ascents and parachute descents at Sydney's Bondi Aquarium, entertaining beach-going crowds in those summer months early in 1891. Many people paid for the privilege of being around the balloon when it rose magnificently from the Aquarium. Others gazed up from the steps of their bathing carriages which colourfully dotted the curving Bondi beach. Small boys forgot their ices and pointed, whooping with excitement, to the object in the skies. The brass band at the covered beach-side stand lost the attention of its audience, which gasped as the brave "professor" tumbled from the basket, his parachute streaming behind him.

There was hardly a schoolboy in any Australian capital city, except, perhaps, Brisbane, who had not seen a balloon at least once, and gone back to school on Monday with excited tales to tell. It was an adventurous age. Queen Victoria, Empress of India, still symbolised the power of the throne behind Australia's affairs. But she was ageing, and the federation of the Australian colonies was a foregone conclusion.

Australia's cities teemed with progress and the gaslights of Sydney lit a sophisticated nightlife in the Nineties. The docks at Circular Quay were jammed with tall masts and funnels, and smudges of black smoke betrayed the presence of a swarm of ferries and tugboats upon the varicoloured harbour. Fort Denison had already become a landmark, and the nation had found its voice in such poets as Henry Lawson.

Lawson himself portrayed the roguish, carefree yet caring Australian figure. He was born in a tent on the Grenfell goldfields in 1867, the son of a Norse sailor and a woman from Kent, and he typified the surging, questing Australian. He expressed the pain of the industrialisation of the brown country, a sophistication which allowed the nation a place in aeronautical research.

But such industry and the benefits it brings come slowly. It was almost four centuries since Leonardo da Vinci said "The human bird shall take his first flight, filling the world with amazement, all writings with his fame, and bringing eternal glory to the nest whence he sprang." And it was to be almost exactly four centuries before da Vinci's words would ring true with the faltering, noisy flight of the Wright brothers. The world took little enough notice of the Wrights.

Australia's military men began to grasp the possibilities of balloon flight in the 1880s. Balloons were first used in action by British troops in 1885, in operations in the Sudan which were helped by a tiny contingent from the Australian colonies. Australian troops also saw balloons being used to good effect in the 1899–1902 Boer campaigns of South Africa. A Balloon Unit had been formed by the Royal Engineers in 1890.

The world began its most fantastic era of change with the new century. The past 100 years had been a century of astounding technological progress, but the developments of the industrial revolution were to compound dramatically in the twentieth century. The British Empire, for one thing, began to change in nature: Queen Victoria died after her fifty-five-year reign; some of the colonies became self-governing; and Australia's own Federation came in 1901.

It is as though this period proved too exciting in itself for any major results to be evidenced in Australian aviation. True, some creative forces remained at work. Hargrave continued to lecture to the Royal Society until 1909. But nothing of dramatic consequence was to happen until about 1906.

The Wrights finally and positively proved man's ability to rise in a mechanically-powered, heavier-than-air craft that December day in 1903, but aviation enthusiasm did not grasp Australians until about 1906. From then until the first World War there was a spate of activity. So much activity occurred concurrently that the scene—especially in about 1909—becomes confusing. The race for the first powered aircraft flight in the country ended in a closer finish than the Melbourne Cup. But there were successful glider flights before that, model flights, and powered flying attempts. Even the Government got into the act, at last.

But let us begin with the events of 1906.

Messrs Jacobs, McIntosh, Crooks, and Shulz spent much time tinkering and pottering in a shed in the Geelong suburb of Fyansford, Victoria, until June of 1906 when the *Geelong Advertiser* confided, with great restraint, that "some time ago a Fyansford resident designed a flying machine, which gave what were considered satisfactory trials".

There were little enough details to satisfy the curious. Was it a manned flight? The newspaper did not say.

It continued: "The machine was patented in the names of Messrs. Jacobs, McIntosh and Crooks. Since then the inventor and Mr. E. Shulz, a well-known mechanic, have elaborated on the original design, and a larger model, propelled by a three horse power motor, has successfully stood trials, carried out along the ocean foreshore near Bream Creek."

Only a 3-hp engine, but to the then untutored locals would that in itself signify that the flight was unmanned? It had taken thirteen horsepower to lift Orville Wright off the ground at Kitty Hawk. Still, the three inventors and Mr Shulz had doubtless made significant discoveries regarding the theories of flight. Perhaps they were not new discoveries, but they proved the years of research by the Wrights, Hargrave, and Chanute.

"It is claimed that the flying machine is efficient in all winds and weathers, and the question of future operations is now under consideration" ended the article in the *Geelong Advertiser*. There is no evidence to tell of further work by the men.

Global aviation development moved slowly at this time, considering the major breakthroughs offered by the Wright brothers, Hargrave, and the like. The Wrights offered in 1905 to build a war plane for the United States Army, but their good offices were snubbed. Even the enthusiasm of Orville and Wilbur flagged under such indifference to technological revolution. They stored their Wright Flyer until 1908.

Alberto Santos-Dumont, the son of a rich Brazilian, was not so deterred. He had been primed on earlier successes of his controlled balloon flights which themselves would prove valuable for some thirty years more. He went on to make Europe's first officially-observed powered flight on 23 October 1906. The Brazilian's flight, in fact, caught the world's imagination even more than did the Wrights' flight—perhaps because Santos-Dumont was a much more glamorous personality than the rather dour Americans.

In Australia the new Federal Government had taken some official interest in ballooning. One of the Royal Engineers' balloons had come to Australia for the festivities of Federation in 1901. It took another eight years, though, before the interest was to jell into something concrete.

The balloonist and parachutist "Captain Penfold" (Mr V. P. Taylor) continued to thrill crowds in the meantime. By 1908 he was abroad, crossing the wide San Francisco Bay, once by day and then by night. On the night flight he "bombarded" the United States fleet which was anchored in the bay, perhaps to the annoyance of the men on the ram-prowed battle cruisers moored off the earthquake-scarred city. He came home next year, and still occasionally flew over Sydney suburbs and New South Wales country areas.

The next year, 1909, saw considerable progress in Australian aviation. The Aerial League of Australia was formed, and gained an impressive amount of influence. A now-departed landmark, Sydney's Australia Hotel, was the first meeting place for the League. Major C. Rosenthal and George Taylor were Honorary Secretaries, *pro tempore*, for that gathering.

The tenets of the League were formalised at the meeting on 28 April 1909. They were to watch the latest achievements in aerial engineering; to secure best recognition for Australian efforts in that direction; to awaken public attention to the grave danger in allowing foreign nations to excel in aerial navigation; to join forces with the Aerial League of the British Empire in advocating that the Empire should secure the same supremacy in aerial navigation as it had enjoyed in the command of the sea.

Allen Taylor, Lord Mayor of Sydney, was elected President and five Vice-Presidents were named. One of them was Lawrence Hargrave. George Taylor was named Honorary Secretary, and Rosenthal Honorary Treasurer. Their efforts prompted the formation of a Victorian branch that September.

George Taylor was not idle in those months which followed. He established, in September, the first aircraft factory in the southern hemisphere. It was in Surry Hills, Sydney, now so closely urbanised that it is hard to imagine an aircraft factory there. He built a number of man-lifting kites, presumably under the influence of Hargrave, and then a powered monoplane, with a 30-hp engine and a six-foot aluminium propeller. Apparently it never flew. George Taylor was to become one of the most significant men in Australian aviation development.

The Federal Government, through the army, fed the enthusiasm of the League's members with an announcement on 8 September 1909, that it would offer £5,000 for a "flying machine" to be used for "military purposes". This announcement, though it was to contain many unacceptable requirements, proved to be a major factor in spurring aeronautics.

The machine, as far as was feasible, had to be built in Australia by Australians or Britons, and rise from the ground under its own power. No special starting apparatus was allowed. It must land without injury to any of its structure, be able to carry a pilot and observer, must weigh 350 pounds or more, and must be able to "poise" over any given area. It must top 20 m.p.h. and stay in action for not less than five hours. That list of requirements must have satisfied all members of the committee which pondered the future of flight.

Entry was restricted to "Australians only" or Englishmen naturalised and living in Australia, but world-wide interest was aroused. However, there was no aircraft which could fulfil such advanced specifications. In retrospect it is difficult not to think that the Government had either failed to do its preliminary homework or else it merely wished to stimulate research without having to make an award.

A Victorian John Duigan had been interested in making an aircraft a year or more before the Government announcement, and he built a glider to test his theories of aerodynamics. After this he designed a new and awkward-looking biplane. It was held together with bolts and piano-wire bracing, with no screws, nails, or glue. It had a massive eight-and-a-half-foot propeller, and a four-cylinder, 20-hp engine, weighing

George Taylor aloft on Australia's first flight in a heavier-than-air machine

a hefty 135 pounds. Later he found it necessary to upgrade the power output of the beast.

George Augustine Taylor had of course by now seized upon the competition with enthusiasm. He worked with the army to build his box-kite biplane glider, a craft weighing 102 pounds. With a wingspan of twenty-eight feet it looked remarkably conventional by later standards.

Keith Isaacs, in his *Military Aircraft of Australia: 1909–1918*, says: "The glider was launched by being placed into the wind and let into the air like a kite by means of slip ropes, held by assistants, at each end of the mainplanes . . . When well clear of the ground, the glider would be released, the pilot placing himself prone to give the least air resistance."

For the isolated world of Australian aviation, 1909 was a vintage year. It was particularly exciting for Taylor, whose household heard little talk of anything but kites and aeroplanes. His wife Florence took a keen interest in the projects and was able to discuss flying with Taylor's friends Charles Schulz and his wife and Ed Hallstrom, when they dropped in to yarn about their theories and experiments.

They were all at Narrabeen Heads, now a playground of Sydneysiders, on 5 December 1909. That was the seminal date for Australian aviation, because on that day Taylor launched himself into the air and became the first Australian to fly an untethered heavier-than-air vehicle. He made several flights, of which Isaacs says that the best took him eighteen feet high and about 250 yards along the foreshore, though Ron Gibson believes that his longest flight was only about 110 yards.

Mrs Taylor then had a dash at gliding, and thus became the first Australian woman to fly such a craft. In all, twenty-nine flights were made that day by the three men and two women.

But Florence Taylor became worried by the dozens of flights made by the "boys" during the following weeks and persuaded George to give up experimenting with so frail a craft. His "co-pilot" Ed Hallstrom later became Sir Edward, head of a big refrigeration company and a keen leader of Sydney's Taronga Zoological Park.

The army was not altogether ungrateful for Taylor's efforts: it made him an honorary lieutenant in the Intelligence Corps fifteen days after his first flight. After Christmas, and still enthusiastic over his flights, Taylor built a 20-hp engine, weighing sixty pounds. He planned to place it in the glider. With this, he reasoned, he could take that Government prize.

But the engine did not work properly, so Taylor abandoned the project. But his interest in aeronautics did not abate, and in 1918 he wrote a paper entitled *The Air Age and Its Military Significance*.

The Royal Aeronautical Society of Great Britain now makes a yearly award of the George Taylor Gold Medal for aircraft engineering and design. There have been inspiring recipients of this award, though most of them have been little known or appreciated outside the aircraft industry. It is significant that Taylor's questing nature did not stop with his aircraft. In 1911 he sent the first military wireless message, and in 1912 he developed a remote-controlled model dirigible. For that he was labelled, possibly lightly, a lunatic.

Taylor's continued efforts well earned him the respect which his name still carries. He was, in those pre-war days, *the* authority on flight. He watched Colin Defries flying Mr L. A. Adamson's Wright biplane on 9 December 1909 at Sydney's Victoria Racecourse. And he pronounced it a "no show" because, while Defries was airborne, he showed no ability to be able to manoeuvre his craft.

Taylor's head must have buzzed with the events of flying his own craft at Narrabeen, then watching the Defries bid, and reporting to the League and to the Royal Aeronautical Society. George A. Taylor spent a large part of his personal fortune, earned in commerce, in furthering the cause of aviation. He started his own factory to produce gliders and powered aircraft, and his enquiring mind and immense energy led him to research into radio, photography, and other subjects in which he made a number of new discoveries.

Defries tried again at the racecourse on 18 December. About fifty people watched just as the sun began to warm the open area in the early morning. He lifted the Wright up to two feet, and gradually up to fifteen feet, still moving in a straight line.

The fashion of the day, dictating that "a gentleman always wears a hat", led to Defries' downfall that morning. His headgear started to blow off towards the end of his quarter-mile flight; he grabbed for it, and crashed. The Wright slewed into a ditch and was extensively damaged.

The owner of the aircraft, Mr Adamson, like Mr Coppin of fifty years before, never actually flew. He was headmaster of Wesley College, Melbourne, which must have been a well-paid position unless he had private funds. He sent Defries to England and France to buy the Wright and a Blériot. Neither of them ever recorded "official" flights.

The excitement continued into 1910, mainly as a result of the Government-sponsored competition. It was reported that fourteen "full size" machines were being constructed for the competition by March 1910, and seven were to have Australian-made engines.

In that year the first out-of-work pilot was looking for employment. In the Melbourne *Argus* of 11 March an advertisement read "Aeroplane pilot desires situation. Experience with Bleriot, Farman and Wright machines. Address Argus 131."

There was an epidemic of flying fever. John Duigan continued to work on his designs; a young Queenslander named Bert Hinkler was playing with some aeronautical ideas, and so were a score of others.

The big day, however, was 17 March 1910 when Fred Custance, in Blériot monoplane Number 37, an import from France, made several flights at Bolivar, near Adelaide. He won for South Australia the claim for the first successful powered flight—but was to be robbed of publicity for it for more than fifty years by a flight the very next day of Harry Houdini, the escape artist. Ron Gibson says Custance made a flight of five minutes and twenty-five seconds in the Blériot belonging to F. H. Jones, and writes "During this flight he made three circuits of the paddock estimated to be three miles. He made a second attempt the same morning and attained an estimated height of 50 feet. Custance, then, through inexperience moved the elevators too far and the machine crashed."

There was no blaze of glory for Custance, and only the neighbours for witnesses: Mr R. Winzor and Mr and Mrs Sawyer.

Eric Weiss, alias Harry Houdini, next day made three successful flights in his Voisin aircraft with its 60-hp ENU engine, at Diggers Rest, Victoria. Houdini had more stage presence than Custance. He had nine witnesses sign a certificate attesting his flight. Houdini already had achieved world-wide fame, especially in the United States. He had turned up in Australia in the P & O Line ship *Malwa* in February of 1910 and had left again by 11 May for the United States, where he sold his aircraft. He had also done some flying from Sydney's Rosehill Racecourse in April.

By June the number of machines preparing for the Commonwealth competition had jumped to twenty-three. One of them was Ed Hallstrom's 20-hp monoplane, but that never flew either. A gale destroyed it in the next month. Bert Hinkler, the Queensland boy from the bush, was concurrently tinkering with what he named his *Aviette*, a pair of wings to be slipped on like a coat. But he soon admitted that the *Aviette* was not practicable and abandoned his experiments.

Queensland had its first taste of flying in 1910. In May there was an exhibition of gliders and an "aeromarine monoplane". And now that it had discovered the winged world, the State was all abuzz. A branch of the Aerial League was formed there, and the name of the Queensland Aero Club was dropped. That Aero Club name, though, was to be revived by the Royal Queensland Aero Club some years later.

Probably the first officially-designated Australian aerodrome also came into existence in 1910. The Aerial League picked out a flat paddock near Penrith Railway Station, just west of Sydney, for its landing ground. The field was never really lost to speed enthusiasts; it eventually became the Penrith Speedway.

TOP: The French Voisin aircraft flown by Houdini, the famous escapologist, in Australia in 1910 (*The Melbourne Herald*)

RIGHT: Houdini at the controls of the Voisin (*The Melbourne Herald*)

John Duigan had his aircraft ready on 16 July 1910, at Springs Plains Station, Mia Mia, Victoria. It was the first Australian-built heavier-than-air aircraft which worked. On that day he flew it for only twenty-four feet, but he did not stop with that. Duigan made some improvements to the engine which had been built for him by a Mr Tilley, of Melbourne, eighty miles away. It was a 20-hp, four-cylinder affair.

He tinkered with it, changing the belt drive to the propeller for a chain one, water-cooling the heads, upgrading the compression, and finally adding larger cylinders. He was not new to aeroplane building, for this Farman-style flyer was his third aircraft. The first, in 1909, was merely a pair of wings "which he says were not successful", according to historian Gibson. He had then built a half-size glider from a drawing of a Wright machine, and in this, tethered with 120 yards of fencing wire, he left the ground and rose four or five feet.

His third, powered, machine improved so that his flights eventually lengthened to 100 feet and then 100 yards. Then on 7 October 1910 he flew 196 yards at about twelve feet in the presence of some half-dozen spectators. A second run took him to twenty feet in altitude and 200 yards in distance—at 20 m.p.h. The Duigan biplane is considered probably the most important pioneering aircraft in Australia, and is preserved still by the Science Museum of Victoria.

Duigan did not win the Government prize of £5,000, but it is doubtful whether he was in the show for the money. He joined the Australian Flying Corps when war broke out, and won a commission. He served with distinction, flying an RE8 for 3 Squadron, and was awarded the Military Cross.

How recent and living a thing is manned flight. No dark ages shroud its years nor cloak it with the dust which sucks the life from history. Duigan, for example, lived until 11 July 1951, and many of the men who pioneered Australian aviation still live in suburbs and towns around the nation. Few have been held apart from the populace for sufficient time to wrap them in legend, and so we accept them with easy indifference.

The efforts and successes of Taylor and Duigan and the strong response to the Government competition all combined to inspire plans for a military aviation corps late in 1910, when the Military Board of the Department of Defence considered the proposal.

No doubt the plan would have progressed slowly but for the power of the Press. The Board had, on 17 January 1911, deferred action on the scheme because "even the British Army had not been able to select a suitable type of aircraft", but the newspapers agitated. They, like the people, were excited at the martial prospects of aircraft, for those recent campaigns in the Sudan, South Africa, and China had not served to weary the people of war.

A Blériot monoplane was shipped out by its French makers to tempt the Defence Department, but Pilot Gaston Cugnet failed to make a good first impression. He crashed his pretty little machine in front of a big crowd at Melbourne Cricket Ground on 5 December 1910. The crowd unfortunately contained a number of Government officials. But the people at least behaved better than the crowds who tried to lynch fellow-Frenchman Pierre Maigré when his balloon didn't rise from the Domain about sixty years earlier.

Cugnet had come a long way to try to sell aircraft and he thought it wise to achieve something while in Australia. He joined forces with the Queensland Aerial League Secretary, C. Lindsay Campbell. The Queenslander was so keen to fly that he won semi-official Government backing to go to England for flying lessons. There he began well, winning Royal Aero Club pilot certificate 220, but he died in 1912 when his Bristol-Prier monoplane crashed.

The first aircraft seen by many Australians were those early and flimsy Bristol Box-kites, built by the company which eventually became co-builder of the Concorde supersonic airliner. In 1910 Bristol was as proud as possible over its Box-kite and dispatched two sales teams abroad, each equipped with two aircraft. One mission was to India, the other to Australia. Sydney Smith led the team which left for Australia in December 1910, and he had Joseph J. Hammond and Leslie F. Macdonald as his pilots.

Few sales resulted, though the Australian Flying Corps eventually used a Box-kite for training, but the name J. J. Hammond rocketed to great heights in 1911 as he showed his machine to the crowds in Australian cities. Many of the nation's future flyers were among the spectators, too, at the Bristol demonstrations.

Perhaps the visit did Bristol some good: two Australians, in 1911, went to the company. They were Eric Harrison and Harry Busteed, and the company sent Harrison to act as Chief Instructor for the Spanish Army's air corps, which had been equipped by Bristol. He also went to the Bristol branch at Halberstadt, in Germany, and trained young German officers to fly. No doubt some of them were later to encounter Australian airmen in dogfights over the Western Front.

The Australian Government of 1911 was worried by the stir of public opinion, and sent a cable to the Australian High Commissioner in London, saying "Newspapers here agitating formation of Aviation Corps Can War Office give advice?"

Familiar names were behind this public demand. George Taylor and his friends in the Aerial League of Australia wanted the Defence Department to set up and run an aviation school and an air corps.

Aircraft already had been used for military purposes, during Francisco Madero's revolt against the regime of President Porfirio Díaz of Mexico, in 1910. Hector Worden, an American, flew Curtiss and Wright biplanes for the Mexican Government on reconnaissance missions, and in February 1911 flew above

Top: J. J. Hammond, the first New Zealander to learn to fly, in a Bristol biplane over Melbourne in 1911

Bottom: Many early Australian flyers learnt to fly in the aptly-named Bristol Box-Kite (*RAAF Museum*)

the rebels as they massed at Ciudad Juárez. The world was fascinated by the colour and glamour of this Latin-American revolution, but every farsighted military man was more interested in the fact that Worden could rove further and gain greater intelligence than a dozen squadrons of cavalry.

Mexico has never followed up its pioneer use of military aviation. In 1974 this country with a population of 52 million, many of them impoverished, had only twenty-seven combat aircraft. Its front-line strike jets are ageing de Havilland Vampire fighter-bombers.

W. E. (Billy) Hart, a Parramatta dentist, bought a Bristol Box-kite from J. J. Hammond and won the

Top: The Sopwith Scooter monoplane was among the first to have an enclosed fuselage when it flew in 1913 (*Hawker Aircraft Limited*)

Bottom: This Deperdussin monoplane was used at the Central Flying School, Point Cook, during the first World War (*RAAF Museum*)

first Aerial League of Australia pilot's certificate. But this was not a universally recognised qualification so he proceeded to get the Royal Aero Club Certificate 199. And with all that in hand he won a daredevil reputation which offered far more solace to his questing soul than did the extraction of teeth.

He organised and won the first air race in Australia, pleasing the crowds. But he did not delight the New South Wales Government Railways when on 6 January 1912 he was flying the Box-kite near his home at Mount Druitt. Major C. Rosenthal, that stalwart of the Aerial League, was in the passenger seat. When the engine cut out Billy looked for a handy space among the trees and houses, and the best offering was the cleared land on either side of the railway line. The only problem was that train coming up the tracks, but Billy cut past the Blue Mountains Express and landed safely. The train passengers were thrilled, but maybe Rosenthal was a trifle disconcerted.

Hart's enthusiasm for flying did not wane. When the Empire went to war in 1914, so did Hart, with 1 Squadron, Australian Flying Corps. Oddly enough he served as a ground crewman and did not fly during his service.

The year that Hart got his Bristol, 1911, was the year that Australia's military air arm came into official being. But it only just made the year. The *Commonwealth Gazette* of 30 December called for applications for the posts of two "Competent Mechanists and Aviators", but added that no responsibility would be taken by the Federal authorities for any accidents.

What red-blooded lad could turn down the chance to at least apply for such a post, even if it meant a white lie or so? But the flood of applications included quite enough competent aviators, including Harry Busteed, Henry Petre, C. H. Cresswell, L. Turner, F. Raynham, "Colonel" S. F. Cody, and L. McDonald. All of them were pioneers of flight in England, but Busteed seems to have been the only Australian among them. So it was he and Petre who were chosen, with Cody recommended for the post of engineer and administrator. H. A. Petre, a London barrister, learnt to fly at Brooklands and promptly exchanged his law books for a joystick. Before coming to Australia he had been chief instructor at the Deperdussin Flying School.

Flyers then (as now) tended to be individualistic and occasionally had to be prima donnas. Cody rejected the offer, and Busteed, for unknown reasons, did not stay with the team. He was replaced by another Australian, Eric Harrison. Four mechanics were chosen: R. H. Chester, G. A. Fonteneau, C. V. Heath, and A. E. Shortland.

It took until August 1912 to induct Petre as Australia's first aviator, and on 22 October Military Order Number 570 officially established the Australian Flying Corps. By this time Harrison had joined as a pilot. An order was sent to the Royal Aircraft Factory in England for two BE2a tractor-prop biplanes and two British-built, French-designed Deperdussin single-seaters, and in December a two-seater Box-kite trainer was ordered from Bristols. Petre and Harrison were made "honorary" lieutenants in the army, perhaps because officials thought that a flyer's life was likely to be too brief to warrant a permanent commission.

The army wanted the Aviation Instructional School to be embodied in the new Military College at Duntroon, near Canberra, but the two pilots said that the site was "too high" and had "too many hills" for safe flying. The land near Melbourne was better. How about something in Victoria, perhaps at Langwarren, Cribb Point, Western Port, or Altona? Eventually the choice was narrowed down to Point Cooke, near Werribee, and 734 acres of land were bought. By 7 March 1913 the Central Flying School and the Aviation Corps were established there. But it took a year of organisational red tape and problems before the first flight was made there on 1 March 1914, and the first four pupils did not start training until later in the year. By that time the "e" had dropped from Point Cook, and the back-and-forth tide of paperwork had changed the name from Australian Aviation Corps to the Australian Flying Corps, though this title was never officially promulgated.

The first aircraft to arrive had been offloaded at Sydney, and had to be reshipped to Melbourne Ordnance Stores. When the crates were opened it was found that some of the aircraft fabric had been rotted away by tropical mildew.

By now Australia's flying corps was way behind the air arms of the European nations. Germany, particularly, was ready to use her aircraft for military purposes, and France, Britain, and Italy also had operational aircraft.

By this time there was a small but regular flow of Australian aviators or would-be aviators to and from Britain, the place "where the action was" in the building and development of aircraft so far as Australians were concerned. One was Harry Hawker, who spent the years 1911 to 1913 in England, first learning to fly and then as chief test pilot for the recently-formed Sopwith Aviation Company.

Hawker was one of those pioneers who seem to have been born especially to launch the Age of Aviation. He was a natural, skilled, and daring pilot, and in 1913 demonstrated the Sopwith Tabloid, perhaps the most advanced aircraft of its era, at Hendon near London. It flew at a fantastic 92 m.p.h. and climbed at 1,200 feet a minute.

When Hawker made a brief return visit to Australia in January 1914 he brought a Tabloid with him and offered it to the Defence Department. But in those sleepy summer days when the world was at peace he was told that the Australian Flying Corps only required training aircraft.

Another of the Aviation Age Australians who went

Norman Brearley in a Maurice Farman Longhorn biplane at Thetford, England, in December 1915 (*Sir Norman Brearley*)

to England in that period was Horrie Miller, a Victorian youngster who was fascinated first by motor-cars and then by aircraft. The exploits of Taylor, Duigan, and Houdini inspired him to make model aircraft, which actually flew.

As though fate had determined that he was to be a flyer, one of these models attracted the attention of Maurice Shmith, a director of Tarrant Motors Co., who gave him a job. This brought him into contact with three other youngsters interested in both automobiles and aviation: Harry Hawker, Harry Busteed, and Harry Kauper.

This trio went to England together and all made their name in aviation. Kauper invented the Kauper gun gear which enabled a machine-gun to fire through a whirling air-screw.

Horrie Miller became a hire-car chauffeur, and fate's next gentle nudge was to give him the job of chauffeuring Major H. A. Petre and Captain Eric Harrison as they sought a site for the Australian Flying Corps' flying school. He was with them when they decided on Point Cooke.

Horrie reached England in 1913 and made straight for Hendon, and for two guineas had his first flight—in a Bristol Box-kite with an engine missing on one cylinder. Within a few days he had a job in the Sopwith factory. He soon became acquainted with many of the early greats in aviation, and witnessed the terrifying experiment in which Harry Hawker proved that the way to coax an aircraft out of a spin was to push the joystick forward, not pull it back—the instinctive reaction that had killed all the pilots trapped into doing so.

A few months later the world was at war. The 1st Australian Imperial Force, consisting of 20,000 men of all arms who had been brought to a high state of readiness and training by the dedicated pre-war work of such men as Bridges, Blamey, Monash, and White, was ready to sail overseas within five weeks of the declaration of war. The first Australian troops were in action on 11 September 1914, against the Germans in New Guinea.

But the Australian Flying Corps was far from being ready for combat. With only a handful of men and a few training aircraft, it had to stay behind while the Diggers went cheering overseas to death or glory at Gallipoli.

2

Australians Fly to War

Australians proved apt pupils in a new way of warfare and joined the great air aces of the Eastern and Western fronts

All the countries of the British Empire plunged into the first World War with an enthusiasm which did not falter even after the Allies had suffered such reverses as the retreat from Mons and the gigantic defeat of the Second Russian Army, under General Samsonov, by Hindenburg and von François.

To begin with the aircraft of the warring nations played little part. Some generals saw them as a useful aid to reconnaissance; others thought they were a distracting nuisance. Now and again the airmen experimented by firing rifles, revolvers, and even sporting arms at each other, but with little success.

By the end of 1914 the generals had mauled each other's regular armies almost into extinction. Like prehistoric monsters enmired in the asphalt pits they squatted in the mud while they waited for fresh troops to be trained, and wondered what to do next.

But aviation enthusiasts of all the combatant nations, whether they were pilots themselves or "back-room boys", had now found opportunities to show what aircraft could do.

The Wright brothers had believed that aircraft would put an end to war, because they thought that every soldier or noncombatant would be exposed to bombardment by "aerial artillery" and that such a frightful prospect would cause every nation to turn its swords into ploughshares.

They were proved tragically wrong by Second Lieutenant Giulio Gavotti, of the Flotta Aereo Italiano, who on 1 November 1911 threw four 2-kilogram grenades from his aircraft on to Turkish troops in North Africa, when Italy was fighting Turkey for the possession of Tripoli.

Britain and America had carried out bombing trials but, true to form, their potential had been ignored by the diehard military men of both countries.

The French, however, had been more practical. They built four or five squadrons of Voisin aircraft capable of carrying bombs, and used these on short-range raids on German formations.

In Germany, Major Wilhelm Siegert had long since developed a startling new concept. It was based on a simple theory. Germany had for years been preparing for war on a scale hitherto undreamed of. Even her railways were planned for their strategic as well as commercial value. Her heavy industry had immediate military potential. When war began, the military might of Germany was to burst like a thunderstorm, deluging her opponents.

But—why restrict warfare to military men? Those who worked behind the lines were equally important, all the way back to clerks in offices and workers in steel mills. Why not kill them, too?

With traditional weapons this was impossible. But Count Ferdinand von Zeppelin had been working on navigable airships since early in the century. In 1909 he founded the world's first airline: the *Deutsche Luftschiffahrts-Aktiengesellschaft* (German Aeronautical Limited Liability Company). Between 1910 and 1914, Count Zeppelin's company flew a total of 170,000 miles, with 35,000 paying passengers, between various German cities.

In 1913 one of the "Zeppelins" took part in the German Army's summer manoeuvres, and the General Staff, which until then had been deaf to Count Zeppelin's plea that his *Luftschiffe* should be considered for military purposes, suddenly became enthusiastic.

So, reasoned Major Siegert, why not use the *Luftschiffe* to dump loads of bombs on England, destroy her industries, terrorise her population, and drive her out of the war?

The theory was fine, but like so many theories it failed to work because the practical means at hand were not sufficiently powerful. The Zeppelin bombing of London began in 1915 and caused plenty of civilian deaths and damage, but its principal result was to hand the Allies a marvellous propaganda weapon about "Hun frightfulness".

And the huge, slow-moving Zeppelins were about the most vulnerable military targets since the Spanish Armada. The British ringed London with anti-aircraft batteries, and such dashing pilots as Warneford soon mastered the techniques of night-fighting. In a kind of preview of the events of 1940–41, Londoners cheered as they saw Zeppelins break up and burn, with their trapped crews faced with the choice of roasting in mid-air or plunging a few thousand feet to earth.

But the Germans are nothing if not determined. The aircraft works at Gotha in Thuringia were developing a powerful twin-engined bomber, capable of flying to London and back from airfields in Belgium. On 25 May 1917 the world's first daylight terror raid was launched on civilians by Gotha aircraft of *Kampfgeschwader Drei* (War Squadron Three) and these raids continued until May 1918. By that time the Germans were using a new Gotha design: the *Riesenflugzeug* (Gigantic aircraft).

The Gothas caused more civilian deaths and damage, but their effect on the outcome of the war was zero. The main result of their raids was to cause an outcry for the British to kill German civilians in return. When the war ended the RAF was busily equipping a squadron of Handley Page V/1500 bombers to fly to Berlin.

Major Siegert's theory of total war had failed in this instance, but his compatriots would make it work in 1940. While the Germans were practising for it in 1915, the Allies were developing the principle of strategic bombing.

Probably the Royal Naval Air Service carried out the world's first air raids against strategic installations as opposed to the French raids on forces in the field. In 1914 pilots of the RNAS bombed Zeppelin sheds and supply depots on the island of Heligoland, with minimal destruction but with the effect of convincing staff officers that such operations were worthwhile. A wing of aircraft for strategic bombing was formed, but was broken up to reinforce the decimated British squadrons on the Western Front.

Oddly enough it was the Russians, whose armies suffered so badly from staff and supply failures, who really led the way in large-scale bombing. They happened to have an aviation genius: Igor Sikorsky, who escaped to the USA after the Russian Revolution and continued to develop radical new aircraft.

He designed what was then the enormous *Ilya Mourometz* bomber, which could carry a bomb load of nearly a ton. It had four engines, a crew of six, and three gun positions, and was in many ways the prototype of the bombers of the second World War. The first raid by these giants was on German military installations in Poland on 15 February 1915 and, in all, seventy-three were built before the Russian collapse. But so profound was the chaos of Russian military organisation that only about half of the "Flying Ships" were ever used operationally, and this great weapon which could well have changed the course of the war seems to have been ignored by Russia's allies.

The Italians, also, had built three-engined bombers: the Caproni. These flew their first long-range missions on 20 August 1915, quite soon after the Italians entered the war.

The British long-range bomber, the twin-engined Handley Page O/100, was actually on the drawing board in 1914 but did not become operational until October 1917. They could carry 1,800 pounds of bombs for 200 miles, drop them with reasonable accuracy, and return to base.

The O/100, with a wing span of 100 feet, was a monster for its day but was soon joined by the improved model, the O/400. They were used to bomb cities in south Germany within range of British airfields on the Western Front. The war ended before their "big brothers", the four-engined V/1500, could take off for Berlin with their load of a specially-designed 3,300-pound bomb.

The Handley Page O/400 bomber, seen here in the desert with two day-bombers to show its size, was the "Flying Fortress" of the first World War although it was developed too late to have any significant effect upon Allied aerial strategy (*Australian War Memorial*)

This painting by David E. Hammond was based on an episode described in the Australian War Memorial book *Military Aircraft of Australia 1909–1918* by Wing Commander Keith Isaacs. On 7 April 1917, the Australian-born Lieutenant R. A. Little singlehandedly engaged eleven Albatros scouts while flying his Sopwith triplane "Blimp" N.5493 of 8 Squadron RNAS. Little preferred "in-fighting" and fired only at point-blank range. He was credited with forty-seven kills and was awarded the D.S.O. and Bar, D.S.C. and Bar, and the Croix de Guerre before his death in action in May 1918. (*Australian War Memorial*)

TOP: Another painting based on an episode from Wing Commander Isaacs' book. It shows Sopwith Camels of 4 Squadron AFC, led by Captain A. H. Cobby, diving upon Fokker and Albatros scouts of Richthofen's Circus early on 21 March 1918. (*David Hammond: Australian War Memorial*) LEFT: A model of the Vickers Vimy flown by Ross and Keith Smith and their crew in the London–Australia Air Race, 1919

The BE2a, manufactured by the Royal Aircraft Factory at Farnborough, England, was one of the first aircraft to fly at the Australian Flying Corps Central Flying School at Point Cook, Victoria, in 1914 (*Australian War Memorial*)

But the bomber pilots were never the "glamour boys" of the first World War, and the role of strategic bombing was not thoroughly appreciated although by the end of the war both sides were using a large number of single-engined, two-seater fighter-bombers of various types to attack airfields, gun emplacements, munition dumps, army camps, railway yards, and similar targets.

Even so these tended to be in the nature of "hit-and-run" raids and there were few signs of using the bomber as a truly strategic weapon, launching repetitive mass attacks which could break up enemy formations before they got into the front line or cut them up when they began to advance. Such work was still left to long-range artillery—which, of course, could deliver a heavy weight of metal with great speed on to a distant target. The drawback was that the gunners could not switch to "targets of opportunity" as quickly and flexibly as a marauding mass of fighter-bombers—for example the Mosquitos of the second World War. Many of the artillery shells simply churned the battle-fields into a worse shambles than before without having any decisive effect on operations.

A large part of British staff thinking during that war was devoted to the problem of clearing the way for a "cavalry breakthrough", which never came off.

Yet, if they had cared to develop it, they had a much better weapon available than cavalry.

One shining example of appreciation of how aircraft should be used in modern war was given by General John Monash of the 4th Australian Division when he planned the battle of Hamel for 4 July 1918.

A major Western Front problem was that attacking forces usually ran short of ammunition. Bearers sent to follow up with fresh supplies had to stumble slowly through the chaos behind the attack—and often were killed in the process.

Monash decided that this could be solved by using aircraft. Captain L. J. Wackett, of 3 Squadron Australian Flying Corps, devised a simple release device for ammunition boxes slung beneath aircraft. When Monash's advancing machine-gunners ran short of ammunition, they laid out canvas strips as a signal to the aircraft. About 100,000 rounds were dropped in this way.

On the Eastern Front, on 21 September 1918, pilots of the Australian Flying Corps joined with the RAF in what must be regarded as the world's first "blitzkrieg" use of aircraft.

Australian, British, and Indian troops under the British general, Allenby, and Henry George Chauvel who was the Australian commander of the Desert Mounted Corps, were driving a mixed German and Turkish army northwards from Nazareth to Jordan.

On 21 September the flyers caught the retreating enemy on a narrow mountain road. This was the exact moment for the type of repetitive mass attack in which fighter-bombers can excel.

The British and Australian pilots flew back again and again to bomb and machine-gun the trapped columns, until the flyers could hardly bear to return to the slaughter. It went on all day, until nearly 1,000 vehicles had been wrecked, eighty-seven field pieces abandoned, and hundreds of horses and men killed.

The way in which aircraft could outstrip ground troops was demonstrated by the fact that the advance guard of Chauvel's cavalry did not reach the scene of carnage until three days later. Without the air strike the Turks and Germans would have been able to regroup far more effectively to counter the subsequent Allied attacks.

Germans are good at learning lessons taught by war. When their Stukas, Messerschmitts, and Focke-Wulfs carved up the columns of refugees and Allied

TOP: Lieut. R. Williams, later Sir Richard Williams, with an instructor in a BE2a at Point Cook in 1914 (*Australian War Memorial*)

CENTRE: German soldiers prepare to launch an observation balloon (*Imperial War Museum*)

BOTTOM: The DH2 single-seater biplane fighter, propelled by a 100 h.p. Monosoupape "pusher" engine (*Hawker Aircraft Limited*)

troops in France and Belgium in 1940, they were emulating the Australian and British example.

Such usage of fighter-bombers was fairly infrequent in the first World War. For most people, the aerial sector of that war is represented by dogfights over the Western Front.

The actual purpose of these dogfights has to some extent been obscured by the glamour attached to the names of those who became supreme in aerial combat —Richthofen, Mannock, Ball, and so on.

It is easy to gain an idea that the flyers took off as though to fight a duel, and let the best man win, whereas the battles between "scouts", as they were often called, were merely a part of the strategic picture.

The really important aircraft were not the dashing fighters, but the trundling artillery observation planes and the photographic flights—and even the kite balloons which hung motionless in the skies behind the front lines. These could convey important information to the ground commanders.

So a primary task of the fighters was to keep the skies clear so that their own observation aircraft might function without being shot down. And, of course, in their own dawn and evening patrols, they could pick up information about the enemy's movement—what little of it there was on that bogged-down battlefront.

Apart from this, there was a vague kind of feeling that if only they *could* shoot the enemy out of the sky and keep him on the ground, then eventually it would be possible to make a more purposeful use of the air weapon.

This feeling strengthened as the war proceeded, until it became apparent that he who commanded the sky could then begin to make use of the not yet fully realised potential of the fighter-bomber and heavy bomber.

To begin with the Australians took no part in the war in the air, for the simple reason that they had no suitable aircraft. The first Australian airmen did not go to war until 20 April 1915, when the First Half Flight sailed from Melbourne in the liner *Morea.* On 10 February 1915 the Half Flight had been ordered to join the Indian Army in its operations against the Turks in Iraq, then known as Mesopotamia, as part of the British attempt to drive the Turks northwards and relieve the pressure upon the Suez Canal and the precious oil wells near the Persian Gulf.

With some embarrassment the Australian Government had to send a cable on 26 February to the effect that while personnel and ground equipment were available, the Australian Flying Corps did not have any suitable aircraft. Their two operational aircraft had gone to New Guinea, with pilot E. Harrison, for use in Australian operations against the Germans. German resistance in New Guinea collapsed soon after the Australians landed in September 1914, but it seems that the aircraft were still in their crates somewhere in that part of the world. So it was arranged that the

Indian Army would supply aircraft to the Australians.

The Half Flight—known as the Mesopotamia Flight—was well equipped on the ground. It had a Commer truck equipped as a metal workshop, a Daimler equipped for the carpentry which was essential for aircraft of that era, a Garford store-truck, two Berliet transports, a Hupmobile staff car, and a Villiers motorcycle. Four pilots, Captain H. A. Petre and Lieutenants T. W. White, W. H. Treloar, and G. P. Merz, were to do the flying.

The Half Flight landed at Basra on 26 May 1915 and took over the Maurice Farman aircraft provided by the Indian Army. Only five days later Petre and a New Zealander, Captain H. Reilly, made the first reconnaissance flight over the Turkish positions in the valley of the Tigris River, to observe the effect of the Allied attack at Kurna. They brought back valuable information and were keen to get into the air again, but on the next day it was the turn of Reilly and Captain T. White. In their lumbering dual-control Maurice Farman Shorthorn they plugged along for two hours against the *Shamal*, the northerly wind, on their way to Kurna. Here and at Bahran they saw that the Turks were retreating, and sped them on their way with three 20-pound bombs. The *Shamal* helped the Shorthorn to make better time on the homeward flight, and on succeeding days the AFC contingent carried out further successful operations against the Turks. The flimsy Maurice Farman pusher-prop aircraft had almost entirely superseded cavalry for reconnaissance.

The early warplanes were still enjoying relative safety on their flights over enemy territory, because air-to-air combat was still in its infancy. The main danger was in the aircraft themselves—as Captain White discovered on 22 October 1915 when he and Captain Francis Yeats-Brown, his observer, were flying over the Turkish entrenchments at Zeur in Mesopotamia. The Renault engine began to cough and splutter, and the power faded so rapidly that White had no option but to land behind enemy lines. But the engine was still turning over, and White told Yeats-Brown, "We'll taxi her home. You stand up with the rifle and have a crack at any Turks who have a go."

Trailing clouds of dust they rolled through mile after mile of Turkish-held territory, skirting rocky outcrops and thorn bushes. Then they found a road—but straight ahead of them was Kutanieh and a force of 2,000 Turks on camels and horseback. "Keep her going!" was White's only thought, and he gunned the aircraft along the road. With incredible luck they found a stretch running through the trenches and barbed-wire entanglements, and breezed through the front line under the eyes of the startled Turks. Then, after running for fifteen miles, the engine suddenly picked up again and they were able to take off and fly safely home.

White was not so fortunate on a subsequent occasion. He was captured by the Turks, and later told the story in two classic war books: *The Road to Endor* and *Guests of the Unspeakable*.

Back home in Australia the young trainees at Point Cook had watched enviously as the First Half Flight departed. One of them was Second Lieutenant Richard Williams, who in late 1914 had won the first

No. 3 Training Course at Point Cook, 1915. From left are Lieutenants A. W. Ellis, C. D. Merrett, F. H. McNamara, A. L. McNaughton, L. J. Wackett, E. G. Roberts, E. R. Moseley, C. J. Brookes, R. Ross (*RAAF Museum*). Lieutenant Wackett became Sir Lawrence Wackett, one of the great names in Australian aviation

wings presented during the initial course of the Central Flying School. When preparations were made for a second Half Flight to go to Mesopotamia he was chosen to lead a group of three other pilots and some engineers. His hopes were dashed when the group was disbanded before the end of 1915.

But by that time the situation was changing. Thousands of young Australians were eager to get to grips with the Germans, and some of them could not wait for their own country to send them into battle. Those who wanted to join the war in the air were particularly impatient. Scores of them could not wait to be trained on Australia's only two remaining military aircraft, so they sailed for England to join the Royal Flying Corps. Horrie Miller, who was already in England, rushed along to join up—at first in answer to an advertisement for despatch riders who owned their own motor bikes. When he arrived at the recruiting office he found a crowd of 4,000 youngsters waiting there, and after some hours they were told that no more volunteers were required. So he reversed the process and returned to Australia, where he tried to join the AFC. Here again he was told that there were no vacancies and he had to wait his turn. He filled in the time by building an aircraft to his own design—and during its engine trials it jumped the chocks and smashed into the back of the hangar, almost killing his mate Bob Cousins in the process.

Norman Brearley, then aged twenty-four, was one of the young Australians champing at the bit. He was in the final year of his apprenticeship as an engineer in Perth when the war broke out, and as an enthusiast for the early, rowdy motorcycles and an avid reader of *Aero* and other aviation journals of the day he wanted to be in the air with an engine in front of him.

He weighed his chances of getting into the AFC and thought that they were slight. So as soon as his apprenticeship ended he worked his way to England aboard a ramshackle old steamer and joined the RFC.

Scores of young Australians were there before him, and Brearley soon found that the demand for pilots meant that many of the instructors had little more experience than their pupils. In his book *Australian Aviator* he says that, "After my first circuit—the first time I had ever been in the air—I felt vaguely nervous, but apparently not nearly as nervous as my instructor.

"After several more short flights, and feeling I was making no progress, I decided to do or die."

The training system on the Maurice Farman Longhorn could hardly have been cruder. The instructor sat at the controls with the pupil immediately behind him and reaching round him with both arms. The idea was that the pupil would thus be able to get a feeling of what the instructor did with the controls.

Brearley thought that the only way to learn was literally to take the controls into his own hands. "Pressing my instructor's body forward I took a firm grip on the controls, moving them beyond the timid limits he had previously permitted. Wind-blown oaths floated over my shoulder as he knocked my hands aside and quickly came in to land."

The instructor stumbled away from the machine, vowing that he wouldn't fly any more with the mad Australian, and Brearley asked, "Does that mean I'm solo now?"

"Yes, if you want it that way," the trembling instructor replied as he sat down with some other pupils on the grass. At that time, Brearley's total flying time since starting to learn was only an hour and fifty minutes. However he opened up the engine and gingerly took the Longhorn across the grass until she lifted, and then went up to 2,000 feet. He made a cautious 180-degree turn, rose to 4,000 feet, and tried a few more turns. That was enough to start with, and he descended for a successful landing.

Norman Brearley with an Avro biplane in 1919 (*Sir Norman Brearley*)

While Brearley was training with the RFC, the British War Council had been sufficiently impressed by the rush of young men from "the colonies" and the fact that Australia had formed her own embryo air force, to suggest that the dominions should form complete squadrons instead of sending their youngsters to join the Royal Flying Corps or the Royal Naval Air Service. Australia was the first to jump at this suggestion and thus became the first dominion to have its own complete aerial formations in combat. By the end of the war eight Australian squadrons had been formed and sent to help the British.

The twenty-eight officers and 195 non-commissioned officers and aircraftsmen of 1 Squadron AFC left for the first World War early in 1916 aboard the troopship *Orsova*. One of the officers was young Richard Williams who won his wings in 1914, and by September he was a captain. (The present RAAF ranks had not yet been invented. They derive from those established when the RAF was formed.)

The squadron disembarked in Egypt to continue the war against the Turks. They joined the Australian and British troops who were fighting them on the ground. In some ways they were the "forgotten army" of the first World War. After the failure of Gallipoli much of the world's attention was focused on the death struggle on the Western Front, but the Eastern Front was equally important. There the strategy was simple: stop the Turks from cutting the Suez Canal. In those days the Canal was a vital lifeline between Britain and everything "East of Suez", and if ships had been prevented from steaming through it with essential troops and supplies then the effect on Britain's war effort could have been disastrous. Finally the British

The men and machines of 1 Squadron, Australian Flying Corps, photographed in the desert in 1917 (*Australian War Memorial*)

Aircraft of 1 Squadron AFC starting out on an evening patrol in the Palestine theatre of war in 1917 (*RAAF Museum*)

and Australians not only stopped the Turks from cutting the Canal—they rolled them all the way back to Syria.

But that result was a long way off when 1 Squadron landed. In desert warfare the Australians performed an essential service: that of keeping an eye on the enemy in a way which would have been impossible even for the tough, seasoned men of the Australian Light Horse. An aircraft could go as far in a couple of hours as a horse would go in a week.

Besides this, the airmen could harass bodies of troops on the ground, deliver despatches to isolated units, and give a previously unknown mobility to at least one part of warfare.

The airmen certainly did not have it all their own way. German flyers, some of them blooded on the Western Front, were now flying in support of the Turks, and the Australians had to face some Austrian airmen, too. The enemy used their aircraft for the same purposes as the Australians, reconnaissance and general harassment, and in his book *Doctor and the Aborigines* Charles Duguid recalls that the field hospital in which he was serving suffered several night bombing and machine-gunning attacks.

Another hazard was the Arabs. Bands of these fought on both sides, and they fought without mercy. When young Lieutenant G. Merz of 1 Squadron was forced down in his Caudron he was butchered by Arab guerillas, and thus became the first Australian pilot to die in action. And the reputation of the Turks was not too good in this respect, so whenever one of their number had to land in the desert the Australians took tremendous risks to pick him up again.

One of these exploits won the Victoria Cross for Lieutenant F. H. McNamara of 1 Squadron, who came from Rushworth, Victoria.

He was twenty-three when on 20 March 1917 the squadron was flying a sortie against what was known as Junction Station, on the railway supplying the Turkish troops. Their aircraft were carrying bombs made from converted howitzer shells, and six of these were slung below McNamara's Martinsyde G100. He had dropped four successfully, but when he pulled the toggle for the fifth its delayed-action fuse operated prematurely and the bomb exploded below the aircraft. A fragment ripped through the fuselage and into McNamara's thigh, but even through the shock and pain of the wound he noticed that the BE2c flown by Lieutenant Rutherford was making a forced landing near the railway line.

For McNamara, while he still could fly, to leave a comrade was unthinkable. With the blood pouring from his wound he sideslipped his G100 round and eased back on the stick to stall her on to the ground near Rutherford, descending through a hail of Turkish bullets. Before he landed he could see a group of cavalry galloping towards them.

Rutherford raced across the 200 yards between the two machines and flung himself on to the lower wing and across the engine cowling, but when McNamara

Lieutenant George Merz. He was one of the first four trainee pilots for the AFC at Point Cook, and the first Australian airman to be killed in action: in Mesopotamia 1915 (*RAAF Museum*)

tried to control the overloaded aircraft his injured leg betrayed him and the G100 slewed, tipped, and turned over. Somehow the two pilots scrambled out. They remembered to set the aircraft on fire, and Rutherford helped McNamara to hobble across to the BE2c. With its broken stay-wires, damaged fuselage, and a tyre off one wheel it seemed like a pretty insubstantial hope, but at least they could use its machine-gun to hold off the Turkish cavalry. A few short bursts made the enemy think twice about an easy capture, and as they reined in the two Australians decided that an attempt to get the BE2c into the air was better than a meek surrender.

With McNamara in the pilot's seat, and getting weaker every moment, Rutherford dashed round to swing the prop. The Turks renewed their charge towards them, but the engine caught and Rutherford scrambled into the observer's seat as McNamara lurched the aircraft forward.

With agonising slowness the aircraft trundled forward as the Turks galloped nearer, its flat tyre flapping over the soft ground, and then as McNamara eased back on the stick its nose lifted and they were off the ground—and flying just above the heads of the Turks.

The airstrip at Kilo 143 was seventy miles away, and Rutherford hung on grimly as McNamara struggled for consciousness and the fragile little aircraft dipped and swayed. But they made it, and McNamara slumped into unconsciousness as the aircraft rolled to a stop. He was repatriated to Australia and went on with his career in aviation which culminated in the rank of air vice-marshal in the second World War.

Rutherford's luck did not hold. On 1 May 1918 he had to force-land his Bristol Fighter and another Bristol landed to pick up Rutherford and his observer. They clambered on to the wings and held on to the struts, but the weight of four men was too much even for one of the gallant Bristols. The gear collapsed and all four became prisoners of war.

There were several similar exploits, one involving Richard Williams of 1 Squadron who was earning a solid reputation for verve and leadership. On 21 April 1917 he saw Lieutenant A. T. Cole's Martinsyde falter down for a forced landing, and in his BE2e Williams followed him down. Just as he landed he saw Cole fire his Very pistol into the Martinsyde to deny it to the Turks who were closing in fast, and then toss the pistol away as he ran towards Williams. "Darn me—you've left your pistol! Don't you know we're short of them?" Williams roared, and Cole sprinted back to pick it up again. He still owned it fifty years later.

Apparently the whole episode so confused the Turks that they forgot to open fire, and Williams and Cole escaped.

Among the Australians flying in the desert war were young Ross Smith, later to achieve fame with his brother for their England–Australia flight, and Hudson Fysh whose name has become synonymous with Qantas. Fysh was occasionally on the carpet before Williams, who had become commander of 1 Squadron, and he recalls that, "I usually came off best, but not always."

Fysh was rather proud of the fact that he had resisted the temptation to drink hard liquor. He had carried a flask of brandy through the Gallipoli campaign and then through the desert war until a spell of exhaustion made him take his first sip. After the war he proudly told this to Williams, who said, "You know, I can go one better than that. I left Australia with a flask of brandy, carried it all through the war, and returned with it unopened."

This paints rather a different picture from the conventional one of the hard-drinking pilots of the first World War. But they were all very young men, and perhaps their youth was intoxicating enough.

Williams rose from captain to major to lieutenant-colonel: from command of a flight to command of 1 Squadron and then to command of 40th Wing, comprising the four squadrons which played the leading part in the war against the Turks. The Wing gained absolute domination of the desert air and the British and Australian flyers were able to carry out their "blitz" on the retreating Turks, on 21 September 1918, absolutely unopposed.

Williams and his men flew a great variety of aircraft as the huge programme of aircraft building and development inspired by the war put one type after another into the skies. The combatant aircraft at the end of the war had only one resemblance to those with which the war began: the fact that they flew. In speed, manoeuvrability, and overall deadliness the four years of war had brought progress to military aviation which in peacetime would have taken decades. And of course the lessons learnt in wartime flying were applied to peacetime aviation from 1919 onwards.

In December 1916, 1 Squadron received their first Bristol Scout C aircraft, one to each of the three flights. They were the first to arrive in the desert with machine-guns synchronised to fire through the propellers, but otherwise their performance was somewhat disappointing in the heat-thinned air of the desert. But the Australians made good use of this and subsequent models of the Bristols. Lieutenants C. S. Paul and W. J. Weir painted one of them yellow and used it to such effect that Turkish orders noted: "All ranks are to take immediate cover on approach of the yellow English aeroplane."

Ross Smith, back in the air after being grounded with a wound, used a Bristol to fly Lawrence of Arabia to join his Arab irregulars, and subsequently worked with him on several occasions.

Some of the most important work by Australians on the Eastern Front was done when the final attack on the Turks was being prepared in late 1918. The enemy tried desperately to discover the direction from which the attack could be expected, but the German reconnaissance aircraft were shot out of the sky by the Australians. When the attack was launched the Australians hit hard at the ground troops, and were aided in this by their only Handley Page bomber, flown by Captain Ross Smith.

On the Western Front, Australians in both RFC and AFC squadrons were making a name for themselves. Sixty-one of them became accredited aces, having shot down the requisite number of more than five enemy aircraft. The total of their recorded "kills" was 644.

The highest-scoring Australian ace was Captain Robert Little from Windsor, Victoria. He was only nineteen in 1914, and when he was unable to get into the AFC he paid his own fare to England in 1915 and then paid for his own flying instruction. By January 1916 he had his pilot's certificate and was accepted into the Royal Naval Air Service.

Though this was part of the navy, aircraft carriers were still in the planning stage and to begin with the naval airmen were used mainly in spotting for submarines in the North Sea and English Channel. The constant drain of aircraft on the Western Front resulted in some RNAS squadrons being sent to serve there, and by October 1916 Little was flying a Sopwith Ships' Pup in that theatre. His flight was under the command of another Australian, only four years senior to Little, Major S. J. Goble of Croydon, Victoria. With seven accredited "kills" he too became a first World War ace and eventually became an air vice-marshal in the RAF.

The Sopwith Ships' Pup was a diminutive radial-engined aircraft, but Little handled it in so masterly a fashion that he shot down nine Germans. When his squadron was re-equipped with the new Sopwith Triplanes, known as "Tripes", he really began to make his presence felt. He was undaunted by the Germans and in April 1917 he tackled eleven Albatros scouts single-handed, and out-manoeuvred them for a full half-hour. Baron Manfred von Richthofen, the famous "Red Baron", observed this fight from the ground, and in July 1917 saw Little in action from the air. He watched Little in his "Tripe" outfly and outfight all comers, including the deadly Albatros DVII, and reported these successes to Anthony Fokker the Dutch aircraft designer. They made Fokker hurry to his drawing board with a new set of ideas.

By 1918 Little had forty-seven victories to his credit. By then he was flying the Sopwith Camel, one of the aerial terrors of the Western Front, and maybe his successes had made him over-daring. He was then a member of the newly-formed Royal Air Force which was slowly gaining supremacy in the sky, and one of its tasks was to tackle the giant twin-engined Gotha bombers as they flew over France on their way to bomb London.

The Gothas had abandoned daylight raids because

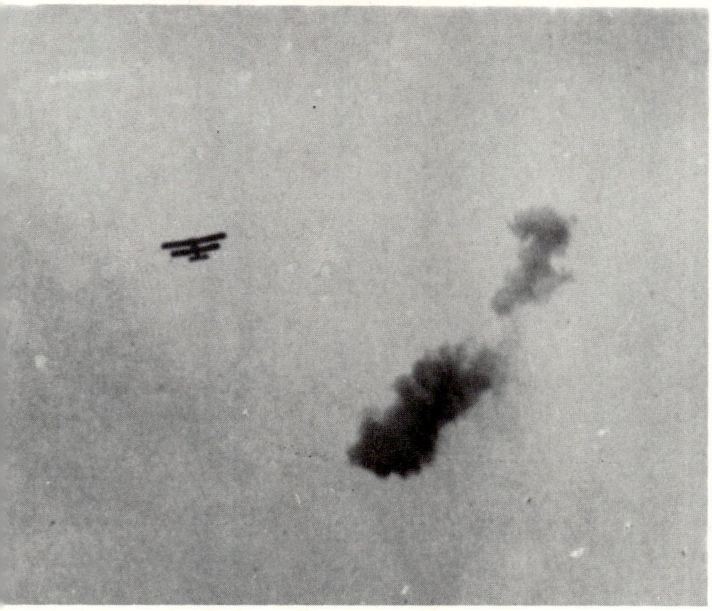

TOP: The Sopwith Pup, Series 1, was one of the speedy lightweight fighters developed during the first World War and used by the AFC (*RAAF Museum*)

CENTRE: A Sopwith Camel high above the Western Front (*RAAF Museum*)

BOTTOM LEFT: "Archie" explodes near a patrolling AFC scout (*Australian War Memorial*)

BOTTOM RIGHT: Norman Brearley's DH2, No. 5961, in which he flew in the first World War (*Sir Norman Brearley*)

TOP LEFT: "Bombing up" an RAF RE8 night bomber in 1918 (*Australian War Memorial*)

TOP RIGHT: The wreck of a German aircraft on an AFC airfield, 1917 (*Australian War Memorial*)

CENTRE: Men of an unnamed AFC unit pose with a Sopwith Snipe in France, 1918 (*Australian War Memorial*)

BOTTOM LEFT: General W. Birdwood, commander of 1 Anzac Corps, is the observer in this Avro 504 (*Australian War Memorial*)

BOTTOM RIGHT: Horrie Miller, later one of the founders of MacRobertson Miller Airways, about to fly from England to France in 1917 (*H. C. Miller*)

A sleek Albatros DIII fighter in captivity on an AFC airfield in 1918 (*Australian War Memorial*)

they suffered too heavily. The British had to attack them as they flew through the night skies: a difficult task in those pre-radar days and with nothing to guide the attacker but the flare of the bomber's exhausts or the beams of searchlights from the ground. But Little hurled his Camel into the air and picked out his target—just as a searchlight beam swung towards him instead of on to the bomber. He was momentarily blinded—and his aircraft was illuminated just long enough for the Gotha's rear-gunner to loose off a quick burst. Bullets ripped through the fabric fuselage and into Little's groin and stomach.

He crashed near the village of Norviz and was buried near Major McCudden, VC, another great ace of the first World War. Aged twenty-two, Robert Alexander Little left a wife and child. Perhaps they found some consolation in the Distinguished Service Order and bar, Distinguished Service Cross and bar, and Croix de Guerre which had been awarded to him during his meteoric career.

Exactly the same decorations were bestowed upon Lieutenant Dallas from Mount Stanley, Queensland. He too enlisted in the RNAS, and in his Nieuport II *Bébé* he hacked his first kill from the sky on 12 May 1916. It was a two-seater Aviatik.

He notched up two more kills about a week later, and went on to a total of thirty-nine official victories. But the cryptic official note on Dallas' record says: "Did not always submit tally claims." So he may have more to his credit, but the facts will never be known. The friendly Queenslander known to his friends as "The Admiral" tackled three Fokker DrI triplanes, and lost.

The third-highest Australian score went to Captain A. H. Cobby of the AFC: thirty-two confirmed victories. Among other feats he led the patrol which on 21 March 1918 shot down four of Richthofen's "Flying Circus" in a four-minute dogfight.

And Cobby led the way in a different kind of aerial warfare. The Staff was at last beginning to realise the potential importance of aircraft in tactical and strategic bombing. In August 1918 he led 4 Squadron in what for those days was a "multi-bomber" raid: an attack by over 100 aircraft of 80th Wing on German aerodromes in the Lille area. It was during the great final Allied offensive of the war.

Earlier that year, when the Germans broke through the Allied front in their March offensive, Australian airmen of 3 Squadron had been engaged in "beating up" the advancing enemy columns. At first they were kept busy in attacking enemy aircraft, but when these had been cleared out of the way they bombed and machine-gunned the German infantry and transport floundering along the muddy, jampacked roads. As soon as each pilot had dropped his bombs and fired off his ammunition he returned to stock up again and within minutes was swooping down on the enemy once more—sometimes with his wheels almost touching their helmets.

In July 1916 the British War Office took a most unusual step: it actually invited Australians to join the Royal Flying Corps. Usually a man could not transfer from the Australian to the British forces or vice versa, but the British had been so impressed by the achievements of Australians in the RFC that they wanted more of them. It seems that the cocky and dashing attitude of the Australian servicemen, so much deplored by most British officers as being "undisciplined" was just what was needed for flying a fighter in the first World War.

Of the 197 men who applied for transfer from the Australian Imperial Force 183 were accepted. As well as these, a good many Australians went direct from civil life into the RFC.

Equipped with RE8 aircraft for artillery observation, 3 Squadron were the first to reach France, followed

by 2 Squadron with DH5 fighters and 4 Squadron with Sopwith Camels. Leading 2 Squadron was Major W. O. Watt, already a veteran because he had flown with the French Air Force as early as 1914.

All the Australians were in action in November and December 1917, and 2 Squadron was particularly heavily engaged as the Germans struck back after their defeat at Cambrai. They bombed and machine-gunned the Germans from treetop height.

The three top-scoring Australian aces of that war, Little, Dallas, and Cobby, all flew with incredible verve and brilliance—though Cobby records that when he went to the Gosport School of Special Flying for an instructor's course his eyes were opened by one Major Norman Brearley. He had thought the instructor's course would be easy, but, as he said in his book *High Adventure*, "I was sadly disillusioned. It took something like twenty-four hours to discover how little I knew about the accurate flying of an aeroplane . . . the manner in which he [Brearley] put the crate down into tiny fields surrounded by tall trees and got it out again was astonishing."

By that time Brearley had survived being shot down on the Western Front. He was shot through the lungs and crash-landed in no-man's-land, but after being invalided home to Australia he returned to Britain where he became an instructor. It was all part of his progress towards founding Australia's first airline: Western Australian Airways.

The names of many Australian flyers of the first World War are remembered only on war memorials. Others flew their stint, perhaps made their kill or two or perhaps never managed to catch a German in their sights before the war ended and they returned to civil life.

Others made their names on the fighting front, and then came back to Australia to make a further impact, however small, on the world of aviation. Lieutenant Len Taplin, for instance. A slow-speaking South Australian, he seemed to saunter through the war and accept the good with the bad. Flying a BE2a and then a Sopwith Camel, he notched up twelve kills. Home in Australia again he flew the airmail routes through the nor'-west of Western Australia, and eventually settled in Port Hedland long before the smell of iron-ore dust had settled on the town. He left WA Airways and virtually took over Port Hedland, becoming the town butcher, undertaker, and electricity supplier.

Captain Edgar Johnston from East Perth, Western Australia, was credited with nine-and-a-half victories. After the war he slipped naturally into the new Civil Aviation Branch of the Defence Department, founded by Colonel Brinsmead. When Brinsmead died Johnston succeeded him as controller.

Lieutenants Keith V. Anderson and Charles E. Kingsford Smith notched five kills apiece, but it

TOP: This Armstrong-Whitworth crashed in a tree while trying to land in mist in England, 1917 (*RAAF Museum*)

BOTTOM: An air training diagram of the first World War (*Hawker Aircraft Limited*)

was not only that which bound their destinies. They became close friends and after the war were two of the first pilots in Brearley's WA Airways. Charles Kingsford Smith had his first victory in a Spad VII on 14 July 1917. The next four came exactly within a month, and on 14 August he was shot down himself but escaped alive.

Horrie Miller went to France with 3 Squadron in August 1917, and in his book *Early Birds* he says, "Having a strong instinct for personal survival, I did what was required of me without taking unnecessary risks or finding anything glorious—except the sheer relief of coming out alive—in sending other airmen to their deaths. In fact, I had a sneaking regard for those brave, efficient, often gallant German pilots with whom we were obliged to engage in mortal combat and in this I was not alone."

This feeling certainly was displayed when Richthofen was shot down in 3 Squadron area and was given a hero's burial by Australian troops and airmen. On 16 and 21 March, 4 Squadron AFC already had tackled the "Flying Circus", and had shown that Australians were a match for these picked German flyers. The score for the first encounter was three Germans, one Australian; for the second, four Germans and no Australians.

From this distance it is easy to discuss the first war in the air as though it was a great sporting event, with all the fear and anguish departed, but there is no doubt that many of the pilots on both sides had a recklessness which gave it a sporting atmosphere. In a period when from your own slow-flying aircraft you could actually see the goggled features of your victim or assailant, and if he was flying a two-seater could see his observer standing up to shoot at you with a machine-gun sliding along a ringed track, there was a personal flavour which departed as aircraft began to fly higher and faster. There was more room for individualists—like Lieutenant A. King Cowper, an Australian with the RFC, who forced a Pfalz DIII to land intact on an Allied airfield. But then the Pfalz was really no match for Cowper's SE5a.

Apart from the men who came home to make subsequent careers in aviation, plenty of Australians made their names over the Western Front but after demobilisation returned to their peacetime occupations. Men like Captain Roby Louis Manuel, DFC, the young farmer from Kerang, Victoria, who ended the war as OC, C Flight, 2 Squadron AFC. With 43 and 80 Squadrons RAF the Australians, in SE5a fighters, were keeping the Germans effectively under control in their sector of the front, and under Major Murray Jones were trained in teamwork which defeated the Germans' "circus" tactics. By manoeuvring his flight as a unit Manuel was able to shoot down seven aircraft between 1 and 12 June 1918.

Sopwith aircraft of the first World War. TOP: Type B1 bomber. BOTTOM: The Sopwith Baby Seaplane used between 1909 and 1918 (*Hawker Aircraft Limited*)

The combat reports of such encounters leave much to the imagination. "While flying east over Bussy I saw two Pfalz scouts attacking a Camel. I dived on the one nearest to me, firing about 150 rounds at 100 yards range. Machine fell in flames near wood to E. of Bussy", was all he wrote after an action on 2 June.

Hundreds of combat reports of the AFC flyers are preserved in the War Diaries section of the Australian War Memorial. They make fascinating reading: typed out, presumably by an orderly, in the large-character typing of that era, with the ink in which they were signed now faded to a rusty brown. Some of the signatures are those of well-known flyers; many are by men whose names did not become known to the public.

On 16 September, in a dogfight near Lille, Manuel shot down a Fokker with his last ten rounds of ammunition at ten yards' range. In that fight, eleven SE5s took on fifteen Fokkers and destroyed seven of them.

With machines equipped with the new Wolseley

Top: Air training posters. Left: "Decoyed," showing an AFC machine trapped by three Germans. Right: "Looping" to escape an attack from the rear (*RAAF Museum*)

Bottom: This training poster dated May 1918 shows how aircraft in V-echelon could concentrate fire on a target. The caption is "Strafed" (*RAAF Museum*)

Viper engines, and improved gun-synchronising gear, the Australians were beginning to have it all their own way—though Len Taplin had a remarkable escape.

He crashed his Camel on take-off when she was loaded with a twenty-pound phosphorus bomb, a twenty-five-pound high-explosive bomb, hundreds of rounds of ammunition, a batch of Very light cartridges, and a full tank of petrol. The whole collection exploded, but Taplin escaped with no more than superficial burns. When the war ended, he had shot down fourteen German aircraft.

Led by Murray Jones, 2 Squadron pasted the Germans on the ground as well as in the air. On 16 and 17 August 1918 their total of nineteen machines joined the armada of sixty-five which bombed the German airfields at Harbourdin and Lomme, and put paid to the enemy forces commanded by von Leutzer and Bolle.

For some airmen over the Western Front life tended to be very short, but strangely enough they were safer in combat than they were in training. According to *Aircraft*, a death rate of sixty per cent in the training squadrons was not uncommon, whereas the death rate on active service was about thirty per cent.

TOP: "The Last Loop" warns against stunting after a dogfight, because the victor's machine might have been damaged in the combat (*RAAF Museum*)

BOTTOM: The training poster captioned "Trapped" advised caution when attacking an enemy kite balloon (*RAAF Museum*)

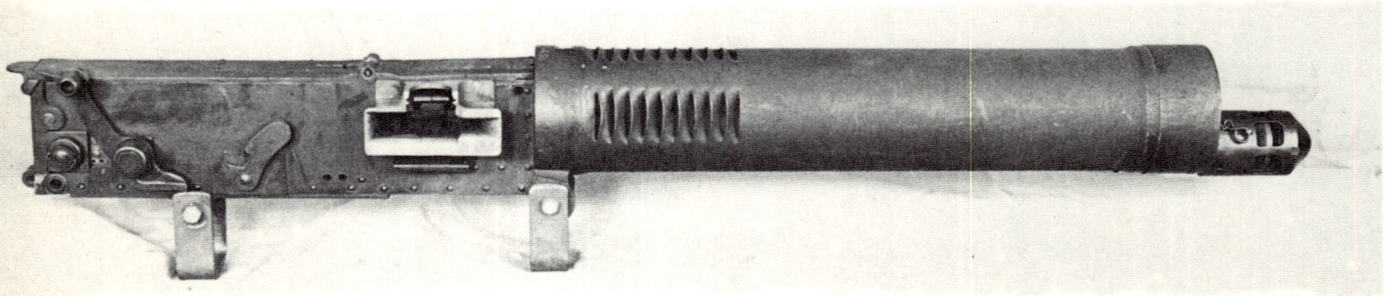

TOP: The body of Lieutenant Potts, killed in aerial combat, lies beside his RE8 after it crashed behind German lines (*RAAF Museum*)

CENTRE AND BOTTOM: .303 calibre Lewis aircooled and Vickers watercooled machineguns used on AFC aircraft during the first World War (*RAAF Museum*)

J. S. Bendrodt, a Canadian ex-pilot who settled in Brisbane after the war, implied that this heavy loss in training was due to poor supervision of the trainee pilots when they were not flying. He contrasted the British and Australian training loss with that of the Canadians, which was less than one per cent, and attributes this to rigorous physical conditioning, "Foreign Legion type" discipline, and careful class-room instruction. The result was shown by such Canadian aces as Bishop, Collishaw, Barker, and McElroy, who accounted for 258 enemy aircraft between them.

The early aircraft gave plenty of scope for individual ingenuity, especially from such men as Lieutenant L. J. Wackett who later became Australia's best-known aircraft designer. Wackett and Brearley both learned their trade in the old BE2c, for which few pilots had any kind words. Faced with such aircraft as the Fokker scouts and two-seater Aviatiks they were flying coffins—except that coffins have sides and the BE2cs did not. One of its great disadvantages was that its machine-gun was difficult to train on an attacker.

So Wackett, with 1 Squadron in Egypt, designed a new gun mounting for his BE2c and fitted it to the top centre-section of the wings. On the next bombing raid on Beersheba, with five lumbering BE2cs escorted by one Martinsyde G100, a Fokker and an Aviatik took off in expectation of another victory. They made the mistake of picking on

TOP: Lieutenant F. H. McNamara, 1 Squadron AFC, winning the Victoria Cross in Palestine, 20 March 1917. This incident is described in the text. (*painting by H. Septimus Power: Australian War Memorial*) BOTTOM: A Handley Page O/400 bomber, C.9681, was used to airlift supplies to the Bristol fighters of 1 Squadron AFC operating with T. E. Lawrence's Arab guerillas. It was flown by Captain Ross Smith. The painting shows a landing at Lawrence's camp at Um es Surab. The Arabs were amazed at the size of the O/400 and believed that the Bristol fighters were its foals. (*painting by Stuart Reid: Australian War Memorial*)

TOP: On 19 October 1918 two Bristol fighters of 1 Squadron AFC forced a German DFW CV reconnaissance biplane to land in the Palestine desert. Captain Ross Smith, flying Bristol fighter B.1229, with Lieutenant A. V. McCann, landed near the German and set her afire with a Very cartridge while her crew surrendered. *(painting by Stuart Reid: Australian War Memorial)* BOTTOM: The death of Captain Baron von Richthofen. Gunners Ridgway (right) and Langier, 29th Australian Field Artillery, were repairing telephone lines on a hill near Vaux-sur-Somme when they saw Richthofen's Fokker Dr I triplane struck by bullets fired from the position occupied by Sergeant C. Popkins, firing a post-mounted Vickers machinegun. *(painting by Keith Swain: Australian War Memorial)*

Wackett, who turned on them with withering effectiveness.

The war ended at last. As part of the British Army of Occupation, 4 Squadron stayed at Euskirchen, Germany, but most of the other Australian flyers were grounded and for months had nothing to do but wait for a troopship to carry them home.

Most of them heard about the prize of £10,000 offered by the Australian Government for the first flight from England to Australia. Most of them shrugged their shoulders. It was hard enough to fly the English Channel—and having come through the war in one piece they were not about to take any more risks.

But to some others, including the Smith brothers and Ray Parer and John McIntosh, it was a challenging temptation. In true Australian style they prepared to "give it a go".

The end of the war meant the sudden run-down of the powerful air forces built up by the combatants. At the end of 1918 Britain had about 5000 serviceable aircraft, but before long they had mostly been scrapped or sold at disposals prices.

TOP AND CENTRE: SE5a Training Scout used by the AFC in England and an Airco DH6 used for War Loan propaganda (*Australian War Memorial*)

BOTTOM: This DH6 used as a trainer at Point Cook came to an undignified end after the war, when it was used as an advertising vehicle (*RAAF Museum*)

The next few years would in any case render most of them obsolete.

The same applied to most of the men. The skills of thousands of trained flyers, all over the world, rusted into disuse because civil aviation simply was not ready for them. But for a select group of Australians the war had been a launching pad into lifetime careers in aviation.

This group included Lieutenant-Colonel Richard Williams, commander of 40th Wing of what was now the Royal Air Force. As 1 Squadron prepared to board the *Port Sydney* on 5 March 1919, bound for Australia and demobilisation, he told them, "The victory gained in Palestine and Syria has been one of the greatest in the war . . . you gained for us absolute supremacy in the air."

At thirty-one, Williams became Australia's first Chief of Air Staff. On his promotion to air marshal, Australia's first, he was knighted, and as Air Marshal Sir Richard Williams, KBE, CB, CBE, OBE, DSO, Order of El Nhada of the Hedjaz, he retired from the RAAF and in 1946 became Director-General of Civil Aviation, a post which he held until 1955.

Six more of Australia's airmen of the first World War have at various times been knighted for their services to aviation: Sir Ross Smith, Sir Hudson Fysh, Sir Lawrence Wackett, Sir Peter Drummond, Sir Norman Brearley, and Sir Charles Kingsford Smith. Their stories will form part of the developing history.

And, of course, there were the men who did not receive the formal recognition of knighthood but who received the acclaim of Australians in that glamorous period between the wars when flying was something very special. Men like Hinkler, Ulm, and Horrie Miller. Their stories also will be told.

In an indirect way, one Australian flyer's name became known to hundreds of thousands of people. He was Sidney Cotton, who invented the Sidcot flying suit. Countless folk who never got closer to a fighting aircraft than the early "Biggles" stories by Captain W. E. Johns, first published in the British boys' magazine *Modern Boy* in the late 1920s, have heard about Sidcot flying suits but few know anything about their origin or their inventor. One of them is still preserved in the Australian War Memorial.

Early flyers suffered badly from the cold. Usually they flew in their service uniforms with as much as they could cram beneath or on top of it in the way of underwear, jerseys, scarves, leather jackets, and so on. Flying helmets were developed quite early, but when a man was up in the open cockpit of an aircraft above the snow-clad Western Front it seemed impossible to prevent the draughts from whistling around his shrinking torso.

According to T. G. Jones, in an article in *Australian Flying* of November 1973, Sidney Cotton hit upon the idea of his flying suit almost by accident. The flyer, who was born near Proserpine, Queensland, in 1897, was serving with the Royal Naval Air Service in 1916. One particularly cold day he was working on his aircraft when the squadron was ordered into the air in what would later be called a "scramble". Cotton had no time to change out of his greasy overalls, and when he landed again he realised that he had not felt the cold as badly as usual. After a little thought he found the clue: the oil and grease on his overalls had kept his body heat inside them in much the same way that a "wetsuit" does for a modern skindiver.

He took the idea further by designing various types of flying suits and having them made up by a clothing manufacturer. The one which worked best was a fur-lined suit with an outer skin of waterproof fabric and an insulating layer of airproof silk. The idea was soon adapted by flyers of many nations and when von Richthofen was shot down he was wearing a British Sidcot suit.

Sidney Cotton survived the war and tried to settle down to farming in Australia, but the call of the air was too strong and he returned to the northern hemisphere. Between 1919 and his death in 1969 he made a name for himself as a "trailbreaker" of some Canadian air routes and a pioneer of aerial photography.

Few Australians even knew that Cotton was one of their countrymen. He was like another famous pioneer airman whom many Australians do not realise was one of their own: Harry George Hawker from Moorabbin, Victoria, the blacksmith's son who founded one of the world's most famous aircraft factories. It was to build the Hawker Hurricane which with the Spitfire smashed the German air armadas of 1940–41. Hawker formed the company but was never to see its progress towards such achievements as Hawker Siddeley, the British aerospace manufacturer; Hawker de Havilland Aviation; and the Hawker Siddeley Harrier, a vertical take-off and landing (VTOL) fighter of revolutionary capabilities. He died on 12 July 1921 in his Nieuport Goshawk G-EASK, but the ideal of world-beating aircraft which he pursued in the Sopwith factory, and which manifested itself in such fighters as the Sopwith Pups and Camels, still lives on.

For each of the Australian flyers of the first World War whose name became well known to the public for his exploits during or after the war, there were a score who became redoubtable combat pilots. They were a special breed of men, characterised by Captain Harry Busteed who, according to a writer in the British magazine *The Aeroplane* on 28 September 1938, "had a motto which runs, 'Get up every morning with a clear conscience. Look every man straight in the eye. And tell him to go to Hell.'"

3

The Great Pioneers

Harry Hawker, Lawrence Wackett, and others of their breed blazed aviation trails for Australia and the world

Many of the early flyers "learnt by the seats of their pants", using a special kind of instinct to find out just what an aircraft would and would not do. Some died in the process, but a number helped to blaze the aviation trails which were to be followed by others. Some, like Max Immelmann the German ace, had their discoveries named after them. "I did an Immelmann turn," a flyer of the first World War would say when describing a manoeuvre which had first been performed successfully by the German pilot.

Other flyers were not content with simply flying the aircraft. They had such a passionate involvement in the whole theory and practice of flight that they wanted to design and build their aircraft as well. Again, some succeeded and some died in the process. Harry Hawker did both.

Hawker was one of that group of Australians who went to England, then one of the few cradles of the aircraft industry, shortly before the first World War. They included Harry Busteed, Harry Kauper, Horrie Miller, and Bob Cousins. Of them all the name of Harry Hawker is best remembered today simply because his name prefixes the titles of the air industry giants Hawker Siddeley and Hawker de Havilland.

Kauper got a job in Tom Sopwith's factory at Kingston-upon-Thames and soon persuaded Sopwith to give Hawker a chance. As Horrie Miller says, "Hawker soon made it clear that he was to be more than just a cog in the wheel. Before long Sopwith had agreed to let him learn to fly and within seven weeks of his first flight he had won a trophy and £500 by breaking the all-British duration record in a Sopwith-Wright machine."

This was an early example of the Australian instinct for flying: the innate ability which made the British War Office anxious to add Australian flyers to the RFC.

Hawker's next attempt at a record-breaking flight ended in disaster. The aircraft was unable to take off with the load of fuel required for a duration attempt, and he crashed. This sent him to hospital for a fortnight, but when doctors told him that a recurrent spinal pain might be a warning of eventual paralysis he just shrugged it off. Like Cecil Rhodes, he had "so little time, so much to do".

He broke Geoffrey de Havilland's height record by 800 feet when he took Tom Sopwith's "passenger bus" up to 11,450 feet, and won another prize: £50. After this the London *Daily Mail* offered £10,000 for the first flyer to complete a circuit of Britain by seaplane. Sopwith supplied the seaplane, and Hawker took off with Harry Kauper as mechanic.

A flyer of those days had to take the full blast of engine noise and exhaust fumes, despite the rush of air past the open cockpit—or even the open seat on which he perched among the struts and wires of a kite-like fuselage. In this case, the exhaust fumes proved too much for Hawker and he had to abandon the round-Britain attempt.

Nevertheless he had established a new seaplane record: 1,043 miles in twenty-one hours and forty-four minutes. The *Daily Mail* gave him a consolation prize of £1,000.

Hawker now worked with the Sopwith design

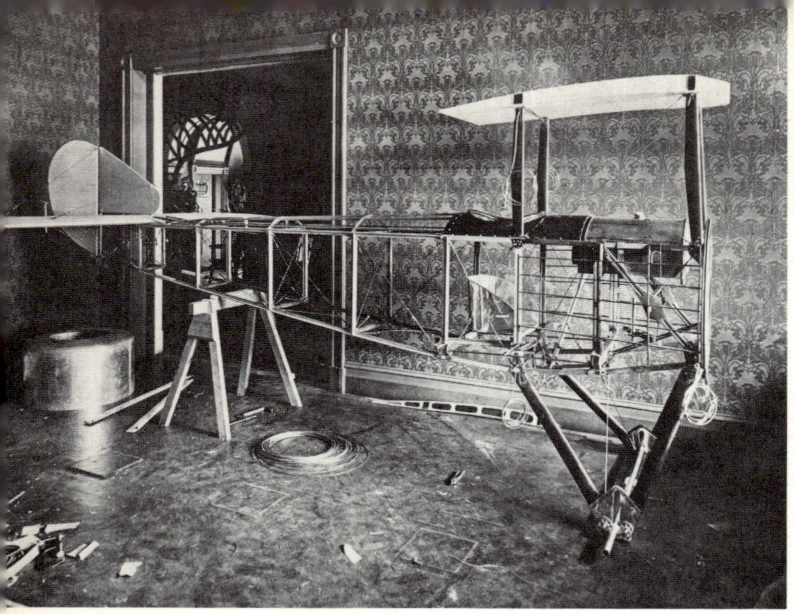

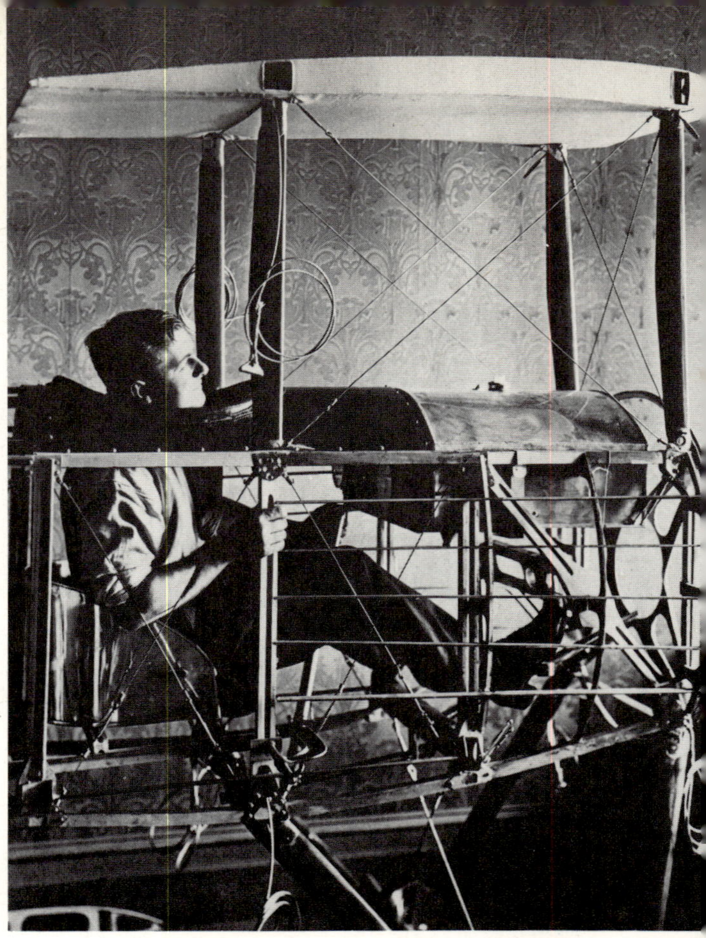

In 1916, Basil Watson of Melbourne built a biplane in his living room. He put the wings on in the garden and later got the machine off the ground, but crashed fatally at Point Cook on 28 March 1917 (*RAAF Museum*)

team in planning a new aircraft which they hoped would be adopted by the Royal Flying Corps: the Sopwith Tabloid. When work was finished, Hawker asked Sopwith for permission to take a Tabloid back to Australia to demonstrate to the Commonwealth Government.

Horrie Miller says that "Sopwith agreed, realising that it would advertise his aircraft and gratify Hawker's wish to show his folks how far he had come in the aviation world. In fact to disagree would probably have lost Sopwith his most valuable man."

The Sopwith flights in Australia were unsuccessful as a sales pitch but attracted enormous attention. Hawker put on a spectacular show at Caulfield Racecourse but the field was only just large enough. When he landed, the aircraft came to a stop only just before hitting the fence—and Hawker's uncle, rushing to help, received a bloody nose from the Tabloid.

Hawker returned to England, and on his first day back found that Harry Barnwell from the Bristol Aircraft factory was "looping the loop" in a Tabloid. This manoeuvre had only recently been "discovered" by the Russian pilot, Nesterov. Hawker immediately climbed into a Tabloid and took her up for a perfect loop. He was the darling of the crowd.

Mascot Aerodrome in the 1920s. The aircraft in foreground is a DH9. Above her are an American Eagle, a modified Curtiss Jenny, a Curtiss Meteor and a DH60 (*McGrath Collection*)

When the first World War broke out, Hawker's back injury barred him from entry into the Royal Flying Corps. For the British aircraft industry this was a blessing in disguise. He was far more valuable as a designer, test pilot, and aviation innovator than as another pilot over the Western Front . . . although his work for the Sopwith factory was far more gruelling and just as dangerous as the life of a combat pilot.

He became chief test pilot for Sopwiths, but to begin with the machines which they developed were too sophisticated for the raw youngsters being trained as pilots. Training losses from crashes caused by pilot error were staggering. So many of the new aircraft had to be tucked away in hangars until enough skilled pilots were available to fly them.

Aircraft from the Sopwith stable played a major part in gaining Allied air superiority in the wartime skies. In the late winter of 1918 an air armada swept across southern England in an awesome display of manoeuvres to celebrate the Armistice. They were followed by a lone civilian flyer, Harry Hawker, and his performance made the superb RAF display look like an amateur show.

With the war over, flyers began to look for new challenges. One was the England–Australia flight. The other, closer to home so far as the British were concerned, was the trans-Atlantic flight. As a matter of pride, the Americans had to be beaten to this feat.

Naturally, Tom Sopwith chose Hawker for the attempt. It was to be made in a biplane powered by a single 350-hp Rolls-Royce Eagle engine—and Hawker came up with the idea of dropping the undercarriage to eliminate weight and drag. Thus the idea of the retractable undercarriage was born.

With Lieutenant-Commander Mackenzie Grieve as his navigator, Hawker took the aircraft to Newfoundland to make the flight in the west–east direction because that is the direction of the prevailing Atlantic winds. They had to carry 400 gallons of fuel, so the aircraft had to be coaxed into the air but took off sweetly and reached 12,000 feet by nightfall.

They droned steadily along over the black Atlantic, but even for a Rolls-Royce engine the hours of continuous flying were too much. It began to overheat, and Hawker passed a note to Mackenzie Grieve telling him this and saying, "May have to alter course to shipping lanes. Have course ready."

Hawker then began some hair-raising attempts to cool the engine, by cutting it off, gliding downwards, switching on again, and climbing once more. But each time he did this he lost height which could not be regained, and when dawn broke they were uncomfortably close to the cold grey Atlantic.

But luck was with them and Hawker could see a wisp of smoke and then the masts and funnel of a small tramp steamer. He steered towards her, and saw crewmen racing to clear away a lifeboat just before the aircraft splashed down.

The ship was the *Mary,* a small freighter not equipped with wireless. So the world thought that Hawker and his navigator had been lost at sea, and when the *Mary* was able to signal to a lighthouse that she had the two aviators aboard Britain went wild with delight.

When the two men reached London they were mobbed by excited crowds. Tom Sopwith tried to drive them through in his Rolls-Royce, but the crowd climbed on to the car and he had to stop. The only way that Hawker could escape was by climbing on to the back of a policeman's horse.

Even though Hawker had not flown the Atlantic, he had got more than halfway. The *Daily Mail* gave him £5,000, and he was awarded the Air Force Cross.

After this exploit, Sopwith, Hawker, and F. Sigrist decided to exploit Hawker's fame and prestige. They formed the H. G. Hawker Engineering Company, and it was this company which became the Hawker Siddeley Group.

But by this time the effects of Hawker's first crash were beginning to cripple him. In July 1921 he was testing a racing aircraft at Hendon. Suddenly the aircraft slammed into the ground at full speed, and it is thought that his disability may have caused him to lose control.

Ground crews rushed to the wreckage in which Hawker lay. He was shockingly injured and barely conscious. His last words were, "Did I get her down all right?"

Back in Australia, another flyer was experimenting with new aircraft. He was Lawrence James Wackett, a Queenslander who had been one of the earliest cadets to graduate from the Royal Military College at Duntroon. Early in the first World War he had transferred to the Australian Flying Corps and had won his wings in the frail Bristol Box-kite trainers at Point Cook.

As a subaltern pilot he flew with 1 Squadron AFC over the deserts and date palms of the Middle East, and gained a reputation for innovation. This soon led him into meetings with staff officers in Egypt, and he was given permission to go to Britain. Promoted to captain he visited Royal Flying Corps establishments to learn as much as he could about aircraft design and manufacture. In 1918 he went to France to join the Australian forces then being prepared by General Monash for the battle of Hamel which was to be fought on 4 July. In his memoirs *Aircraft Pioneer* he recalls that "General Monash had heard of me from Duntrooners and sent for me to say he was planning offensive operations and

TOP: Oswald Watt served first as an officer in the French Air Force, then transferred to the AFC. The Oswald Watt Medal is now prized by Australian aviators (*RAAF Museum*)

BOTTOM: The Avro 504K was widely used in both peace and war (*Hawker Aircraft Limited*)

wished to reduce casualties by having ammunition conveyed by air to troops in the front line during battle."

A few days later Wackett came up with the answer: a device by which aircraft could drop forty-pound boxes of machine-gun ammunition to the advancing troops. Wackett was paid £300 as a "reward to inventors".

He ended the first World War as a major with an Air Force Cross, a Distinguished Flying Cross, and a Mention in Despatches. But like many regular soldiers he soon had to drop in rank, and as Lieutenant Wackett was given command of an almost non-existent military base at Townsville, his old home town.

With his new wife he could have settled down to the undemanding life of a soldier between the two World Wars. But soon he was plucked from obscurity by Lieutenant-Colonel Richard Williams who was re-forming the old AFC into the Royal Australian Air Force. Wackett, with all his experience in flying and modifying aircraft, and with a keenly inventive mind, seemed to Williams to be the ideal man to maintain Australia's military aircraft at Point Cook.

This was still not enough for Wackett. He devoted every moment of his spare time to study, and within a year had a Bachelor of Science degree. During this period he was developing the belief from which he never wavered: that Australia should plan and build her own aircraft. It is a belief for which he has had to fight a lifelong battle; it has not yet been won. Today's aircraft industry has been plagued by the stupidity of politicians, the vested interests of various people, and the ignorance of those who just do not want to know.

Wackett was to encounter those who apparently were happy to see Australia devoid of a viable aircraft industry in times of military crisis. Others were content to buy only foreign machines, often unsuitable to local needs. Many could not see the economic benefits of the industry to Australia.

To begin with, however, Wackett's enthusiasm met with equal keenness on the part of the Government. When he heard that a war-surplus workshop in the Sydney suburb of Randwick was to be sold, he sent a direct request to the Minister of Defence for permission to convert it into an experimental station. Permission was granted and Wackett built up a workforce of civilians and airmen. He set them to work on repairing air force aircraft, and also on a design of his own.

This was the Widgeon I, a seaplane. He took the completed Widgeon on to Botany Bay for a trial run, but as she began to taxi along the surface, her nose dipped beneath the swell and she sank. However, this was more the result of bad luck than bad design and Wackett rebuilt the aircraft. He took her up and gave a perfect performance, and recalls that "Congratulations came from all over Australia. At last I was an aeronautical engineer in embryo. Nothing succeeds like success. Soon I was authorised to improve the Widgeon by installing a larger, more powerful engine and to develop an amphibian retractable landing wheel undercarriage. I was also authorised to proceed with the design and construction of Widgeon II, a more advanced amphibian."

The Randwick Experimental Station now had a staff of eighty, and Wackett and his family lived there in one of the two attached cottages. When some RAF flying-boats visited Australia, Wackett

An Avro 504K at Point Cook shortly after the First World War (*RAAF Museum*)

Top: A 1921 photograph of a DH9a with its characteristic "motor car" radiator (*RAAF Museum*)

Centre left: SE5s and DH9s were among the aircraft given by Britain to Australia after the first World War as a nucleus of the RAAF (*McGrath Collection*)

Centre right: Lawrence Wackett designed and built the Warbler, with a three-and-a-half h.p. engine, to enter the 1924 Light Plane Competition (*H. C. Miller*)

Bottom: The Wackett Widgeon II amphibian flew well until it crashed in Hobsons Bay and killed the crew of three (*RAAF Museum*)

Sir Hudson Fysh, co-founder of Qantas, in the pilot's garb he wore in 1920 when he flew the first Qantas aircraft: an Avro 504K. The vintage biplane (above) was restored by Qantas personnel in Sydney as a replica of the first Qantas passenger aircraft. The replica, on permanent loan from the Australian War Memorial, will eventually become the centrepiece of the Qantas Aviation Museum in Sydney. (*Qantas Airways photograph*)

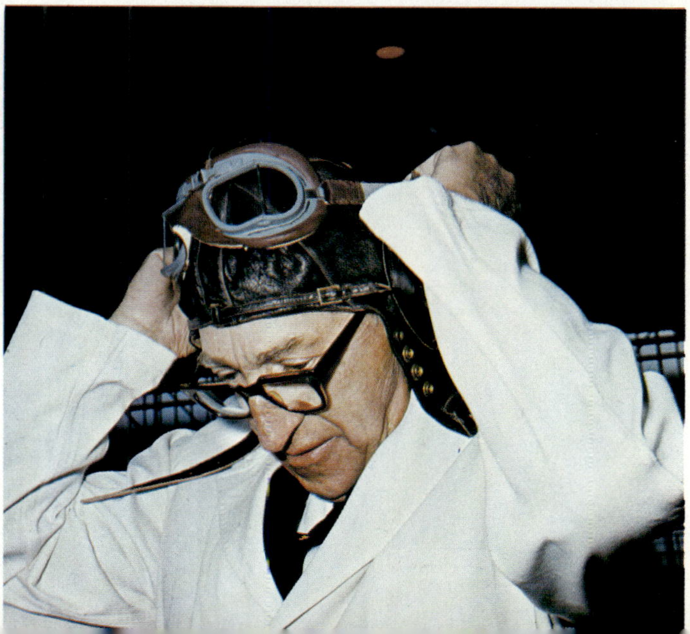

On 17 December 1917, Lieutenant L. J. Sandy, Sergeant H. F. Hughes as observer, was flying an RE8, A.3816, of 3 Squadron Australian Flying Corps, on a corps reconnaissance mission over the Western Front. The RE8 was attacked by six Albatros DVa scouts, one of which was damaged in the fight and forced to land. It was captured by AIF troops and is now on display in the Australian War Memorial. An armour-piercing bullet fired by one of the German scouts passed through Hughes' left lung and into Sandy's head and killed both men instantly. The RE8 continued to fly itself until it crash-landed in a field about fifty miles from the scene of aerial combat. (*painting by Keith Swain: Australian War Memorial*)

A Vought-Sikorsky OS2U-3 Kingfisher seaplane of the RAAF on patrol over HMAS *Hobart* (foreground) and HMAS *Australia* in September 1942. (*painting by Dennis Adams: Australian War Memorial*)

in his Widgeon II joined them in Darwin and flew with them down the coast of Western Australia. He returned home across the desert. He had learned much from designing, building, and flying his new aircraft, and his aeronautical skill and inventiveness grew rapidly.

He made many friends but an occasional enemy. C. G. Grey, that notable editor of Britain's famous magazine *The Aeroplane*, had kind words to say about Norman Brearley and Hudson Fysh who were then running Australia's first airlines. But, Wackett says, "Every now and then *The Aeroplane* published a venomous article condemning my work." As Wackett points out, Brearley and Fysh bought British aircraft whereas he was proving that Australia could build her own.

Wackett is a big man with a thrusting chin and broad shoulders, and with great self-assurance: not the type of man to suffer fools gladly. Maybe this was why, when the Australian economy began to falter in the late 1920s, the Randwick Experimental Station was eliminated by the Treasury. In protest, Wackett resigned. The decision cost him heavily, and he began to design racing boats in order to keep his family. But he could not abandon Australian aircraft so easily, and soon came up with a proposal to work on aircraft at the Cockatoo Dockyard in Sydney. Richard Williams, then Chief of Air Staff, retained his confidence in Wackett and backed him by sending some RAAF overhaul work to the new establishment. It lasted for three years until the Great Depression was at its worst. Cockatoo Dockyard was leased to a firm of gelatine makers, and Wackett like millions of others lost his job. But he soon found work with a small airline and at home worked late at night on designs for a new passenger aircraft. Eventually a German, whom Wackett does not name, financed the Tugan Aircraft Company and Wackett soon built an aircraft which was bought by the RAAF.

By 1933 Wackett was Aeronautical Superintendent of C. G. Robinson's New England Airways. It was the period in which some people were beginning to wake up to the fact that the Germans, Italians, and Japanese were not just playing with words but were building and equipping huge military forces as fast as they could. When the inevitable war broke out the spearhead of these forces would be aircraft. In Abyssinia (now Ethiopia) and then in Spain the devastating power of modern aircraft was demonstrated for anyone who wanted to learn the lesson.

In Australia, some men could at least glimpse the threat of the future. Wackett recalls that "The Minister of Defence began to make enquiries as to whether aeroplanes could be made in Australia. Someone suggested the name of Wackett, and I was sent for." This meeting was the genesis of the Commonwealth Aircraft Corporation Pty Ltd, which

The DH60 Cirrus Moth as a seaplane (*RAAF Museum*) and as a wheeled aircraft (*McGrath Collection*). The latter flew from 1926 well into the 1940s in Australia

was to become an entity greater than the sum of its many members, but was also to suffer from political indifference.

The initial planning led to a meeting between Wackett and L. J. Hartnett, head of General Motors-Holden's Ltd Australia, in late 1935. Richard Williams had persuaded Hartnett to talk with Wackett. He was impressed, and immediately organised further meetings with Essington Lewis and Frank Mitchell, the chief general manager and secretary of The Broken Hill Proprietary Co. Ltd; Harold G. Darling from the same organisation; and Sir Colin Fraser and M. L. Baillieu from Broken Hill Associated Smelters.

They agreed to form a consortium to buy the Tugan Aircraft Company and provide "massive" finance. Wackett and a "small mission" would go to Britain, Germany, Czechoslovakia, and the United States to see what the competition was doing. Wackett returned pessimistic about Germany's intentions but full of admiration for her aircraft industry. He also came home with a licence to build the North American NA16 two-seater monoplane, an ideal starter for Australia's new large-scale aircraft industry. It began a long association between North American and the firm which became known as CAC, of which Wackett was manager, Harold

Darling was first chairman, and Essington Lewis and the others were on the board. T. W. Air, who was to contribute much to the Australian aircraft industry, joined Wackett as the CAC factory was being built at Melbourne.

Williams and the RAAF did all that they could to encourage the new venture, but bureaucrats used delaying tactics against the import of essential machine tools and other equipment. Like the British aircraft manufacturers, they believed that Australia should buy all her aircraft from the "mother-country".

In his efforts to get Australian-built aircraft off the ground, Wackett spent much time in the United States. He shuttled between Los Angeles, the home of North American, and East Hartford, Connecticut, home of the famous Pratt and Whitney aircraft engines. His unstinted efforts rapidly brought CAC to the stage in which Pratt and Whitney radial engines were being forged and machined in Melbourne. They were the nine-cylinder, 600-hp Wasp S1H1-G engines, whose sound still snarls in the memory of many ex-RAAF pilots. Only one engine a week was produced in those early days, but eventually single-row and twin-row Wasps glistened in long rows on the CAC assembly lines.

Two new shareholders were providing finance: Imperial Chemical Industries of Australia and New Zealand, and the Orient Steam Navigation Company. The paid-up capital was £1 million, of which General Motors-Holden's had subscribed ten per cent. Later it sold its interest to Rolls-Royce of England.

The first Australian-built NA16 was test-flown on 27 March 1939, less than six months before Hitler marched into Poland. The modified NA16, known as the Wirraway, was to see much service with the RAAF.

Lawrence Wackett's achievements were eventually recognised with a knighthood, but in October 1970 the irony of fate struck him down. This man who had flown so many air miles in test aircraft slipped from the roof of his house, and the fall of only a few feet left him a quadraplegic.

He was not the only Australian to dream of a national aircraft industry—and to strive mightily to make his dream come true. There must be others who for one reason or another, especially finance, have made little impact upon history, and certainly there is one whose name is little known by Australians. He is Nigel Borland Love, who was born at Kurrajong, New South Wales, in 1892.

His father John Love was a gentle student of the classics, but a prosperous merchant for all that, and he schooled young Nigel in the family business. In 1913 Nigel made a business trip to England, and returned to Australia to join the flood of youngsters yearning to serve the Empire. He joined the AFC and was taught to fly at the Richmond Flying School, near Sydney: a State institution apparently having no connection with the Central Flying School at Point Cook. Its graduates were awarded civil pilots' certificates, and Love's was one of the first twenty. No. 1 had gone to Billy Hart, some years before.

Love was taught to fly by Billy Stutt, an Australian who had been a test pilot for the Royal Aircraft Establishment at Farnborough. Love describes him as "a splendid chap and a born pilot".

The first World War gave Love a couple of years of extended flying training and experience in both Australia and Britain, and when the Armistice was declared he had determined to make a future in aviation. Love, in the wait to be repatriated to Australia, opened negotiations with A. V. Roe of Manchester for the manufacture and sale of Avro machines in Australia. The agreement with Roe provided for Love to establish an assembly factory in Sydney, create an airfield, promote the sale of aircraft in Australia, and further civil aviation as an industry. To help him in these aims he had an invaluable assistant: Harry E. Broadsmith, who had been Avro's chief engineer since 1912.

The homeward journey, aboard the P & O liner *Kaiser-I-Hind*, must have been an impatient one. As soon as Love landed he began driving around Sydney in search of an ideal airfield. Finally he called on the real estate firm of Raine and Horne in Pitt Street, Sydney. They asked him if he had thought of "the bullock paddock" at Mascot.

Love says, "I replied that I had never heard of it, so there and then we arranged an inspection, and lo and behold here it was a readymade aerodrome and the real answer to a maiden's prayer."

The ground was so flat that Love thought it must be reclaimed land. To the south lay Cooks River with what in those days were grassy green banks. Across the river were Bonnie Doon Golflinks. Ascot Racecourse ran along the northern edges of the "bullock paddock" and "Chinamen's Gardens" lay to the east. All these areas gave clear approaches, and in another direction lay the broad stretches of Botany Bay.

Love's immediate decision to sign the lease meant that Mascot airfield was born. It was his private airfield until 1923.

With his partners Harry Broadsmith and W. J. Warneford, Love registered the Australian Aircraft & Engineering Company Limited at 12 Bridge Street, Sydney, in late 1919. The address was that of Love's father's business.

Love soon became one of the best-known aviators in Sydney and his name appeared frequently in the *Sydney Morning Herald.* When Ross and Keith Smith were in Australia on the last leg of their epic England to Australia flight, Love sent a telegram to them at Cloncurry, Queensland, saying, "Should you decide to honour Sydney with a visit on your way south

TOP: The DH50 was one of the "workhorses" of Australian aviation in the 1920s and 1930s. This photograph shows one as an RAAF seaplane in about 1923 (*RAAF Museum*)

BOTTOM: A civilian DH50 of the same era, photographed at Point Cook in 1925 (*RAAF Museum*)

[to Melbourne] this company wishes to extend an invitation to land at its aerodrome at Mascot, position extreme northern corner of Botany Bay landing ground marked with a white circle."

After a number of delays the Smith brothers flew in on 11 February 1920. In his Avro 504K biplane, Love flew out to escort them in from over the Blue Mountains to Mascot.

In April 1920 Nigel Love carried the first paying passenger on the Sydney–Melbourne run. He was John Gibson, of Foy & Gibson, the big Melbourne department store. On 18 April the *Sunday Times* reported, "The total flying time was nine hours although the trip occupied some days. This was the first flight of a commercial character which has been undertaken between the two capital cities."

Pioneer airliner. This eight-passenger ANEC111 biplane, powered by a single Rolls Royce Eagle engine, was redesigned from the unsuccessful Martinsyde Monoplane and used on the Melbourne to Hay service in the 1920s (*McGrath Collection*)

However, it was not the first paying flight carried out by the new company—although they did not cash in on the proceeds. In March, Harry Broadsmith had fitted an Avro with floats and moored it on the Cooks River. He made frequent flights to Manly to take up joyriders, and raised funds for the Manly District Hospital.

No doubt it was a useful piece of public relations in making Australians conscious of the potential of air transport. Love also undertook a number of extensive country trips to promote aviation.

In June 1921 the company began work on Australia's first commercial aircraft building project: six Avros for the air force. The Sydney *Daily Telegraph* said "it is a fact of tremendous significance" and quoted Nigel Love as saying that the first machine would be ready in about four months. The company's work was checked by the Director of Arsenals.

The official maiden flight of the first Avro 504K built for the air force was a great event. It was attended by Dame Mary Hughes, wife of Prime Minister "Billy" Hughes. The crowd watched Love put the machine through such a startling display of aerobatics over Mascot that his fiancée's mother thought she would lose her prospective son-in-law before the wedding.

But Love did not receive much more Government support after delivery of the six Avros, even though the Australian Aircraft & Engineering Co. had designed and built a new five-seater commercial aircraft under Broadsmith's direction and took it up on successful test flights. Love was ahead of his time, and when he was forced into liquidation in 1923 the Government paid him a pittance for his stock and equipment, including the new five-seater. He is one major Australian aviator whose life has gone largely unrecorded, and it is certain that most of those who now pass through Mascot airfield have never heard of him.

Another airman who was well known in the period between the wars, but now is little remembered, was Captain F. S. Briggs. As an RFC pilot he served from 1916 to 1918, and after the Armistice was given the job of flying various delegates from one part of Europe to another as they hammered out the Peace Treaty. He flew the first draft from Paris to London for signature by King George V.

Back in Australia he made a number of "first flights": from Sydney to Melbourne in one day, Brisbane to Melbourne in one day, Melbourne to Adelaide and return in one day. First flights from Melbourne to Perth and back to Melbourne and Sydney; first flight from Melbourne to Alice Springs and back.

He carried the first intercity airmails Sydney–Melbourne–Adelaide, and altogether flew sixty-seven different types of aircraft, logging 4,000 hours in the air before joining Shell Company as an Aviation Officer. His memoirs, published in 1935 under the title *Joysticks and Fiddlesticks*, give an exciting account of the great aviation era in which he played an important part.

4

Barnstormers and the Great Air Race

Young men back from "the war to end war" laid the foundations of Australian civil aviation. Some of them became world-famous in the process

When the first World War ended a good many young Australian flyers tried to find some future in aviation. Apart from anything else, it was difficult for them to settle down to a nine to five routine after they had known the thrill of hurling their tiny aircraft over the mud of the Western Front or the deserts of the Middle East—with death always waiting for the unwary or unskilled pilot.

Some of these pilots picked up an aircraft or two when the Government's auctioneers were getting rid of the enormous accumulation of wartime equipment. The machines went virtually for pennies. All four squadrons of the Australian Flying Corps were disbanded in 1919, and most of their aircraft were auctioned at Point Cook. Hopeful little aircraft companies operated by ex-servicemen sprang up in empty paddocks. Most of them foundered within weeks. The average Australian simply did not fancy the idea of trusting himself to roaring, rickety little machines flown by devil-may-care youngsters. But a few of these pioneers of Australian civil aviation did succeed.

In 1919 Captain Harry Butler flew from Adelaide across St Vincent Gulf to Minlaton on the Yorke Peninsula, carrying the State's first airmail. He died a couple of years later in an air crash at Adelaide, but his tiny aircraft "The Red Devil" is preserved at Minlaton in a memorial building containing other relics of his life.

Also in 1919 Captain H. N. Wrigley and Sergeant A. W. Murphy made the first south-to-north crossing of Australia by air. They took off from Point Cook on 16 November in their BE2e and landed in Darwin on 12 December. The purpose of the flight was to survey the route to be followed by competitors for the prize offered by the Australian Government for the first flight from England to Australia.

Wrigley and Murphy barely completed the survey in time. In fact, Keith and Ross Smith drifted their twin-engined Vickers Vimy bomber on to the flats near Darwin on 10 December, two days before the survey ended.

Major Norman Brearley in Western Australia was steadily forging ahead with his own project. Despite the entreaties of "Boom" Trenchard, father of the RAF, to stay with the new service, Brearley had gone home to Perth to prepare for the arrival of two Avro 504J biplanes which he had ordered in London. In Perth in 1919 he was a war hero and local-boy-made-good. And as a handsome, dapper young man with a beautiful wife on his arm, he looked the part. This hero's welcome was all very well for a while, but he wasn't one to rest on his laurels. When the crates containing his biplanes were delivered to a Perth warehouse, he and a Danish-born technical warrant-officer, Peter Hansen, enthusiastically began to assemble the two Avros.

Even at this stage he began to show his essential difference from the would-be civil aviators. Some of them were binding their kites together with baling wire but Brearley was paying particular attention to detail. And he was one of the first to consider insuring his aircraft. He had taken care of that in London and even the West Australian Insurance Company knew about his superb flying record. They said, "We have taken into consideration the

TOP: Sergeant A. W. Murphy and Captain H. N. Wrigley (*A.W.A. Ltd*). BOTTOM: Brearley's Avro 504J being towed to Carnarvon racecourse for assembly and take off (*Sir Norman Brearley*)

very excellent record we have before us of your flying experience." They charged him £25 for a three months' cover of £1,000.

While Brearley was fitting his Avros together two other young airmen were contemplating the possibilities of an airline which would speed mail, freight, and passengers at such astonishing speeds as 100 m.p.h. above the Australian bush. They were Paul McGinness and Hudson Fysh, away on the other side of the continent from Brearley. Their problem was finance, and there seemed no way to solve it until they met Fergus McMaster in a Brisbane hotel. McMaster was a grazier from the Cloncurry area, and McGinness had once helped him when his car was in trouble. Maybe Fysh and McGinness had been wondering about some way in which they could make use of McMaster's contacts in the world of ready cash, because when they met him in Brisbane they outlined their proposals for an airline. They spoke convincingly, and McMaster agreed to invest in the new company and to hunt around for other investors.

Soon enough cash had flowed in for Fysh and McGinness to order two Avro 504 Dyak biplanes from Nigel Love in Sydney. They received only one of them because Love was having production problems, but they picked up a spare aircraft closer to home. A stock agent at Longreach had just bought a BE2e and he agreed to sell it to them. Their first name for the new company was the Western Queensland Auto Aerial Service Limited, which was rapidly changed to Australian Trans-continental Aerial Services, and finally to Queensland And Northern Territory Aerial Services Limited—now, of course, known as Qantas.

But none of these grand titles meant very much until the company could begin its first regular air service, and that did not happen until 2 November 1922 when McGinness flew an Armstrong Whitworth FK8 from Longreach to Charleville. Next day, the first fare-paying passenger climbed aboard to fly from Longreach to Cloncurry. He was eighty-one-year-old Alexander Kennedy. But while Qantas was being established, Brearley's pilots were already making the first flights of WA Airways, which in 1921 won Australia's first airmail contract. Other men were also flying, with varying success. Some were simply aviation addicts to whom the smell of burning castor oil used as an engine lubricant in those days, and of the "dope" made of celluloid dissolved in acetate used for treating aircraft fabric, was truly intoxicating. A light came into their eyes when they saw the graceless constructions of taut fabric, struts, and wires, or when they heard the thin growl of a Gnome rotary engine. For them, comfort lay in being able to wear a cotton overall soaked with oil or in feeling the cracked leather of their flying-helmet straps flutter against their chins as they squinted out into the slipstream.

Such men as these were still barnstorming around the country towns, and making little more than enough to keep their aircraft flying. Others were notching up new records, like Briggs and de Garis who on 9 July 1920 flew from Sydney to Melbourne and back in one day in a Boulton Paul P9. Records were being scored with remarkable speed, if only because there was a clean slate on which to write them. A DH4 made the first east–west flight in November 1920, and flew home in a respectable twenty-one-and-a-half hours' flying time.

But the fun was beginning to grow a little stale, especially when the Government passed the Air Navigation Act in December 1920. This cut a swathe into the ranks of the barnstormers because it demanded that pilots, engineers, and aircraft should be licensed. A number of the old-timers and their aircraft failed to make the grade, but forty-five pilots, 142 ground engineers, and thirty-eight aircraft were registered. For a number of years the aircraft were registered in British style with the prefix "G". Pilot licence No. 1 was never allotted, but No. 2 went to Norman Brearley.

Despite the new regulations many of the flyers

Top left: One of the six Bristol aircraft used to inaugurate West Australian Airways, the nation's first regular airline service, in 1921 (*Sir Norman Brearley*)

Top right: Major Brearley's first aircraft at the unofficial airstrip at Belmont, near Perth, in 1919 (*Sir Norman Brearley*)

Centre: The first Qantas aircraft, an Avro 504K flown from Sydney to Longreach by P. J. McGinness in January 1921 (*Qantas Airways photograph*)

Bottom: The second Qantas aircraft, an RAF war surplus BE2E purchased by Charles Knight of Longreach in 1921 and later sold to the airline (*Qantas Airways photograph*)

Major Brearley (centre) with the pilots who flew Australia's first scheduled air services. Brearley's West Australian Airways flew northwards from Perth. From left: Charles Kingsford Smith, Bob Fawcett, Len Taplin, Val Abbott (*Sir Norman Brearley*)

were still nonconformists and individualists. Men like Syd Marshall, of Bankstown, near Sydney. If you coaxed him, Syd would tell you how he fell out of a Curtiss Jenny over Mascot in 1927 or 1928, and wait for the laughter of disbelief to die away before he continued.

He was out at Mascot in the days when grass still grew on the airfield and smog-free breezes blew off Botany Bay. A friend asked Syd if he'd like to fly up front while he took the Jenny for a flip, so Syd slipped into the open front cockpit and felt his usual pleasure at the sight of the struts and taut wires which held the wings in harmony. Perhaps this is why he felt no urge to fasten the safety harness.

The Jenny took off sweetly and climbed high above the widely-spaced houses of south-eastern Sydney. The air bit sharply into his exposed cheeks; he felt alive.

The pilot did a few stall-turns, then pushed the stick forward. Wind hummed through the wires. Then the stick went right back into the pilot's stomach, and Syd's own stomach slewed towards his toes. God, no, not a loop. With no harness. He saw sky, tilting clouds, then lurching, sickening, unreal brown horizon. He grasped frantically for something to hold, but there was nothing. His fingers slipped off the small ridge inside the cockpit. And the loop was not perfect: the Jenny hung almost silently upside down, instead of nosing over immediately.

Marshall began to fall. He could feel the seat of his pants lift from the warm comfort of the cushioned seat. It was unreal, but he was falling from the Jenny. He hit something: the inverted upper wing. He grasped at the wires and a strut as the pilot coaxed the nose over. The aircraft began to level off, and Marshall tumbled back into the cockpit.

If Marshall had fallen to the ground instead of into the cockpit he would not have lived to slog his way through the mountains of New Guinea and play about with aircraft in the pioneering days of that island. Nor would he have been one of the last to wave farewell to Amelia Earhart and listen to her last crackling wireless messages. Nor would he have returned to Sydney to become a kind of stubble-chinned demigod to aircraft buffs who admired his cheeky disregard for authority and his life of adventure. He was living aviation history: a man who had actually seen Houdini and Hawker and had known Billy Hart.

Marshall's tales are priceless and yet he is seldom

Left top: P. J. McGinness with an aerodrome Inspector at Longreach in 1921, in front of the Qantas Avro 504K (*Qantas Airways photograph*)

Left bottom: Qantas bought this Avro Triplane for airline service but she proved unairworthy and was broken up before going into service (*Qantas Airways photograph*)

Right top: Brearley and Len Taplin watch as Australia's first scheduled airmail freight is entered on the aircraft manifest, 1921 (*Engineering Division, PMG's Department, Western Australia*)

Right bottom: Arthur Kennedy, a bearded eighty-five-year-old pioneer, was the first Qantas passenger. He flew from Longreach to Cloncurry on 3 November 1922. Here, on a later trip, he sits in the rear cockpit while his baggage is loaded (*Qantas Airways photograph*)

Below: This group posed at Longreach when the Armstrong-Whitworth FK8 arrived on the inaugural flight of the Qantas airmail service in 1922, on the Charleville-Cloncurry route (*Qantas Airways photograph*)

referred to in the chronicles of Australia's early history. It has always been an honour to be invited to sit down in his Bankstown hangar while he or his mates boiled the billy, and to look around at his collection of veteran aircraft and other souvenirs. There was an old silver de Havilland Puss Moth, a shining Lockheed 10 Electra of pre-war vintage, a Bf 109G Messerschmitt, and a DC2—a rare example of the forerunner of the DC3, both of which flew so long and faithfully over Australian air routes. At one time he owned three of the DC2 airliners, and he sold the best of them to a Texan, Richard M. Rosser, who had it overhauled in the Hawker de Havilland workshops at Bankstown before flying it home. The aircraft had not flown since 1957 and was hangared until the early 1970s. It had logged no more than a modest 9,000 flying hours.

A number of historic aircraft still survive in Australia—to an aviation addict there is nothing so precious as these veteran machines. They include Kingsford Smith's *Southern Cross* and Bert Hinkler's Avro 581E (G-EBOV) in which he established his solo record from Croydon, England, to Darwin in 1928. Both are in Brisbane. In the RAAF museum at Point Cook is an early Deperdussin; Ross and Keith Smith's Vickers Vimy is in a special building at Adelaide airport. The Duigan biplane, the first successful aircraft to be designed and built in Australia, and the Blériot monoplane believed to have been used by Guillaux on the first Melbourne to Sydney airmail run on 16–18 July 1914, still survive. The DH9, registration number D-EAQM, in which Ray Parer flew from England to Australia in 1920 is at Parer's old school, St Stanislaus Vincentian College, at Bathurst, New South Wales. And there are others, though many historic airplanes went the way of all flesh and fabric before they could be preserved.

While men like Marshall were doing all that they could to keep Australian aviation going, the Australian Government showed little enthusiasm. The Australian Air Force was formed on 31 March 1921 and was granted the prefix Royal in June 1921 by King George V, but had only twenty officers and 120 airmen under the command of Richard Williams at Point Cook, Victoria. Oddly enough, the RAAF had exactly as many aircraft as airmen. In 1920 Britain gave Australia 120 free aircraft: remnants of the huge fleet built up by the RAF during the first World War. But most of them stayed in their crates until 1925—when they were rapidly becoming obsolete.

Organisation of the RAAF's aircraft only began in 1919, with the "A" series. Each aircraft type had an "A" number. The de Havilland 9A, for instance, was A1, the SE5a was A2, the Avro 504K was A3, and so on. The first "A" series ended in 1935 with the A12, the Bristol Bulldog MkIIA.

TOP: A DH9 of the RAAF photographed in the 1920s (*RAAF Museum*)

TOP CENTRE: A 1926 photograph of an RAAF Supermarine Seagull III flying boat (*RAAF Museum*)

BOTTOM CENTRE: A Fairey IIID seaplane of the type which made the first round-Australia flight (*RAAF Museum*)

BOTTOM: A Supermarine Southampton flying boat photographed in 1928 (*RAAF Museum*)

A5 was officially the Westland Wapiti Mark One and Mark Two. The number was originally allocated to Ross and Keith Smith's Vickers Vimy IV (G-EAOU), but was never taken up.

The RAAF also had four of Lawrence Wackett's early aircraft, but they were never given "A" numbers. Still, the Wackett name was well known in the air force, with the Wackett Widgeon Marks One and Two, and the Wackett Warrigal Marks One and Two.

Tiny as it was, the RAAF had a proud tradition to maintain and fortunately for Australia in later years there were dedicated men to keep it alive. For the average Australian, flying was still a dangerous enterprise and its danger made it glamorous. Consequently, anything to do with aviation made good newspaper copy. Journalists gave so much attention to everything to do with flying that, probably unintentionally, they helped to promote its development in Australia.

But at least one journalist was an aviation enthusiast too. He was Edward J. Hart, an Englishman whose connections with aviation went back to 1906. As a boy he had helped to push the kite-like aircraft of the Brazilian pioneer Alberto Santos-Dumont when he made the first officially-observed powered flight in Europe on 23 October 1906.

Despite his interest in aviation, Hart slogged through the first World War as a soldier with the AIF but when the war ended he became managing editor of the magazine *Sea, Land and Air*, a journal published in Sydney and bearing the notation "official journal of the Australian Aero Club". There was already an Australian aviation magazine: *Aircraft*, published in Melbourne. When Ted Hart called a meeting on Friday evening 23 May 1919, in the Elizabeth Street chambers of the Royal Society in Sydney in order to form a New South Wales branch of the Aero Club, *Aircraft* wrote: "Such a club was badly needed, if for no other purpose than to stir up the governments, Commonwealth and State, as to the necessity of drafting legislation to control civil and commercial aviation. The first president is Mr H. C. Macfie, FCP, FCIS. Mr 'Ted' Hart was unanimously elected honorary secretary and treasurer, and the club could not have chosen a better man. He is as full of enthusiasm as the proverbial egg is of meat."

Aircraft was an influential magazine in the tiny aviation world of Australia in that period. The noted aviation journalist Stanley Brogden, of Victoria, searched through many files to discover the name of the first editor of *Aircraft*. The man remains anonymous despite the mark he left. The early issues were filled with enthusiastic adjectives which must have played their part in inspiring those who believed in the future of aviation.

Less than a year after the end of the first World War, *Aircraft*'s anonymous first editor dreamed: "In a few months there will be hundreds of aeroplanes in Australia. There will be thousands of men capable of teaching others and hundreds of them will be making handsome incomes by doing so." But such a forecast was premature in a country in which even the ownership of motor cars was still fairly restricted.

His first issue was a paper foundation-stone for one of Australia's aviation landmarks, because *Aircraft* still appears each month and is believed to be the oldest aviation monthly in the world. The oldest aviation weekly is London's *Flight*. Also, *Aircraft* is probably the oldest aviation institution of any kind still functioning in Australia: older than the RAAF, the Department of Civil Aviation, and Qantas.

After three years as editor of *Sea, Land and Air* Hart went to Melbourne and edited the short-lived aviation journal *Slipstream*. In March 1921 he amalgamated it with *Aircraft* when he was appointed editor of that magazine, then accredited as the official journal of the Victorian section of the Aero Club.

Aircraft was first published on 1 July 1919 and was owned by Attkins McQuitty Limited of 16 Pitt Street, Sydney. Disregarding *Sea, Land and Air* they called it "Australia's first and only aviation journal". Later they put it under the auspices of their Motor Press of Australia Ltd which also produced a motoring journal.

The circulation of an aviation magazine in that era was somewhat small, and Hart became "sole proprietor" in July 1921. When he took it over he said in an editorial that "the new ownership of *Aircraft* is entirely free from 'Trade' influence. No member of any aircraft company or allied industry has the least voice in its control." But the sixpenny magazine suddenly ceased to appear in December 1921, and for three months there was editorial silence until Hart published "Vol. 1 No. 1 New Series" in March 1922.

Hart wrote, thundered, bullied, and praised in *Aircraft* until 1 October 1930, when he decided to return to England. Alan Moyle persuaded Keith Murdoch, chief of the Herald and Weekly Times Limited of Melbourne, to buy the magazine, and Moyle edited it for a year or so. He was followed by Jack Bellew, and in September 1933 the editorial chair was taken over by Douglas Robertson. He is still editor of *Aircraft* and his knowledge of Australian and world aviation is all-embracing.

He is a quiet, modest man who nevertheless exerts a strong will and influence in the aviation industry through the pages of the magazine.

But it was Ted Hart who had the honour of writing about the pioneer airmen who were blazing the trails. These were not only the "first-ever" flights, like those of Arthur Long who made the first aerial crossing of the Bass Strait from Melbourne to

Tasmania on 17 December 1919, but also the administrative changes being made on the ground. If Australian aviation was ever to be properly established, then these were equally important.

To begin with, civil aviation was under the firm hand of Senator G. F. Pearce, Minister of Defence from 10 January 1918 until 21 December 1921. On 16 December 1920 the Civil Aviation Branch of the Defence Department was formed, and remained in control of non-military flying until 1936 when the Civil Aviation Board was established, still under the Defence Minister. In 1939 the Board was abolished, and the Department of Civil Aviation was set up, with its own Minister. (In 1973 this department was incorporated in the Department of Transport.)

The first Controller of the Civil Aviation Branch was Lieutenant-Colonel Horace Brinsmead. He was not a pilot himself, but had transferred to the staff of the AFC after having been so badly wounded as an infantry officer on the Western Front that he was unfit for further active service. *Aircraft* applauded his appointment with the words, "It is a happy augury for the future of civil aviation in Australia." Norman Brearley describes him as "an alert and friendly man with a keen sense of humour".

Brearley met him soon after his appointment, and was to have frequent contacts with him. He "had an effective and diplomatic way of cutting red tape", Brearley writes, and Brinsmead was of great assistance to Brearley when the latter was nursing his WA Airways through its earliest years. Brearley was to have numerous brushes with bureaucracy, and recounts one incident in which the public servants insisted upon him using totally unsuitable lighting equipment for night-landing operations. "Some years later I saw it [the equipment] covered in dust, lying in a small store shed," Brearley says. It had never been used, because Brinsmead had authorised Brearley to install lighting which would do the job properly.

The whole aviation picture was changing rapidly, especially when Brearley was pioneering airmail flights along the rugged coast of north-west Western Australia. Brinsmead and Brearley were a well-matched pair: both of them intent on doing a job properly and safely instead of arguing about the way in which it should be done.

While Brearley was planning Australia's first airline, two more young Australian pilots were making one of the flights which holds pride of place in aviation history. They were the brothers Ross and Keith Smith.

It all began when the Australian Government offered a prize of £10,000 for the first flight from England to Australia. This was a huge amount of money in those days, but the risks were equally great. And the conditions, laid down by the Royal Aero Club of Great Britain, were fairly stringent.

Colonel H. C. Brinsmead, appointed Controller of the Civil Aviation Branch of the Defence Department in 1920. He was a vital force in Australian aviation until his death in 1934

These stated that the aircrew had to be Australian, that the flight must be completed within 720 hours, in an aircraft built entirely within the British Empire and with a range of at least 500 miles, and that competitors must complete the flight in the aircraft which took off from England. Five sections of the machine and five parts of the engine must be marked, and at least two parts of each set must still be in position when the aircraft was checked at the Singapore control point and when landing in Melbourne. A competent navigator, who could be the pilot, must be in the aircrew. On top of all this each entrant had to pay £100 entry fee.

When the Australian pilots waiting for repatriation from England heard about the competition, several of them decided to give it a go. They included Captain Ross Smith and Lieutenant Keith Smith; Ray Parer and John McIntosh; Captain Howell and Tom Fraser; Captain Matthews and Tom Kay; and Lieutenant Williams, Valentine Rendle, Garnsey Potts, and Sir Hubert Wilkins. The last four flew a twin-engined Blackburn Kangaroo, and Parer asked for the job of piloting this aircraft. But it was

Sir Keith Smith (left) and his brother Sir Ross Smith made one of the world's epic flights and gained international acclaim when they won the England–Australia air race in 1919, flying a Vickers Vimy bomber (*Elder Smith Goldsbrough Mort Ltd*)

given instead to Charles Kingsford Smith, who had to withdraw when for various reasons the flight was delayed. In the event, only two aircraft made it to Australia. They were the Vickers Vimy bomber flown by Ross and Keith Smith, with Sergeants Jim M. Bennett and Wally H. Shiers; and the DH9 two-seater day bomber flown by Parer and McIntosh. But Parer and McIntosh did not arrive in Australia until nearly nine months after the contest closed.

The French pilot, Poulet, decided to make an independent attempt to seize the record, and left Paris for Darwin in a Caudron a month before the Smiths left England.

The Vickers Vimy bomber, reconditioned after a fire, may still be seen at Adelaide airport and by comparison with the huge jets which take off for the mere 400 miles of flying between Adelaide and Melbourne it seems an almost laughable machine in which to embark upon a flight of 11,340 miles. It has open cockpits in which the flyers had to sit for hour after hour, and two huge Rolls-Royce Eagle engines which deafened them with their steady roar. But in 1919 it was one of the biggest and most modern aircraft available. Its aircrew already had some experience in long-distance flying. After the Armistice Ross Smith piloted a Handley Page bomber on the first flight from Cairo to Calcutta, together with Sergeants Bennett and Shiers. After this, with Major-General A. E. Borton, he began the ground survey of a proposed air route from India to Australia, but fell sick at Timor and had to abandon this project.

When he and his brother decided to enter the contest they picked the Vickers Vimy bomber as the most suitable type of aircraft, because it carried 865 gallons of fuel giving it a cruising range of 2,400 miles. Vickers Ltd agreed to provide an aircraft for the flight.

They left Hounslow airfield, near London, at 8.30 a.m. on 12 November 1919, and after flying across the English Channel in bitter, squally weather they hopped across Europe via Lyons, Pisa, Rome, and Taranto. This sector took five days, and on 17 November they flew from Taranto to Suda Bay on the island of Crete. They arrived at Heliopolis airfield, Cairo, on 18 November. The seemingly

TOP: The winning crew in the Great Air Race of 1919. Wally Shiers, Keith Smith, Ross Smith, Jim Bennett (*Wally Shiers*)

BOTTOM: The winning crew stand before the Vickers Vimy at Hounslow, England, shortly before take off

A letter flown in the Vickers Vimy from Bandar Abbas, Persia, to Australia, franked with Indian stamps and signed by Keith Smith (*H. R. Harmer and Stamp Magazine, London*)

slow progress was because they flew only during the day. In those days, before the introduction of more sophisticated navigation equipment, long-distance visual night-flying had many dangers; it still has, and now cross-country night-flying is generally done with instrument back-up.

There was an RAF base at Heliopolis, and the skilled mechanics gave the Vimy a rapid check before the Australians took off for Damascus on 19 November. The next stop was to be Baghdad, but as they droned above the desert the Vimy had to bore into head winds. It soon became clear that they would not reach Baghdad before dark, and they decided to land at Ramadi. The 10th Indian Lancers were based there, and they gave the flyers a hearty welcome. The Australians were now exhausted, but their hope for a peaceful night was to be frustrated by the wind which had opposed them all day. It rose to a gale and caused the Vimy to strain wildly at its moorings. With the help of fifty soldiers they had to spend the night fighting to prevent the aircraft from being turned over and smashed. In the calm of the morning they topped up the fuel tanks and checked over the Vimy. Departure was at noon. They decided to bypass Baghdad and fly straight to Basra, then make another hop of 630 miles to Bandar Abbas in Persia. Thirteen days after leaving the green pastures of Hounslow they flew 730 miles from Bandar Abbas to Karachi.

It took the four Australians nearly seven hours to fly the 730 miles; hours which contained more tension than any other part of the journey. They were flying over one of the world's most savage landscapes, with range after range of jagged waterless hills and vast stretches of desert. A forced landing would have given them little hope of survival.

But they landed safely at Karachi, flew on to Delhi, then arrived at Allahabad on 27 November. The next stop was Calcutta, where they received an excited welcome and the news that Poulet's tiny Caudron aircraft had left only hours earlier. The Smiths decided to press on immediately to Sittwe (then called Akyab) in Burma.

Poulet now had little chance of beating them to Darwin unless something happened to the Vimy, which was faster than his small machine. The Australians made the comparatively short hop to Rangoon and then had to face one of the trickiest pieces of flying of the whole trip. The next control point was Bangkok, beyond mountains which clawed up to 7,000 feet. Slowly they coaxed the laden Vimy up into the thick clouds, never quite sure whether one of them hid a jungle-clad peak. Eventually, however, they left the moisture and cold and descended towards Bangkok.

On the next stage they encountered what Ross Smith described as the worst weather he had ever experienced. On 1 December as they were flying towards Singora the Vimy was lashed by a continuous series of heavy rain squalls. Visibility dropped almost to zero and they seemed to be flying through solid sheets of water. But Singora was reached more or less on schedule, and they flew on to Singapore where they arrived on 4 December.

Now they had to face 2,500 miles of flying over the jungle-clad islands between Singapore and Darwin. Any fault in the aircraft would leave them with only two choices: splashing down in the sea or trying to find a landing spot among the swamps, jungle, and mountains of the Indonesian (then East Indian) archipelago. The Vimy was checked with painstaking care before the take-off for the nine-hour flight to Kalidjati, Java. They were beginning to feel that they

The French pilot Etienne Poulet gave the Vickers Vimy some stiff competition during the Great Air Race but he was forced to abandon the flight in Burma

might make it to Australia but there was still plenty of time for things to go wrong.

The first real setback was at Surabaja (then Soerabaya) on 6 December. Tropical rains had turned the airstrip to slush and they became hopelessly bogged. The four had to spend the night persuading the locals to lay a track of bamboo mats ahead of the aircraft, and next morning they lumbered slowly along this, tense until the Vimy left the ground.

They were now headed for Bima on the island of Sumbawa: the last stop before Darwin. Ahead of them lay 470 miles of the Arafura Sea: their longest flight over water. A morning fog seemed likely to delay them, but by noon it had drifted away and they headed for Darwin. Somewhere in the Arafura Sea was the cruiser HMAS *Sydney*, which might have been able to come to their rescue if they came down in the sea.

The lookouts on the *Sydney* were only a few of the many eyes straining to see the Vimy as she neared Australia. Those who had prepared for the flyers' arrival included Hudson Fysh and "Ginty" McGinness, who had slogged through Queensland and the Northern Territory getting landing sites ready for the Vimy. In his book *Qantas Rising* Fysh recalled, "We anxious watchers at Fannie Bay at last saw the Vimy appear as a little speck away towards the Timor Sea and then grow larger and larger as it circled the ground and landed. It was one of the most moving sights I can remember—the termination of one of the greatest flights, if not the greatest, in the history of aviation."

As the flyers jumped from the Vimy on to the hard-packed red earth, Fysh greeted his wartime acquaintance Ross Smith and handed him a bundle of fifty welcoming telegrams. Smith's first words on the completion of the epic flight were, "Well we're here. I didn't recognise you at first. Where's your uniform?"

The flight had taken them just under twenty-eight days of which 135 hours were spent in the air. They flew on to Melbourne and received an uproarious welcome at every intermediate stop. The prize of £10,000 was divided equally among the crew, the Smith brothers were knighted, and the two sergeants were promoted to lieutenants and awarded the Air Force Medal.

Vickers Ltd gave the Vimy to Australia and for some years after that it was on exhibition in the Australian War Memorial, Canberra. Eventually it was moved out to make room for other exhibits, and was almost forgotten until 1955 when it was presented to South Australia. The building in which it now stands was erected with the help of a legacy from Sir Keith Smith, who died in 1955. He had joined Vickers Ltd as their Australian representative and became well known in the Australian commercial world as the director of a number of companies associated with Vickers.

The Vickers Vimy on a Northern Territory airstrip after arrival in Australia. The aircraft now stands in a specially-constructed memorial building at West Beach Airport, Adelaide

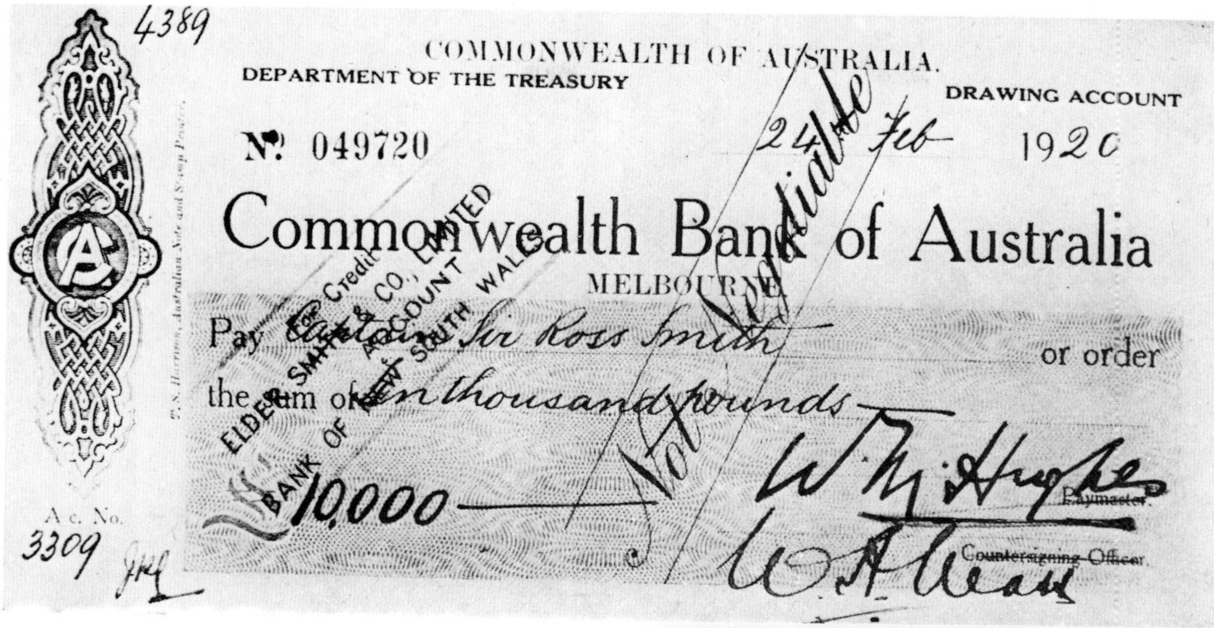

Top: The cheque for £10,000, signed by Prime Minister W. M. "Billy" Hughes, paid to Sir Ross Smith as the prize money for the Great Air Race (*Elder Smith Goldsbrough Mort Ltd*)

Right: Great Air Race medals. Top: The 1969 Fiftieth Anniversary Medal (*Pacific Promotions, Adelaide*). Centre: 1920 Aero Club de France Medals. Bottom: 1920 Royal Aero Club of London Medal (*Art Gallery of South Australia*)

The career of Sir Ross Smith was cut short on 13 April 1922. In 1920 he wrote *The First Aeroplane Voyage from England to Australia* and then lectured on the flight in Australia and England. Early in 1922 he planned a flight around the world in a Vickers Viking amphibian, but on its trial flight the machine crashed. With him perished Lieutenant J. M. Bennett. A few weeks after Smith's death, his book *14,000 Miles Through the Air* was published.

But while the crew of the Vickers Vimy were receiving the acclaim of the Australian public, Ray Parer and John McIntosh were still plugging doggedly towards Australia. These two had not found it easy to obtain an aircraft for the flight, and Ross and Keith Smith had already reached Karachi before they even raised the money to buy one. It was given to them by Peter Dawson, a Scottish whisky millionaire who was a friend of "Billy" Hughes, the Australian Prime Minister. Dawson had entertained hundreds of Australian servicemen during the war.

Before Parer and McIntosh even took off from England they heard that the Smiths had reached Darwin. Dawson none the less urged them to carry on with the flight. The Royal Aero Club returned their entry fee and Australian headquarters in London forbade them to fly to Australia, saying that the whole project was irresponsible. But they defied

orders and took off from Hounslow at 10.30 a.m. on 8 January 1920.

Everything went wrong all the way to Australia. They had a forced landing in France soon after leaving England, and wrecked one of the wheels.

This was followed by further forced landings at Bron airfield, near Lyons, and in northern Italy. As they flew over Italy the engine caught fire and Parer had to put it out by diving 2,000 feet until the flames were extinguished. By the time they reached Rome they had only thirty shillings between them and had to cable Peter Dawson for more money.

An unwary flight over Mount Vesuvius, to look into the crater of the volcano, almost ended in disaster when they were caught by a vicious down-draught and nearly crashed into the bubbling mass of lava. In Naples they met Colonel Horace Brinsmead, who was on his way home from his post as Staff Aviation Officer of the AFC in England. He encouraged them to press on, and another officer told them that the authorities had decided to ignore their unofficial flight.

Most of their landings were unexpected and almost disastrous. They had to put down on a beach in Italy, in Brindisi, and then after hopping slowly through southern Europe made another beach landing in North Africa. After reaching Heliopolis, where their aircraft—which they called the PD after Peter Dawson—was reconditioned by the RAF, they began the long flight across the Middle East, hampered by a forced landing in the desert. They were attacked by bedouin but managed to take off again and reach Baghdad on 28 February—after completing the first crossing of the Syrian Desert in a single-engined aircraft.

They continued slowly towards Karachi, with the PD looking more dilapidated all the time and having to be patched up with improvised repairs, and then pressed on through India. By the time they reached Calcutta they were so broke that they had to put on a flying demonstration, drop advertising leaflets over the city, and have the PD repainted with the words SHELL in large red letters on the underside of the mainplane, FIRESTONE TYRES underneath the fuselage, and MILTON on the base of the fuselage. These advertising efforts earned them £2,000 and they received another £1,000 from Peter Dawson, but the PD was still very cranky when they left Calcutta for Burma. As they flew over the Arakan mountains, the engine cut out at a height of 4,000 feet. With a mixture of good luck and skilful flying, Parer managed to land the PD on a sandy island in a river. It was only 200 yards long: barely enough for a landing and take-off.

With the help of some friendly natives they patched up the PD and improvised an airstrip, from which they took off for a safe flight to Rangoon. The next disaster came when they were flying south down

Top: Ray Parer learns to fly in a 1914 Bristol Box-Kite (*Mrs R. Parer*)

Centre: Ray Parer (left) and John McIntosh at Darwin after their hazardous flight across the Timor Sea (*Brisbane Sunday Truth*)

Bottom: The PD on arrival in Calcutta

the coast of Burma. The engine began to misfire, an explosion blew a hole in the exhaust pipe, flames were everywhere, and their gravity-feed petrol tank began to overflow. Petrol was sprayed over both the airmen.

It seemed that they had no hope of survival, and Parer was preparing to land in the sea when the engine started again. Far ahead, Parer could see Moulmein, and he coaxed the aircraft down towards a landing. This should have been successful but as he prepared to touch down on the racecourse a huge crowd swarmed on to the turf, and he had to choose between ploughing into them and crash-landing the aircraft. He chose the latter, and by some miracle he and McIntosh were unhurt. But the PD looked like a write-off. The propeller was broken, the undercarriage wrecked, the petrol and oil tanks were split, and the engine, radiator, and fuselage had all suffered. It seemed like the end of the flight, but after seven weeks' delay during which they adapted a propeller supplied by some Italian flyers, and virtually rebuilt the whole machine, they took off for Penang.

By now it seemed that the whole venture was operating under the rule of what is popularly known as Sod's Law: "Everything that can possibly go wrong will do so." En route for Penang the oil pressure suddenly dropped to zero, but there was nowhere to land so there was nothing to do but press on. With the engine shuddering on nearly dry bearings Parer spotted an invitingly open stretch of turf on Penang. The only problem was that it was occupied by galloping horsemen involved in a polo game. He landed in the middle of it.

It took him and McIntosh five days to patch up the PD again, but as soon as they got into the air on their way to Singapore the exhaust pipe broke loose. So they landed at Prai, on the mainland of what is now Malaysia—and snapped off their tail-skid in doing so.

But they made it to Singapore, where they had to land on the golf course when the oil pressure disappeared again. Wearne, an Australian engineer, virtually reconditioned the aircraft for them—but when they flew out of Singapore the PD still had a car radiator up front. It had been one of their Moulmein improvisations and so far they had not been able to find anything better.

Because the race was over so far as the Australian Government was concerned, there would be no rescue ship stationed between Darwin and Timor and all the petrol and other stores were being removed from the airstrips between Darwin and Melbourne. But the two flyers decided that having got as far as Singapore they might as well press on.

Fate was still against them. They left Singapore on 20 July 1920, and ran straight into a gigantic storm which forced them to return. Next day they tried again, and reached Djakarta (then Batavia) two days

TOP: Parer and McIntosh with the PD in Melbourne (*Australian Post*)

BOTTOM: The PD with admiring welcomers at Darwin (*Weekly Times*)

later. Now they faced the long stretch through the islands, and on 23 July they headed east. Soon two cylinders of the six-cylinder engine stopped firing and they had to land on a sand-bank near a sugar plantation. One of the magnetos was giving trouble but was soon fixed up, and next morning they took off along an airstrip of bamboo mats laid down by the plantation workers.

Next stop: Surabaja in eastern Java. And it was almost the last stop because the landing ground was in such bad condition that the PD lurched into a ditch. Apart from some minor damage, the undercarriage was broken and the propeller smashed. Parer was about ready to give up, but the Dutch air force offered to help and in four days the PD was ready to go again.

They reached Timor on 29 July after a comparatively uneventful flight through the rest of the island chain, and on the next day made elaborate preparations for the flight to Darwin. Among other things they built an emergency raft out of a petrol tank fastened to a bamboo framework. They tried to take off but just before they left the ground the engine was stopped by a blocked feed-pipe. They cleared it, tried again, and managed to climb to 2,000 feet. The day was clear and the sea was calm, but it was potentially the most dangerous stretch of their flight. They had just enough petrol for about eight hours' flying, but when they had been six-and-a-half hours in the air they had still not seen Australia. There were no landmarks and a minimum of navigation instruments; there was always the chance that they had somehow turned off course and were simply flying on and on across the ocean.

But at 4.30 p.m. they sighted Bathurst Island, and at 6.25 p.m. on Monday 2 August 1920 they landed at Fannie Bay on the airfield which had been prepared for the Great Air Race. Their petrol tanks were dry.

They were greeted almost as rapturously as the Smith brothers had been, and later made easygoing progress from Darwin to Rockhampton, where they had a civic reception, and then to Brisbane. In the Queensland capital they were met by the mayor and given a ticker-tape welcome, followed by an official lunch and a cheque for £100 from the State Government.

By this time the PD was almost falling to pieces, and when they had to make yet another forced landing at Tweed Heads Parer patched up a rip in the fabric with brown paper. They had yet another forced landing near Newcastle, but finally made it to Sydney. As they circled above Mascot they saw an enormous crowd waiting for them and as soon as they landed they were mobbed. They were carried shoulder high to an official marquee but souvenir hunters almost tore their clothes off before they got there. Ross and Keith Smith were among the welcoming committee, and the mob followed Parer and McIntosh all the way to their Sydney hotel. The cheering crowd literally put their money where their mouths were and donated hundreds of pounds in collections.

Next they were to fly to Melbourne via Bathurst, where Parer had gone to school. Once more, they barely made it. The PD stammered and stuttered over the Blue Mountains and Parer almost ran off the end of the airstrip when he landed.

When they took off again the PD began to fall to pieces in the air. The starboard lower wing was working loose, and when they were near Albury the engine started to act up again. Parer landed in a ploughed field and the PD turned over. Once more the two flyers were unhurt but the aircraft was now a wreck. They knew that a big reception was planned in Melbourne so the poor old PD was carried there by rail while Parer and McIntosh flew to Melbourne in a DH9A provided by the Defence Department.

The reception, on Flemington Racecourse, was a gigantic affair. The public had to pay for admission but got their money's worth. First they watched an aerial display, and at 2.30 p.m. Parer and McIntosh flew in the DH9A which was decorated with three long coloured streamers and accompanied by thirteen other aircraft. The welcoming committee was led by "Billy" Hughes and the Premier of Victoria. The battered PD was on display, and on 1 September 1920 the Melbourne *Age* wrote, "There was not a single part of it [the plane] but bore unmistakeable traces of the strenuous journey which it had almost accomplished. An improvised motor-car radiator hangs underneath it; both blades of the propeller are broken off; the coverings of its wings, from which paint is peeling in large flakes, are tattered and torn, and its framework has been tied together with string and wires; its rudder is bent and pierced, and its fittings are suffering from the effects of long exposure."

Next day they were given a State Banquet at Parliament House and presented with £500 apiece by the Commonwealth Government. This was increased by more than £1,000 raised by public subscription in Victoria. They had earned every penny of it with their almost incredible flight which had taken them 237 days although the actual flying time had been only about eight days.

Ray Parer was to have a long and adventurous life in aviation and the rest of his story will be told in this book, but John McIntosh, like so many of those early flyers, did not survive his next ventures into the air. For most of the war he had served as an infantry officer and had not transferred into the RAF until late in 1918 so he had not completed his flying training. It is not known whether he took further instruction, and in those easygoing days it is quite possible that he did not. But he had sufficient confidence to buy a flying school in Western Australia, and to begin teaching other people to fly as well as engaging in charter flights. On 28 March 1921 he flew from Perth to Pithara with a mechanic and a passenger. Over Pithara the aircraft engine failed and it crashed. McIntosh and the mechanic were killed, though the seriously injured passenger recovered.

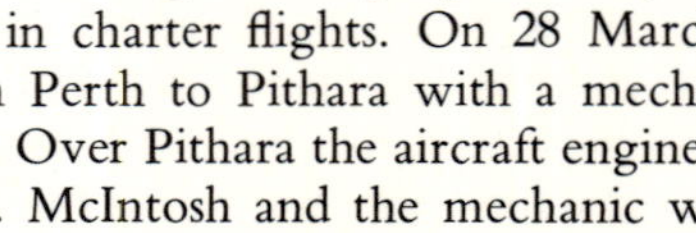

5

Smithy, Ulm and Hinkler

Three of the greatest adventurers in Australian aviation simply couldn't resist a challenge. They survived every adventure but the last

The years from 1920 to 1930 were the real glamour period for Australian aviation. The skies were wide open. Solid citizens like Norman Brearley and Horrie Miller were establishing their own airlines, of which more later, but the public was perhaps more interested in the exploits of such men as Charles Kingsford Smith, Bert Hinkler, and Charles Ulm. And there were others, heroes in their day but almost forgotten now, who also made their mark on Australian aviation—though not all of them were Australians. For example, Major de Pinedo and F. Campanelli flew a Savoia Marchetti from Italy to Australia in 1925, and Alan Cobham with A. Elliot flew a DH50 seaplane from England to Darwin in 1926 and returned in the same year. Cobham—once an advanced flying pupil of Brearley—was knighted and went on to lead Cobham's Flying Circus, which delighted so many people as it flew round English seaside resorts in the summers of the 1930s.

In 1922 Andrew Lang in the flying boat *Seagull* made an air survey of the Australian coast, and in 1924 S. J. Goble and Flight Lieutenant Ivor McIntyre made the first round-Australia flight in a Fairey IIID seaplane, taking ninety flying hours to cover the 8,500 miles.

The flight actually took them forty-four days, and like all the records established by early flyers it involved hard physical labour as well as actual flying time. After the flight Goble said:

"Although 2,500 miles of the journey were made in heavy rain and bad visibility we were always tired by the time we got into the air, owing to the strain of watching the machine, the manual labour involved in doing running repairs and humping drums of petrol. Each day after landing we put in about seven hours work baling floats, cleaning plugs, adjusting magnetos and valve springs, straining petrol through chamois leather, oiling up, filling the radiator etc., in addition to the long and irksome watching to ensure the safety of the machine on the water, and also to avoid being left high and dry. Several times we had to dig her out when the tide played dirty tricks on us."

In 1928 A. Hood and J. R. Moncrieff made the first attempt to fly from Australia to New Zealand, but they and their Ryan aircraft were never heard of again. In 1927 one of the first of the women aviators, Mrs K. Miller, accompanied by W. Lancaster, flew an Avro Avian from England to Australia.

But the darlings of the public and of the newspapers were Hinkler, Ulm, and "Smithy".

Charles Edward Kingsford Smith was born at Brisbane in 1897 and enlisted in the AIF in 1915. He served in Egypt, Gallipoli, and France, and in 1916 was recommended for a commission. While he was with an Officers' Training Unit in England he transferred to the Royal Flying Corps and was commissioned in 1917.

He flew over the Western Front for three months, long enough to win the Military Cross and for a German bullet to whip off three of his toes. Unfit for further active service, he was promoted to captain and made an instructor in the RAF. When the war ended he stayed a while in England, and scratched up a living with an aircraft in which he sold joyrides. When he heard about the £10,000 prize for the air race to Australia, he made two attempts to

Wing Commander Goble and Flight Lieutenant McIntyre won the Britannia Trophy for 1924 when they circumnavigated the Australian mainland in this Fairey IIID seaplane (*McGrath Collection*)

enter it: one as pilot of the Blackburn Kangaroo, and the other with C. Maddocks and V. Rendle. Their entry was rejected because none of them knew enough about aerial navigation.

He was determined to make a name for himself, or at least a living, in aviation. A year in California, stunt-flying for the silent movies, did not take him far towards this ambition. He tried to convince some Americans that he could fly the Pacific to Australia, and that they should finance him. But as he said later "In 1920 the idea of spanning the Pacific by a plane was assured of a reception that a proposal to establish a radio station might have received from Tudor England."

Unknown to him, his future inseparable partner, Charles Ulm, was trying to raise money for the same venture. Smith already knew Ulm, and recalled that "One day in 1921 we met at Mascot Aerodrome, Sydney, for a few fleeting minutes, and parted still in ignorance of the fact that here were two minds thinking along the same channels, and that a fusion of interest might have been helpful."

When Smith returned to Australia in January 1921 he was almost broke. But he found a niche with one of those short-lived companies operated by ex-flyers of the RFC: the Diggers' Aviation Company, which sold joyrides to Australians who wanted to taste the dizzy thrill of getting off the ground. For some of

Flight Lieutenant I. E. McIntyre (pilot) and Wing Commander S. J. Goble (navigator) of the Fairey IIID, photographed during their flight (*RAAF Museum*)

Norman Brearley stands beneath the propeller boss of his aircraft in this group assembled on 4 December 1921, the day on which he began Australia's scheduled airmail services. Kingsford Smith's famous smile is seen on the right (*Sir Norman Brearley*)

them the thrill was a great deal more than they had bargained for. Smith was up with a load of passengers when a wing collapsed, but with incredible skill he landed them all unhurt.

Like the wing, Diggers' Aviation collapsed and Smith had to look for another job.

He found it at Point Cook in August 1921, when Brearley was selecting pilots for his new WA Airways. Brearley told the magazine *Aircraft* that "We only started here yesterday afternoon, but I have already satisfied myself that they include some really 'pretty' pilots, possessing exceptional ability, together with that wonderful judgement that makes flying what it should be."

In his book *Australian Aviator*, Brearley has described his first meeting with Smith. He says, "One young man named Smith was quite outstanding, and as soon as we were in the air I knew he was well above average. When we came down, and I was noting personal details for my own records, I asked him his full name and service background.

"'Charles Edward Kingsford Smith.'

"'And how did you win your Military Cross?'

"'For various acts of foolishness.'"

Smith accompanied Brearley and Peter Hansen when they surveyed the air route from Perth to Port Hedland in December 1921 before Brearley began his first commercial flights. They had some adventur-

TOP: This DH50 was one of three built under licence by West Australian Airways in Perth (*Sir Norman Brearley*)

BOTTOM: Signing for the mails on an early Qantas service. The aircraft is a DH4 and the pilot (centre) is Hudson Fysh (*Qantas Airways photograph*)

ous landings on the primitive airfields, and a forced landing on the beach south of Broome.

Smith flew with and for Brearley until 1924, when with other pilots he joined in a "pilots' strike" in which they requested substantial improvements in their pay and conditions. Brearley believed that their discontent was not really due to their dissatisfaction with their wages, which were the same as those of pilots in British airlines, and with only two or three days' flying a week they were hardly overworked. They had flown most satisfactorily and had helped to establish the fine reputation of WA Airways, but Brearley thought that they still yearned for the adventurous flying of their wartime careers. Smith was only twenty-seven and the others were in the same age-group, and it was obvious to Brearley that Smith was looking for something more exciting than a shuttle service up and down the coast of Western Australia.

So they parted good friends, and Kingsford Smith with Keith Anderson planned to attempt some long-distance trans-oceanic flights. Brearley recalls that Smith wanted to fly the Pacific and Anderson wanted to fly the Indian Ocean.

First of all, Smith and Anderson had to make some money. They opened a trucking and transport business in Carnarvon, sold it at a profit, and moved to Sydney where they joined Charles Ulm and set up an aviation company.

TOP: This DH50J was one of three used by Qantas from 1928 for mail and charter work in outback Queensland. Qantas built them under licence at Longreach (*Qantas Airways photograph*)

CENTRE: S. M. Bruce, in white suit, then Prime Minister of Australia, chats to Hudson Fysh in 1923, during the first airborne electioneering tour made by a prime minister (*Qantas Airways photograph*)

BOTTOM: Local people are intrigued by a Gypsy Moth when the Albury (NSW) airfield opens in 1928

As so often in the history of flying, it seems that fate had taken a hand and was determined to bring Smith and Ulm together. Although they dreamed the same dream, they were physically very different people. Smith was a short, jaunty man, limping slightly from his war wound but effusive and self-confident. Early photographs show his typical challenging stance as though he was prepared to take on anyone or anything—including the Pacific Ocean.

Charles T. P. Ulm was taller than his friend and captain-to-be. He was of French blood but of Australian birth, having been born in Melbourne in 1897. At seventeen he enlisted in the AIF under the name of Charles Jackson, served in Egypt and Gallipoli, and with two wounds was sent back to Australia with a discharge because he was under age. He re-enlisted in 1917 and went to the Western Front, where he was wounded again and recommended for a commission. At the Officers' Training Unit he made a few flights as a passenger and later learnt to fly.

When he was discharged he tried to establish an aviation company in New South Wales but this, like so many others, did not succeed.

As soon as Smith and Ulm met again they began to reveal to each other their ideas about flying the Pacific. The bond was forged immediately, and they no longer talked about *whether* they would do it, but *how* and *when* it could be done. They told the story in the book *Story of the "Southern Cross" Trans-Pacific Flight 1928* which they wrote when their dream had become a reality.

The first problem, as usual, was finance. Australians in 1927 were just as aloof to the idea of flying the Pacific as Americans had been in 1920, so Smith and Ulm decided on a shorter but record-breaking flight to prove their mettle and make their names.

They would beat the record set by Captain E. J. Jones, of the Civil Aviation Branch. He had taken twenty-two days and eleven hours to fly around Australia: they planned to do it in ten-and-a-half days.

H. Campbell-Jones, managing editor of Sun Newspapers Limited, believed that they could do it and backed them to the hilt. They used one of the two Bristol Tourers (which they had bought from Norman Brearley) with a Siddeley Puma engine which already had logged 1,100 hours.

The round-Australia record was broken with what seemed to be ease, though Smith and Ulm put much careful planning and work into the effort. Overnight at Broome, in north-west Australia, they even carried out a top overhaul of the engine. And wherever they landed, exhausted after twelve hours of continuous flying, their first task was to top up fuel tanks and prepare the aircraft for take-off next morning.

From Broome to Perth they were flying over home territory, and rarely rose higher than 150 feet above the arid country as they pressed the airframe for maximum performance. They wasted no time in climbing high. The engine sang sweetly all the way.

By the time they reached Melbourne Smith and Ulm knew they had the record in their pocket, and relaxed enough to talk about the Pacific flight. They knew that it was possible. Aircraft had progressed far from their flimsy beginnings, and Colonel Charles Lindbergh had made his daring Atlantic dash in the *Spirit of St Louis*. All that Smith and Ulm had to do was convince others that the flight was feasible.

When they landed in Sydney they had created a new round-Australia record: ten days and five hours. "Sydney gave us a really wonderful welcome, and we derived new courage for seeking to pursue our Pacific flight ambition," they wrote later.

Sun Newspapers gave the two flyers a luncheon to which most of Sydney's leading businessmen had been invited, and Smith and Ulm seized the opportunity for a bold public relations manoeuvre. "We asked the gentlemen round the table and the people of Australia to supply us with the necessary money."

They had struck at an opportune time, and the money started to flow in. A gift from Sidney Myer, of the big Melbourne department store, a public subscription, and a guarantee by Premier Jack Lang of New South Wales that the State Government would give £3,500, seemed enough to start them on their way.

They asked Smith's partner, Keith Anderson, if he would join them as navigator, and he jumped at the chance. On 14 July 1927 the three men, each in their thirtieth year of age, boarded the liner *Tahiti* for San Francisco. They made good use of the voyage by picking the brains of the ships' officers for hints and instruction on navigation.

Almost the first to meet them in San Francisco was Herbert L. Dickie, the local Vacuum Oil Company

Aircraft silhouettes became more streamlined during the 1920s. This RAAF Hawker Demon was photographed in about 1929 (*RAAF Museum*)

representative. He had a startling proposition: that they should take part in the race from San Francisco to Honolulu, to take off in the second week of August 1927, for a prize of $35,000 donated by the huge Dole pineapple empire of Hawaii.

Dickie even had an option on an aircraft for Smith to fly in the Dole derby: an International biplane, with a single radial engine, named *Miss Hollywood*. The whole city was agog with excitement over the coming race, and for a little while the three Australians were caught up in the whirl. But, "It took us only a few hours to realise that we could not organise efficiently for a flight of 2,400 miles over an ocean in three or four days."

It was a wise decision. Fifty aircraft entered the Dole derby, but only two survived. The race was won by the Arthur Goebel–William Davis team, but the race itself was generally regarded as a fiasco. This outcome made the three Australians think deeply. "The . . . disasters placed a new complexion on long distance flying over oceans," Smith said. "They brought into bold relief those two comparatively new angles to flying—heavy loading and blind flying."

Despite the Dole derby disasters, it had been proved that properly prepared crews and aircraft could make lengthy overseas flights with success and safety. On 28 June 1927 two US Army lieutenants, Lester J. Maitland and Albert F. Hegenberger, flew their *Bird of Paradise* from Oakland airfield near San Francisco to Wheeler Field in Hawaii. They took twenty-five hours and fifty minutes. On 14 July 1927 Ernie Smith made the first civilian crossing. In the same year the first regular trans-America airmail and passenger service began flying in from the eastern States to Oakland.

Obviously the Australians had to plan even more carefully for their own much longer trans-Pacific flight. The first need was a suitable aircraft, and when they heard that the airframe of the Fokker FVII B/3M, used by Sir Hubert Wilkins over the Arctic, was in Seattle, Washington, they decided that it would suit them admirably.

They had just enough to buy the airframe and to begin fitting the three massive Wright Whirlwind radial engines, but funds were running out. The Lang Government was voted out of office in New South Wales, and his successors would not pay the £3,500 which he had promised.

The temptation to abandon the project was almost irresistible, but the Returned Sailors' and Soldiers' Imperial League of Australia cabled them to "Keep on keeping on". They did. Then they changed their minds. Then they decided to keep going. Anderson at last had to go back to Australia, but just after he sailed Captain G. Allan Hancock bought the aircraft, relieving Smith and Ulm of their financial worries. He loaned it to them for the flight, but later was to present it to them outright.

Though the Australian public seemed to have abandoned them, Smith and Ulm struck a chord in the warm-hearted Californians. They won financial support and local fame, and Smith and an American pilot even had a crack at the world endurance flying record. They circled around for forty-nine hours, but just failed to set a new record.

It was all useful experience for the great test looming ahead, and they prepared themselves unsparingly. "We occasionally tested ourselves by driving an automobile for, say, twelve or fifteen hours, then flying three or four hours, then running for an hour or two, then flying some more, and then more automobile driving, etc, for a total of thirty-five or forty hours on end, until we knew the 'human element' would not let us down," they wrote later.

They had cabled Anderson to ask him to rejoin them, but he decided against it. So they took on two American crewmen: Lieutenant Harry Lyon Jr as navigator and James Warner as radio operator. They were skilled, enthusiastic technicians who worked well with the Australians.

By 31 May 1928 the aircraft which Smith and Ulm had named *Southern Cross* was as ready as she ever would be. The radio equipment was the finest available, built and fitted by a San Francisco firm. All the aids to blind flying known to that period were in the instrument panel. For maximum comfort on

RAAF machines in 1929. TOP: A Hawker Fury (*Hawker Aircraft Limited*). BOTTOM: A Westland Wapiti of the type which saw many years of service (*RAAF Museum*)

Two famous Australian pioneer aviators photographed in 1928. Norman Brearley (left) with Sir Keith Smith. (*The West Australian*)

the long trip, deep wicker chairs had been installed for pilot and co-pilot.

The *Southern Cross* sat ready at Oakland airport, which was then less than a year old and had been opened with ceremonies including the landing of Lindbergh in his *Spirit of St Louis* in 1927. It was later to become the aerial home of Amelia Earhart, one of America's great women pilots, and swashbuckling Howard Hughes would flip in and out on visits from Hollywood. Phillip Mumma, press officer of Oakland airport at the time of writing, said: "Perhaps the greatest flight ever made was that flight by the *Southern Cross*, and probably the greatest flier of all time was Kingsford Smith." But apart from men like Mumma only a handful of Americans now know anything about "Smithy".

It was different in May 1928. Smith, Ulm, and the two young Americans were heroes. At dawn they began checking the aircraft inside and out: checking maps, radio, flying suits; packing rations for the long trip. People began to mill around the hangar: reporters, officials, and about a hundred onlookers who had missed their breakfast so that they could watch the beginning of the great flight. Some of them were girls who had come to know the flyers during their months in the San Francisco Bay area.

The inevitable trails of fog were beginning to roll through the Golden Gate and around Alcatraz Island. They clung to the waterline, hid the piers jutting into the bay, crept around the stately Fairmont Hotel and up Nob Hill until all that could be seen were the Victorian mansions on Telegraph and Russian hills. The early-morning fog-horns wailed like sea-lions.

A light offshore wind drifted across the bay and over the city, favouring the take-off. That would be the critical moment. The *Southern Cross* had to lift a total of seven tons of men, fuel, and gear into the air; possibly the heaviest load ever carried by a winged aircraft up to that time.

"Well, nothing has been left to chance," Smith said as though to reassure himself, and he said later "Heroics, we felt, were never less appropriate. We did not look to the venture ahead as being in any sense a dicing against death. Reflections of that kind did not oppress any of the four of us."

Portrait of an Australian aviator: Air Commodore Sir Charles Kingsford Smith, K.B.E., M.C. (*The National Library of Australia*)

In their own minds they had the best possible aircraft: the sturdy Fokker airframe fitted with the mighty Wright engines. Their course had been plotted for them by a veteran navigator, and his wife, Mrs Voortmeyer, was there to accompany the flyers from the administration building to the *Southern Cross*. She and tiny Margaret Fort, apparently an actress, kissed Smith and Ulm good-bye. Tall, serious Ulm had to stoop for Margaret's kiss.

Among the last to grip Smith's hand was the mother of Lieutenant Albert Eichwaldt, who had died after taking off from Oakland a few months earlier to search for Miss Doran, one of those who disappeared in the Dole race. On to Smith's finger she slipped a ring which her son had made from a franc coin while serving in France during the first World War.

The reporter from the *Oakland Tribune* was scribbling his impressions of the scene and when the newspaper was published that afternoon the lead stories were all about the flight of the *Southern Cross*. The reporter mentioned that eighteen vessels would be along the route from Oakland to Honolulu, that a "fleet of defense planes will dash out to meet" the *Southern Cross* as it approached the Hawaiian Islands, and that the Pacific fleet of 100 US warships was engaged in manoeuvres off Pearl Harbor. The newspaper also outlined thirteen previous attempts on the California to Hawaii route, and while it was on the subject of aviation mentioned that two Spanish flyers had been forced to land in Iraq during an attempt to beat the world's long-distance flight record.

At 8.30 a.m. one after another of the Wright Whirlwinds coughed into life. Puffs of white smoke drifted away from their exhausts as the propellers whirled into shining, translucent discs. Smith listened carefully as the engines settled down into an even chorus, and the *Southern Cross* began to roll towards the take-off point almost before the crowd realised that the adventure had begun.

At 8.54 a.m. Smith pushed forward the three throttle control levers which fitted into the callouses of his right hand. The engine roar was deafening as each of the Whirlwinds rose to 1,800 revolutions. The *Southern Cross* rolled forward and her wheels lifted slowly but steadily off the ground. At seventy-six knots, she climbed at less than 200 feet a minute until she reached the speed of eighty knots at 1,000 feet. As she turned west over "the stone and concrete canyons of San Francisco" the flyers looked down and saw that the mists still swirled around the bay. Smith wrote later "Our last picture of 'Frisco, the City where we had doubted and hoped, but had not despaired, was, as it seemed to stand, baseless and serene like a magic city hanging in the clouds." And as they flew south-west along the radio beam out to sea "The Farrallone Islands were grey with the mist wreaths. They swept under us and glided astern, dropping down into the mantle of cloud that enfolded the coast. That dim grey bulk was the last land we were to see for twenty seven hours . . . we seemed to be leaving the world for a world of our own."

The four flyers settled down for the long flight. Jim Warner tuned his radios to thirty-three metres for voice operation, and 600 metres, the distress frequency, for Morse transmission. They could not smoke or talk during the flight and the only way they could communicate was by writing. Later, Smith regretted that the notes which they passed to each other had not been preserved. "The good-humoured banter embellished at times with colourful adjectives would not have looked as amusing in cold, unsympathetic print," he wrote, but the pencilled quips eased some of the tension on the flight.

Apart from these internal communications, Warner sent out a stream of radio messages. Mostly these were to friends in America and Honolulu. Harry Lyon sent one to Wallace R. Farrington, Governor of Hawaii, reading "A brother Beta a'coming to see you—on the *Southern Cross*. Harry Lyon." Both men were members of the same American college fraternity, Phi Beta.

Later, the *Oakland Tribune* claimed that the *Southern Cross* radioed, "Tell the destroyers [of the Pacific Fleet] to stand by. We may need them." But no record of this message is shown in the "complete log" of the flight later published by the flyers. However, it is certain that both Smith and Ulm were somewhat apprehensive about the fuel consumption rate. Smith tended to be pessimistic, but Ulm's calculations were accurate even though the *Oakland Tribune*'s headline of 1 June 1928 was "SOUTHERN CROSS ARRIVES SAFELY IN HONOLULU. PLANE WITH GASOLINE ALMOST GONE WINS RACE FROM OAKLAND."

The *Southern Cross* landed on Wheeler Field, Hawaii, at 9.20 a.m. and was greeted with deafening cheers from a crowd of 5,000. But as the four flyers climbed out of the aircraft they could not hear the cheers: the steady roar of the engines had temporarily robbed them of their hearing. Smith and Governor Farrington could only shake hands and smile, with the Governor deafened by the crowd and Smith deafened by the Whirlwinds.

From Wheeler Field they flew the *Southern Cross* to Barking Sands, on the Hawaiian island of Kauai. So far they had flown over a stretch which others had flown before them. Now, the major adventure lay ahead: the huge unknown stretch of 3,130 miles to Suva, Fiji.

At 5.12 a.m. on 3 June 1928 Smith started the three engines. Ten minutes later the *Southern Cross* began to roll along the 3,400 feet of firm sand and then lifted slowly but smoothly into the air.

The Pacific Ocean breeds a constant stream of tropical storms; many are small and pass swiftly, but each one of them holds danger. As the dark clouds loomed ahead the *Southern Cross* had to be steered over or around them, which meant that the Whirlwinds guzzled more gasoline to provide extra energy. Ulm and Smith shared the flying though Smith did a little more than his partner. Rain lashed the taut fabric of the aircraft while those within it churned a constant question through their minds: would the fuel last until they reached Fiji?

Thirty-one hours after they had taken off from Kauai, at 1.10 p.m. Smith was dozing in his wicker chair. Ulm held the *Southern Cross* steadily on a compass course which led to the featureless horizon. Suddenly, a tiny brown dome seemed to spring over the horizon, seventy miles away. Ulm swung the aircraft eagerly towards it, and Smith was jerked awake. Before he had time to realise what was happening he swore at Ulm for sleeping at the controls. But Ulm pointed ahead, and a few minutes later "the ocean seemed to sprout a rampart of islands right across our path". Their worries were over and they still had enough petrol for five hours' flying.

At Suva they had to land on Albert Park Sports Oval. It was only 400 yards long, and they had to drop down over an embankment and could not touch down until they were 200 yards along the turf. Trees and a fence came rushing towards them, but at the last possible second Smith trod violently on the rudder control and slewed the tail around. The Fokker ground-looped, faced the way it had come, and stopped.

Men of the local defence corps rushed to place a barrier round the aircraft as thousands of Fijians swarmed towards it. "Men clasped our hands and shouted at us. We grinned back at them, we were afraid, a little stupidly. But we could not hear a sound. We were stone deaf," Smith and Ulm recalled.

For three days the flyers fretted to leave Suva because they had determined to have dinner in Sydney on Saturday night. But they had to accept the overwhelming hospitality of the islanders, and also had to find a suitable airstrip from which to take off again. It would have been impossible for the fully-loaded *Southern Cross* to take off from the sports oval.

They eventually found a perfect runway: Naselai Beach. They took the Fokker there and at 2.52 p.m. on Friday 8 June began the flight of 1,795 miles to Brisbane. This was the shortest but the worst of their ocean flights. They had to fly through a series of huge electrical storms, with the *Southern Cross* bucking and swaying and the static so bad that Warner could not send or receive radio messages. Two hours from Fiji, their earth indicator compass failed and they had nothing left but the far less reliable steering compass. Their only consolation was that they had the whole continent of Australia to aim for instead of a few scattered islands.

The darkness of the storms faded into the darkness of the night, and the *Southern Cross* battled on through turbulence which made the flyers feel as though they were in a boat in rough seas. But as the skies paled into dawn the clouds ahead seemed to be less menacing and slowly the flight grew smoother.

The hours dragged on into morning and the

The *Southern Cross* on a flight from Sydney to Brisbane in about 1930 (*The National Library of Australia*)

Southern Cross roared steadily at seventy-eight knots towards the cloud on the horizon. Gradually the cloud assumed more definite form—it was land!

"We felt disposed to yell" Smith recalled. Lyon and Warner shook hands in the tiny radio cabin. Gradually the horizon turned a deep bronze-green, laced by the white of surf, but obviously they were off course. The cliffs, beaches, and towns ahead were not Brisbane, and the world's first trans-Pacific flight reached Australia above the town of Ballina, 110 miles south of the Queensland capital.

Smith turned northwards and at 10 a.m. on Saturday they looked down upon the corrugated-iron rooftops of south-eastern Brisbane. By this time Warner had been able to radio ahead, and the news had flashed around the city. Seven aircraft escorted the *Southern Cross* down to Eagle Farm airfield, where 15,000 people waited to greet the flyers. They had completed the flight in eighty-three hours and eleven minutes' actual flying time.

They did not dine in Sydney that Saturday night, but Brisbane welcomed them almost ferociously. They were world heroes, and every Australian spoke the name of "Smithy" with a special pride. But for Smith himself it was not enough. His dream of flying the Pacific had now become part of a larger dream: that of flying around the world. Ulm, of course, was to be his co-pilot.

Before they made the next long-distance attempt, with the idea of breaking the record from Australia to London, Smith flew the *Southern Cross* from Point Cook to Perth: the first non-stop flight over that stretch of Australia. That was in August 1928. In September he and the *Southern Cross* made the first crossing of the Tasman from Australia to New Zealand, flying from Richmond, New South Wales, to Christchurch on the South Island. He was greeted rapturously everywhere, and became a dramatically glamorous personality beloved by the newspapers. Naturally this also aroused a certain amount of envy which was to become apparent later.

On 30 March 1929 he and Ulm took off from Richmond on their attempt to break the record to London. With them were H. A. Litchfield as navigator and T. H. McWilliams as radio operator. The first leg of their flight was to be non-stop to Wyndham in north-west Australia, but an hour after leaving Richmond the radio aerial, which in those days trailed beneath an aircraft, was carried away. They could still send messages but could not receive them, so they missed the message telling them that the weather was bad at Wyndham and advising them to return to Richmond.

In northern Western Australia the *Southern Cross* plunged into rain and low cloud. With the primitive navigation instruments of those days, it was easy for a pilot to lose his way when he lost sight of the ground, and Smith was becoming uncertain of his whereabouts when through a gap in the clouds they saw a group of buildings. They dropped a message asking for the direction to Wyndham but this was not seen and the men on the ground thought that the aircraft wanted to land. They scratched out a big arrow pointing west, by which they meant to indicate that

The crew of the *Southern Cross* at one of the receptions given to honour the first trans-Pacific flight. *From left*: Charles Kingsford Smith, Harry Lyon, Charles Ulm, James Warner. Smith and Ulm wear RAAF uniforms of the 1920s, and carry the canes which in those days were favoured by officers on informal occasions (*The National Library of Australia*)

a suitable landing area lay in that direction. But Smith thought that it pointed towards Wyndham and headed westwards.

Soon after this they passed over Port George Mission and dropped a message saying "Place white sheets pointing direction of Wyndham. Mark in large figures number of miles." The results showed them that Wyndham was 250 miles to the east—but by this time there was less than enough petrol in the tanks of the *Southern Cross* for two hours' flying.

Nevertheless Smith kept on flying east until he realised that they had no hope of reaching Wyndham and turned back towards Port George Mission. Luckily, McWilliams had been sending out messages which were received in Sydney and Melbourne, and soon he sent Smith's final message which read, "Have become hopelessly lost in dense bush. Now faced with forced landing at place we believe to be 150 miles from Wyndham in rotten country. Wish us luck. Will communicate again as soon as possible."

Colonel Brinsmead and the staff of the Civil Aviation Branch in Melbourne had seen all of Smith's messages and on 1 April realised that he was lost and asked Norman Brearley to help in the search because the WA Airways airfield at Derby was nearest to the area in which Smith had disappeared.

Brearley responded immediately, and the subsequent search which went on for twelve days probably was the world's first big organised search for a missing aircraft. No effort was spared. Wackett was even asked to fly his new Widgeon amphibian to Wyndham, but it was realised that this was impracticable. The RAAF sent five DH9A aircraft to help, but three of these crashed, without loss of life.

Naturally the disappearance of "Smithy" was red-hot news, and Brearley, in Perth, was pestered by journalists and other people seeking information. The search was co-ordinated by Colonel Mansbridge in Broome, who directed Brearley's pilots Eric Chater and Jimmy Woods. A DH61 airliner named *Canberra*, flown by Les Holden, left Sydney to join the search, and the *West Australian* newspaper chartered a DH50 to be flown north by Bertie Heath with two journalists. Brearley sent the best of his mechanics north to maintain the aircraft, and his Perth workshops ensured that the supporting aircraft and engines were kept in good trim.

The searching aircraft criss-crossed the whole area, but none of them spotted the *Southern Cross* lying on a mud flat about 135 miles north-east of Derby in the valley of the Glenelg River north of the King Leopold Ranges. The flyers had named the place "Coffee Royal", because of the dark brown mud-flats. They had landed without injury, and had dragged the *Southern Cross* out of the mud on to drier ground, but without petrol they had no hope of taking off again.

By this time the search had been joined by Smith's old friend Keith Anderson, who with his mate Bobby

TOP: Menu of the dinner given by the Australian Aero Club to honour Kingsford, Ulm, Lyon, and Warner (*The Crome Collection, National Library of Australia*)

BOTTOM: The *Southern Cross* taking off from Christchurch, New Zealand, on Kingsford Smith's trans-Tasman flight (*The Crome Collection, National Library of Australia*)

TOP: One of the early Consolidated PBY-5 Catalina flying-boats of the RAAF at the RAAF base, Rathmines, New South Wales, in 1942. (*painting by Dennis Adams: Australian War Memorial*) RIGHT: Lockheed Hudson A16-210, of 2 Squadron RAAF stationed at Ambon, attacks Japanese ships and landing forces in 1942. (*painting by Roy Hodgkinson: Australian War Memorial*)

A Short Sunderland flying-boat of 10 Squadron RAAF attacking a German submarine in the Bay of Biscay, 1942. Personnel of 10 Squadron arrived in England in 1939 to pick up Short Sunderland flying-boats ordered by the RAAF, but when the war broke out they remained in the United Kingdom and flew many operations with Coastal Command. (*painting by Frank Norton: Australian War Memorial*)

FRANK NORTON 1942

On 26 December 1942, Flying Officer J. S. Archer and Sergeant J. L. Coulston, flying CAC Wirraway A20-103 of 4 Squadron RAAF, were on a tactical reconnaissance over a Japanese ship wrecked off Gona-Sanananda in the Papua New Guinea area. Archer sighted a Japanese Zero about 1,000 feet below, and without hesitation he put the nose of the Wirraway down and fired a five-second burst with his Browning .303 machineguns. The Zero burst into flames and plunged into the sea. *(painting by Roy Hodgkinson: Australian War Memorial)*

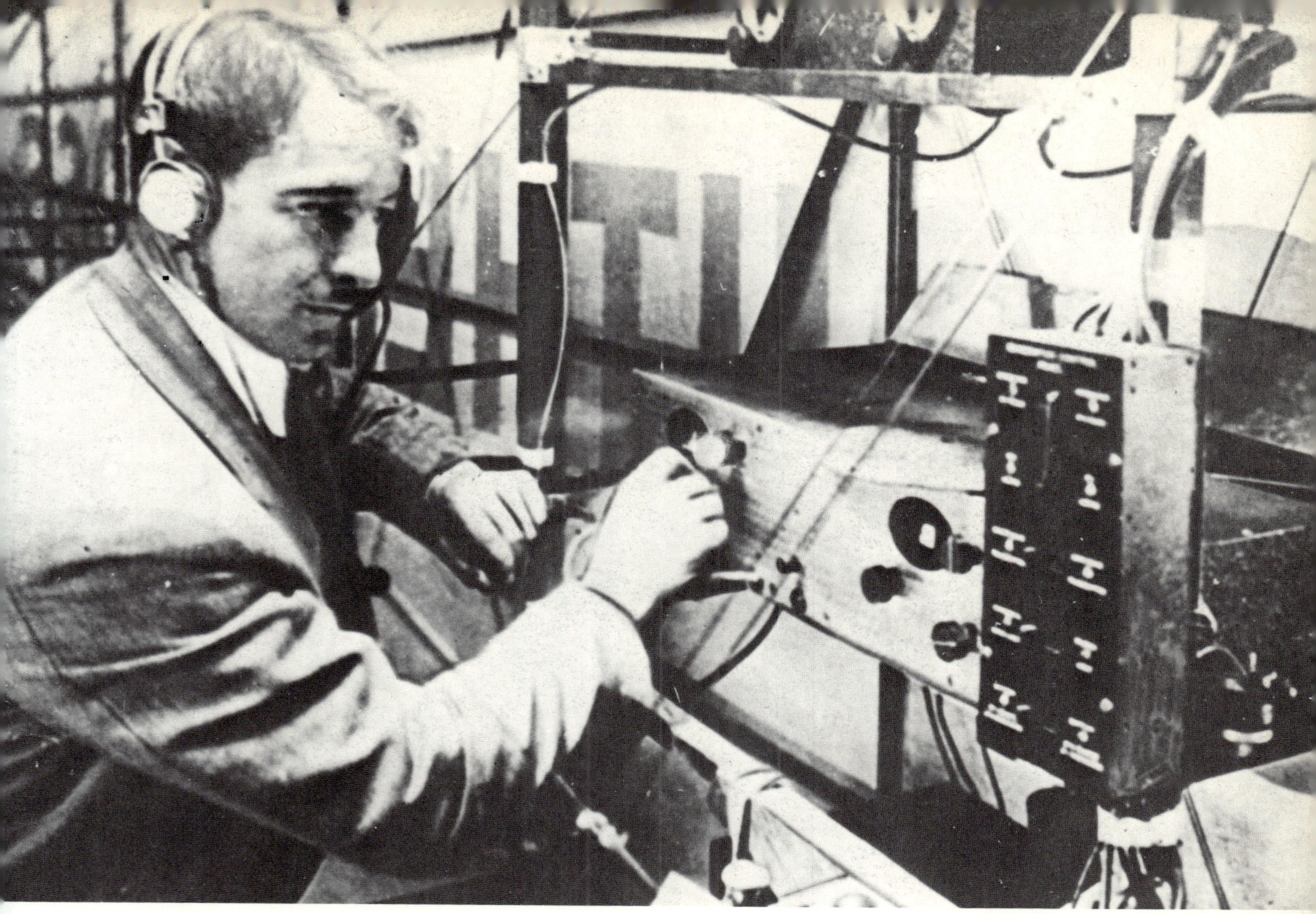

James Warner tests the radio equipment in the *Southern Cross* before the trans-Pacific flight. (*National Library of Australia*)

Hitchcock had set out from Alice Springs in their little aircraft *Kookaburra*. They too disappeared, and were not found until 21 April when Lester Brain, who was to become the first general manager of Trans-Australia Airlines in 1946, spotted the *Kookaburra* in rough country eighty miles south-east of Wave Hill. Three days later a ground party reached the aircraft and found that Anderson and Hitchcock had died of thirst. On the rudder of the aircraft Anderson had written that they had lost their way because of a malfunctioning compass and had been forced to land by an engine fault.

At "Coffee Royal" Smith and Ulm had been trying to make their wireless transmitter work but without success, although by this time they had rigged up a receiving aerial and could listen to news bulletins about the search for them. So at least they had some hope of rescue, but day after day went by without relief. The scanty rations which they carried in the aircraft had soon disappeared and they were living on mud snails. They were sunburnt, smothered with mosquito bites, and parched by the heat. Fortunately there was enough muddy water to keep them alive. From time to time they heard the drone of distant aircraft, but none of the searchers saw them.

Ulm had been keeping a diary, in which he recorded brief details of their predicament and wrote that they saw mirages of good food. On the twelfth day after their forced landing, he wrote "Saved . . . saved . . ."

Les Holden had spotted them from his *Canberra* and had dropped food and messages but could not risk landing his big aircraft on the rough ground. Next day, Bertie Heath landed the DH50 which the *West Australian* had chartered from Norman Brearley, and the first people to talk to the flyers were a Press photographer and journalist. For the flyers this experience was a real anticlimax, especially when the photographer was crass enough to ask the exhausted men to lift Bertie Heath on their shoulders so that he could obtain a good picture.

Like so many other flyers of those early days, the man who spotted Smith and Ulm was to die young. He went to New Guinea and opened Holden Air Transport to fly from the coast to the goldfields, and was killed in a crash.

Litchfield and McWilliams were flown to Derby

by Bertie Heath, and a few days later Smith and Ulm flew the *Southern Cross* to Derby and then to Sydney where they arrived on 27 April. The Commonwealth Government had underwritten the cost of the search, and this amounted to nearly £20,000. Anderson and Hitchcock had died in the attempt to find the *Southern Cross*, and soon the newspapers began to demand an inquiry. Unpleasant rumours began to circulate, and these were dragged into the light when, during the official inquiry which began on 13 May 1929, Kingsford Smith was asked, "Was the forced landing a premeditated publicity stunt?"

He snapped, "That is an absolute, deliberate and malicious lie that affects dead men's reputations. I consider it disgusting!"

Unfortunately, Charles Ulm had made a remark which stuck in people's memories. In 1927, when he and Smith were trying to raise money for their record-breaking flights, he said jokingly that all they needed for some big publicity was to get lost in Central Australia. This remark was repeated in court, but, in defence of the flyers, Norman Brearley's pilots and one of the journalists said that the idea was unthinkable. The journalist said, "It is my emphatic opinion that no sane man would have undergone the rigours of their experience for any sum of money." Brearley, too, has never believed this rumour. He believed that the safe landing of the *Southern Cross* on the mud-flats had been yet another example of Smith's magnificent flying, and thought that if anyone had wanted to carry out such an exploit for publicity he would have chosen a much safer place to land. After the inquiry, he asked Smith, "Tell me, was it a stunt?"

Smith looked hard at him and said, "You know I could never lie to you, Norman. No, it wasn't a stunt."

In fact they were extremely lucky to be found at all. In April 1974 it took thirty-three aircraft, including four RAAF Dakotas and a helicopter, a week to find a Cessna which crashed in north-west New South Wales, and in August of the same year it took five days to locate a Cessna which had force-landed in the Simpson Desert.

The inquiry was held in Sydney by an Air Enquiry Board consisting of Brigadier-General L. C. Wilson as chairman with C. N. McKay and G. Hughes. One of the men who appeared before the Board was Edward Carey Francis, a Perth mining engineer and geologist, who said that he could not understand why the flyers should have suffered from lack of food. He said, "I am in the position of stating definitely that I know the exact spot on which the *Southern Cross* landed.

"I was camped half a mile from the landing place during January, and about five miles away during February last. On my return to Perth on May 16 I was interviewed by the press. In the course of that interview I said that one could not quite understand the difficulty Kingsford Smith and party had in obtaining food of some kind during their stay at Coffee Royal. My reason for making this statement is on account of my knowledge of the natural food resources of that particular part. From my own experience there mudfish are easily obtainable by hand. I have never eaten mudfish, but I know they are esteemed as food by the Tamils of India. Local natives say they are good food . . . there are also in the mangroves, within 400 yards of the *Southern Cross* landing place, quantities of oysters, which are really good food and which I have personally gathered and eaten in that locality. Crabs are very plentiful, and to get them it is only a matter of digging out with your hands their holes at the foot of mangroves along the creeks exposed at ebb tide."

Francis said that goannas were plentiful and that he had "eaten and enjoyed them" though inexperienced men might find it hard to catch them, and that kangaroos, turkeys, and wallabies abounded. He pooh-poohed a published statement that the flyers would have been in danger from dingoes and crocodiles, and said he couldn't understand why they had not set a

The *Southern Cross* in trouble. This photograph was not taken during the "Coffee Royal" incident, from which the aircraft escaped comparatively unscathed, but after a later crash landing (*Hawker Aircraft Limited*)

signal fire because "round the Coffee Royal salt pan are tons of dead mangroves and other wood of good burning qualities, and in spite of wet weather, anyone might reasonably, even without an axe or other instrument, be expected to have no difficulty in making a fire of any size."

Probably the last man to see Anderson and Hitchcock alive was a prospector named Martin G. Krommer. He told the Board that on 9 April he was at Algebuckina, about thirty-six miles south of Oodnadatta. He saw an aircraft pass overhead, return, and then land near by. The Adelaide *Advertiser* of 30 May 1929 reported "He went over to see it, and there were two men, to whom he spoke. One of the men said his name was Anderson, and he called his mate 'Bob'. They told him the exhaust valve was out of order. It knocked, and caused them to land. They also said it was the best landing place they could see from the air. One of the men went to look for a get-off. Mr Hitchcock had only a few tools. He had no spanner to turn the nuts, and had to undo them with a screwdriver and a thing that looked like a corkscrew. Mr Anderson said to him, 'I thought you had a good set of tools,' and Mr Hitchcock replied, 'Someone stole them while I was in hospital.' Mr Hitchcock said he had only come out of hospital when the bus started. Mr Anderson said to him, 'If you had only told me at Broken Hill I would have got you some.' Mr Hitchcock replied that there were some tools in the machine. When they started up the engine two or three pieces of cotton waste came out of the exhaust pipe, and Mr Hitchcock said it was enough to cause the trouble. The machine had landed on soft, spongy ground, and the witness and Mr Hitchcock pushed it until it moved under its own power to higher ground. The airmen told him that they reckoned it would take four hours to reach Alice Springs. They then took to the air, and he did not see them again."

There are always those who like to snipe at heroes, and the rumour clung to Smith for the rest of his life but did not worry him very much. The "Coffee Royal" incident also did not deter him, and in June 1929 he and the *Southern Cross* completed the round-world tour by flying from Australia to London in the record time of twelve days eighteen hours, and in June 1930 he flew from Ireland to New York and on to San Francisco, the place where he and Ulm had started to make their dream come true.

In 1928 Smith and Ulm launched into a venture which in one sense was as risky as flying the Pacific. They started a commercial airline to fly regular passenger services on the Brisbane–Sydney–Melbourne–Tasmania route and called it Australian National Airways Pty Ltd—which had no connection with the later firm of the same name. But despite all the problems of finance and the sidelong way in which many of the public still looked at aviation, the two men had a lot going for them. They were riding high on a crest of fame and so their airline gained an instant reputation, and they could attract some of Australia's best pilots to fly for them—such men as P. G. Taylor, Jim Mollison, and E. Chaseling. And, of course, they themselves flew.

They began the service with the *Southern Cross* and expanded their fleet with four Avro X aircraft, which were Fokkers built under licence by Avro Capable of carrying eight passengers and two pilots, they were named *Southern Sun, Southern Star, Southern Cloud,* and *Southern Moon.*

The very first flight of the newborn Australian National Airways came close to disaster. Smith and Ulm announced a daily Sydney–Brisbane and Brisbane–Sydney service, flown by three-engined aircraft carrying the pilot, co-pilot, and a payload of eight passengers. The fare was £9 13s and the flight took five-and-a-half hours.

Aviation writer T. G. Jones has researched the story of the first flight, and says that Ulm left Brisbane at 8.00 a.m. on 1 January 1930 in the *Southern Sky* while Smith left Sydney at the same moment in the *Southern Cloud,* with his father among the passengers. The two aircraft should have passed about midway between the two cities.

As Ulm crossed the McPherson Range he flew into such severe turbulence that the aircraft became almost uncontrollable, and when he descended through thick cloud he found that he was flying into a valley of the ranges. He had to land the *Southern Sky* in a small paddock among the timber. Miraculously, no one was hurt, and the aircraft suffered so little damage that she was flown out again a few days later.

After this all went fairly well. Although the pilots had no radio and only rudimentary weather forecasting, they were all experts on instrument flying and kept the service going in all but the worst of visibility. People were beginning to think that flying was not all that much more dangerous than driving.

Then, on 21 March 1931, the *Southern Cloud* disappeared. With T. W. Shortridge as captain and C. L. Dunnell as co-pilot she had taken off from Sydney on a routine flight to Melbourne, carrying six passengers: W. O'Reilly, C. Hood, and Miss E. Stokes of Sydney, N. A. Farrell of Caulfield, Victoria, a Mr Margules of Brisbane, and Miss F. Glasgow who had been "booked through Thos. Cook & Sons" so may have been on a world tour.

The *Southern Cloud* should have landed at Melbourne's Essendon airport at 3.30 p.m., but did not even arrive at her refuelling point of Bowser, five miles north of Wangaratta.

Scores of aircraft joined the search. Jim Mollison landed at Melbourne in the *Southern Star* from Tasmania, and within half an hour took off again to scan the thickly-forested mountain ranges on the route between Melbourne and Sydney. G. U.

"Scotty" Allen was in the air, in the *Southern Moon*. Costs were sky-rocketing, and during the search Smith crash-landed the *Southern Sun* at Holbrook.

Hopes faded very slowly. A surprising number of people survived crash landings in those days, when even large aircraft could be let down at low speeds and pilots were skilled at tackling rough-and-ready airstrips. F. H. Stewart, the managing director of ANA, said, "We know the man [Shortridge] and so long as Mr Shortridge could see the timber he would get the aircraft down without serious damage. A portion of the aeroplane would be damaged, but Mr Shortridge could get the plane down in the most difficult conditions and save the passengers."

But it seems that Shortridge had not been able to "see the timber". The search was called off at long last and the *Southern Cloud* remained a mystery until more than twenty years later, when workers on the Snowy Mountains Hydro-Electric Scheme found her remains scattered in dense bush. Apparently the aircraft had ploughed into the mountainside, killing all the passengers on impact.

The disaster came a few months after another crash which received wide publicity. Captain C. H. F. Nesbit, managing director and chief instructor of Wings Limited, died with two "pupil pilots" Haidee Rae and Bill Bell when his de Havilland Puss Moth crashed at Chidlow in the hills behind Perth. Flying was getting a bad name again, the ANA passenger bookings fell away, and this together with the cost of the search and previous financial problems forced Ulm and Smith to put their airline into liquidation.

Kingsford Smith was reduced to making a living by selling joyrides, though the honours continued to flow in. In 1932 he was knighted.

In September 1933 he went to England again, this time with the ambition of establishing a new solo flight record to Australia. Previous records had been set by such famous names as C. W. A. Scott and Jim Mollison, who in 1931, in DH Moth aircraft, had established successive new records which broke Smith's own solo record established in an Avro Avian in 1930. Mollison made it in eight days, nineteen hours, twenty-five minutes. In a Percival Major Gull, Smith set a new record of seven days, four hours. Things began to look up again when the Commonwealth Government gave him £3,000 and he was appointed aviation consultant to the Vacuum Oil Company.

In 1934, after several more flights which kept him in the headlines, he made the first west-to-east crossing of the Pacific—and in a single-engined aircraft to boot. She was a Lockheed Altair which he named *Lady Southern Cross*. His co-pilot was P. G. Taylor.

Taylor also accompanied him, together with J. Stannage, in a flight from Australia to New Zealand with a special airmail cover for the jubilee of King George V, in the old *Southern Cross*. When they were about 600 miles out above the Tasman one of the engines faulted and Taylor performed the astonishing feat of climbing out under the wing, not once but several times, to transfer oil from one engine to another. For this he was awarded the Empire Gallantry Medal, later replaced by the George Cross. "Smithy" was in the headlines once again, because apart from Taylor's courage the flight had been yet another example of Smith's flying skill. Soon after this the *Southern Cross*, which Smith called "the old bus", was sold to the Commonwealth Government for £3,000 and became a national relic.

And Air Commodore Sir Charles Kingsford Smith, KBE, MC, was not merely a national but an international figure. He was older now than the young man who had won the MC for "various acts of foolishness", and was beginning to feel the physical and nervous strain of many years of flying and economic pressures. On 6 November 1935 with J. T. Pethybridge he left England for Australia in the *Lady Southern Cross*. They reached India safely and then took off for Singapore, but were never seen again. One of the aircraft's wheels was washed up on the coast of Burma months later but nobody knows what happened. There have been many theories, including the one that Smith was dazzled by the sun on the sea and flew too low, but it seems hardly likely that this could have happened to a man of so much experience. But somehow, even though his death shocked Australia into national mourning, there is something strangely appropriate about the manner of his end.

It was almost exactly a year since his old friend, Charles Ulm, had disappeared in exactly the same way. Ulm had been managing director of Australian National Airways, and when they had to liquidate this company he bought the *Southern Moon*. He modified her in such ways as lengthening her wings and renamed her *Faith in Australia*, and in this aptly-named aircraft began to make a name which was as well known as Kingsford Smith, though he did not radiate the same kind of glamour as "Smithy". With a crew of such aviators of the day as P. G. (later Sir Gordon) Taylor, G. U. Allen, and R. T. Bolton, he made two flights from England to Australia and in 1933 created a new record of six days, seventeen hours, fifty-six minutes, and in 1934 made three trans-Tasman flights on one of which he carried the first official airmail to New Zealand. His wife and a Miss N. Rogers flew with him as the first air passengers on that route.

In the same year he made the first return airmail flight to New Guinea and on 4 December 1934 he left Oakland with G. M. Littlejohn and J. L. Skilling to fly an Airspeed Envoy to Australia. Contact was lost when they were between San Francisco and Hawaii and they were never seen again. When an air and sea search had failed to find them, the Australian Government voted £5,000 for Mrs Ulm and their son John.

A man who was to the Australian public almost as glamorous as "Smithy" was Bert Hinkler the boy from Bundaberg. He was born there in 1892 and became fascinated by flying almost as soon as he could walk. He studied the flight of ibises and learnt enough from them to build successful model gliders. He tried to find someone who could teach him to fly, but that was difficult in Australia in those days so he went to England and joined the "Sopwith school" where so many young Australians were learning the flying trade. As soon as war broke out he joined the Royal Naval Air Service and flew as what was known as an "aerial gunlayer". He won the Distinguished Service Medal, was promoted to sub-lieutenant, and ended the war in the RAF.

Even before the Australian Government had decided to give a prize for the first England–Australia flight, Hinkler wanted to give it a go in a tiny Sopwith Dove. The authorities thought he was crazy and he could gain no support. So instead of returning to Australia he worked for the A. V. Roe company in Manchester for a couple of years, saved up enough to buy a second-hand Avro Baby and in 1920 made the first non-stop flight across Europe from London to Turin. This won him the Britannia trophy of the Royal Aero Club, and the support of the Avro Company, which shipped an Avro Baby to Australia as a demonstrator for him to use.

In a letter home to Bundaberg in 1919 he had written, "What a day when I land in Gavan Street!" On 11 April 1921 he did just that. He flew the Avro Baby from Sydney to Bundaberg, and incidentally created a new Australian non-stop solo flight record of 650 miles, landed in a reserve, and then taxied along the streets to his parents' home. This exploit in itself was a remarkable example of the way in which the world was changing. The streets were so free from traffic, electric light poles, and other obstructions that an aircraft could taxi along them, yet the whole Age of Flight was bursting from its chrysalis.

Hinkler never sought publicity and this stunt was just for the fun of it. He returned to England and Avro, and for seven years flew for them as a test pilot. His only eccentricity was that of wearing a hat, sometimes a bowler, instead of a flying helmet. But he was making a name for himself and in 1925 he was chosen by the Royal Aero Club to join the Schneider Trophy team going to America. In 1927 he created another long-distance record by flying from London to Riga in Latvia, and later that year made an unsuccessful bid to establish a new England–India record. But his dream was that of flying solo from England to Darwin, a dream which seemed so crazy that he could not even interest a newspaper in financing him or raise enough money to insure the Avro Avian which he would use for the flight. Finally backing came from various firms who would use his name in advertising if he

Bert Hinkler's aerial exploits captivated the Australian public and brought him international fame (*RAAF Museum*)

succeeded, and on 7 February 1928 he went to Croydon aerodrome with his wife Nancy and a man from Avro. In extreme contrast to the uproarious farewells and welcomes given to other flyers of the period, they were the only people apart from the ground crew to watch Hinkler take off.

Aboard the Avian were cigarettes, newspapers, and a few sample bottles of Scotch, together with a *Times* atlas to help him find his way. The liquor and tobacco were for trading en route and proved most useful. The flight itself was comparatively uneventful. Each day for fifteen-and-a-half days he flew steadily onwards, landing at pre-arranged points and carefully checking his engine and airframe before eating or sleeping. Each dawn he took off again, and when he reached Darwin he had broken the previous England–Australia record by twelve-and-a-half days and had made the world's longest solo flight until that time. The retiring and almost unknown airman shot suddenly into the headlines and received a full-blown Australian welcome. He had to endure a continual round of official banquets and speech-making though it left him as quiet and modest as ever. About all he seemed capable of saying in reply to the speeches was, "I'm glad to be home. Thank you." The Queensland Government gave him £500 and the Commonwealth Government gave him £2,000. He was awarded the Air Force Cross and made an honorary squadron leader in the RAAF but he never used the title or wore the uniform.

"Hinkler in his tiny plane dwarfs giant distances" one newspaper headline said, and another: "Hurrying East—Hinkler's Great Flight". London's *Punch* magazine greeted the feat with a cartoon of a kangaroo jumping up and down, Australian flag in paw, greeting the aircraft. The caption read: "Kangaroo to his hero: 'Hinkle, Hinkle, Little Star! Sixteen days and here you are!'"

Later the London *Daily Telegraph* wrote: "Squadron Leader Hinkler's flight will always remain unique, whether his record's broken or otherwise. Not only was he his own mechanic, but for many stages he had no ground organisation. The Australian flight, unlike some tests, is in the highest degree useful. Much knowledge has been accumulated by the pioneers in charting strange regions, in their endeavour to girdle the earth." The Avro Avian 581E, registration G-EBE, in which Hinkler "girdled the earth" now hangs from the ceiling of Queensland Museum, in Brisbane.

He went back to England and set to work to design an amphibian monoplane which would incorporate all that he had learned as a solo flyer of light aircraft. But he was ahead of his time, and neither in England nor America could he interest a manufacturer. Boyhood memories of the days when he had studied birds and dreamed of flying all went into the name which he gave the monoplane: *Ibis*. At last he had to sadly store the prototype in a barn near his Hampshire home, and it stood there through all the years until 1953 when it was brought to light again—and copied by the local aero club which produced a number of similar aircraft.

In further attempts to raise money he made what was then one of the most remarkable flights ever known: from New York to the West Indies, Brazil, West Africa, and then to England. He flew a Canadian-built de Havilland Puss Moth.

The man whom *Aircraft* called "the tough little pilot" took off on 25 November 1931, and the magazine described the rest of his flight. The toughest part was the twenty-two hours of flying from Port Natal, Brazil, to Bathurst in West Africa. "He fought his way through terrific rain and thunderstorms, and the immense cloud formations that overhung the sea.

"Six hours he spent flying a few feet off the water below clouds that raced by just overhead. Six more hours, and those during the night, he battled through cloud banks at heights up to 12,000 feet, his outward vision completely blanketed, and his only guide the two compasses and his own rare instinct for direction.

"When the fuel remaining in the tank would have

Bert Hinkler casts an eye over his Avro Avian in a hangar at Maylands Aerodrome, Perth (*The West Australian*)

Bert Hinkler and his Avian at the centre of an admiring crowd in Perth (*The West Australian*)

been enough only for another two or three hours in the air, Hinkler crossed the African coast 10 miles south of Bathurst, turned north and landed on the Bathurst flying field. The close approximation of the landfall to his predetermined objective is noteworthy after a sea crossing of such distance."

For this exploit Hinkler received a great deal more of the fame which he did not seek but very little of the money which he needed to put the *Ibis* into production. Among the honours heaped on him were the Johnston Memorial Prize of the Guild of Air Pilots, given each year for the best feat of aerial navigation, in memory of Squadron Leader E. L. Johnston, navigator of the ill-fated airship R101.

Among its recipients was Francis Chichester, who as an immigrant New Zealand amateur pilot flew 500 miles over the Pacific by means of solar observation and returned safely. He, of course, gained later fame by his sailing exploits.

Some of the financial proceeds of Hinkler's flight came from allowing his name to be used in advertisements—just as Kingsford Smith allowed his to be used by Avro advertisements headed "KINGSFORD SMITH'S CHOICE". With some lack of originality, further ads said "THE CHOICE OF LIEUTENANT HINKLER".

On 7 January 1933 Hinkler took off in the Puss Moth in an attempt to make a new solo record from England to Australia. By that time C. W. A. Scott had established a new record of eight days, twenty hours, forty-seven minutes. But Hinkler, like Ulm and Smith, disappeared, though unlike them his fate was revealed. In April 1933 mountaineers found Hinkler's body close to the wreckage of the Puss Moth near the Passo Delle Vacche in the Apennines of Italy. The winter snows had preserved his body, which the Italians buried with military honours.

In those dashing days Bert Hinkler's solo record from England to Australia did not last for long. In 1930 Kingsford Smith set out to break it in an Avro Avian single-engined biplane, *Southern Cross Junior*, powered by a DH Gypsy II engine. He was sponsored by the Vacuum Oil Company. Lieutenant C. W. Hill was also making a separate attempt at the record.

Smith made Rangoon, Burma, in seven days. Hill took nine days; Hinkler had taken eleven. All went well until Smith pancaked the Avian at Singapore, but night-long work got her into the air again. Smith and Hill were almost neck-and-neck at Bima, but Smith was leading on time elapsed.

As usual the public waited to acclaim a new record-breaker—especially when it was their beloved "Smithy". The only sour faces were on the Darwin Council. The Mayor, Alderman Brogan, asked his fellow councillors what plans should be made for a civic reception of the aviators "who might arrive within a few days".

"These flyers are becoming as common as potatoes," one councillor cried.

"They're more common," retorted another. "Potatoes are scarce up here."

"Well," chipped in the Mayor again, "it is a bit rough to expect this poverty-stricken town to be running about entertaining them."

"Still," said another, "if they get no welcome at the front door of Australia and have to wait until they reach the back door or the side door, there will be a fuss. Perhaps it would be just as well if the Mayor took a run to the aerodrome and met them; but as for a civic reception in the town—they would rather sleep than listen to us."

The meeting was adjourned when it was decided that the Mayor and two councillors should discuss the matter with the Government Resident. The result is lost in history, but Smith flew triumphantly in to Australia to establish a new record of just under ten days, and this stood until C. W. A. Scott, Jim Mollison, and C. A. Butler established a rapid succession of new solo records in 1931. The airmen were bringing Australia closer to England all the time.

"Cor, look at that!" the small boy in the centre of the picture seems to be saying as a crowd of Perth people watch Hinkler's aerobatics on one of his trips to the West (*The West Australian*)

A formation of Australian-built CAC Wirraways in flight over the New South Wales coast in August 1942, showing the type of camouflage and markings in use at the time. (*painting by Dennis Adams: Australian War Memorial*)

Top: Personnel of 22 Squadron RAAF sheltering under the tailplane of one of the squadron's Douglas Boston twin-engined attack bombers during a tropical downpour. Flight Lieutenant W. E. Newton won a Victoria Cross while flying Bostons of 22 Squadron. (*painting by Alan Moore: Australian War Memorial*) Left: An Avro Lancaster four-engined bomber of the type used by Nos 460, 463, and 467 Squadrons RAAF in the United Kingdom for attacks against targets in Europe 1942–45. (*painting by Colin Colahan: Australian War Memorial*)

6

Aero Clubs, "Aircraft," and the Second Great Race

A magazine editor pushed at politicians to support Australian aviation while the Aero Clubs kept Australian flying alive and well

In 1912 Bert Hinkler had been unable to find anyone to teach him to fly. Twenty-one years later Australian aviation had come of age. The issues of *Aircraft* for 1933 show that despite the Great Depression everything to do with Australian aviation was struggling vigorously for continued existence—though the magazine for that year bore advertisements "inserted by a number of aircraft operators, pilots, and aircraftsmen on behalf of their less fortunate unemployed fellow craftsmen". These advertisements appealed to Members of Parliament to demand "reform for the future, and a Royal Commission for the past" on the spending of money by the Civil Aviation Branch.

It claimed that the money was being "pumped into a few departmental favorites" and that "many branches of aviation, notably gliding, tuition, taxi services, aerial surveying and outback mail routes are being neglected or stifled by the Civil Aviation Branch, and forced out of existence by oversubsidised monopolies".

"It is the duty of all Federal Members to INSIST THAT THE WHOLE OF THE CIVIL AVIATION VOTE IS SPENT EMPLOYING AUSTRALIANS WITHIN AUSTRALIA, AND NOT ON COLORED LABOR BETWEEN DARWIN AND SINGAPORE" the advertisement proclaimed.

Many Australians were near to starvation in that year—including trained pilots and ground crew. Nevertheless the magazine shows that at least the nucleus of Australian aviation was being kept alive in a variety of ways. There were 185 civil aircraft of all types, but principally various de Havillands, on the register on 31 December 1932.

The register of commercial pilots' licences lists 220 names, including such famous identities as Brearley, Ulm, Smith, Miller, Parer, and Fysh. Of these, seventy-six were licensed as flying instructors.

On the list of licensed private pilots there were 356 names, of which twenty-seven were women, a number were doctors, and at least one, the Reverend C. M. Kemmis of Wilcannia, New South Wales, was a clergyman. Perhaps he had found that flying was the best way to cover his outback parish.

By 1973 Australia had 2,257 commercial aircraft and 1,648 private aircraft; 3,251 commercial pilots and 12,701 people holding private licences.

The private and commercial flyers of the early 1930s could buy their aircraft through any of the thirteen Australian representatives of British and American manufacturers. Only one Australian aircraft manufacturer, Larkin Aircraft Co. Pty Ltd, was listed in the magazine's register of manufacturers, and the same firm built gliders, as did two others in Victoria.

If you broke an air-screw then Hodge & Dose, of Marrickville, New South Wales, could repair or replace it, and engines could be purchased from Cirrus, de Havilland, or Bristol branches in Australia.

Nine agencies supplied accessories from leather coats to maps, and if you wanted to book air travel on any of the Australian or international airlines of that period you could do so at forty-six agencies.

A possibly incomplete list totalling twelve "aircraft operators", including Australian National Airways and WA Airways, was published. Five years before, in May 1928, the first mercy flight had been made by

TOP: The Bristol Bulldog was one of the classic aircraft produced between the wars. BOTTOM: Another classic: The Tiger Moth, used by the RAAF for training and general aviation (*both photos RAAF Museum*)

the Australian Inland Mission Aerial Medical Service, which in 1942 had its cumbersome name shortened slightly to Flying Doctor Service of Australia and later to Royal Flying Doctor Service.

Aviation was still glamorous but was becoming somewhat more regimented and businesslike. Men like Ray Parer could still hop into their aircraft and take off for the wild blue yonder, as he was doing in New Guinea, but both he and his aircraft had to be registered and he would be flying under civil aviation regulations—though in the faraway places these sometimes seemed a trifle irrelevant and were treated accordingly.

Yet it seems that the regulations were still insufficient. *Aircraft* lambasted the authorities for allowing "the famous little Australian, Bert Hinkler" to take off without filing a flight plan, which was an almost unheard-of thing in those days.

The magazine also scolded the aviation authorities about "the Whelan affair", in which a pilot had taken off in "an aeroplane with a water cooled engine and only the poorest equipment to fly into the desert in the very hottest season of the year". Whelan crash-landed, and was rescued only after an expensive search.

In the same year Captain Hans Bertram and "Scotty" Allen crashed an overloaded Junkers while trying to take off from Point Cook on an Australia–England attempt. Among the harmful effects of this, the magazine said, was that the public became still more convinced that flying was unsafe and that an Australia–England airmail service was impracticable.

These disastrously-unsupervised attempts were contrasted disapprovingly with the care taken by other flyers, such as Frank Neale and Bert Hussey who spent three months in preparing airstrips and store dumps for the Mackay Aerial Survey expedition into Central Australia in 1930, and Jack Turner who prepared emergency landing grounds before he flew a party of gold prospectors to The Granites, north-west of Alice Springs.

"The majority of flying men will probably think additional regulations unnecessary . . . the operator who fails to take proper precautions does all the damage," the article said.

The pages of *Aircraft* in 1933 and 1934 were able to list a steady series of new records, beginning with Mrs Amy Mollison (Jim Mollison's wife, *née* Amy Johnson) who broke her husband's London–Cape Town record by seven hours. She flew the distance in four days, ten hours, twenty-two minutes.

Such overseas events occupied a good deal of the magazine which served as a world-wide reporter and interpreter of the aviation scene for Australians, though local happenings received full coverage when they were worthy of it. The various branches of the Aero Club were reported on regularly, and it can be seen that these enthusiastic amateurs were keeping the flame burning brightly. And not all of them were amateurs, because pilots who won their wings in France still kept their hands in by weekend flying with the club branches.

Whatever may be thought of Britain's preparedness for war by 1939, *Aircraft* seemed to have little doubt that the RAF led the way in military flying in the early 1930s and there were frequent reports of new aircraft being taken into service. They included troop carriers, flying-boats, bombers, fighters, torpedo-bombers, and seaplanes. Some of these were employed on active service in distant corners of the Empire, as when a bombing squadron cleaned up a fracas between Zeidi tribesmen and men of the Koteibi and Alawi tribes near Aden. The result, *Aircraft* said complacently, proved that air control was "effective, swift in action, inexpensive and humane".

An aerial search for Lasseter's Lost Reef, by Keith Farmer in a Sopwith Gnu, was reported in detail. So were the training activities of the Citizen Air Force units, which then were very much in embryo though they were supported by the Aero Club branches. The Air Board was so bogged down in its

One of the best-known of all aircraft: the DH Gypsy Moth used by so many flyers for so many purposes for more than twenty years. Veterans of this marque are now eagerly sought after by clubs and individuals who restore them to airworthiness, and a number are still flying. The photograph, of an RAAF Gypsy, was taken in 1925 (*RAAF Museum*)

own bureaucracy that it would not even lend training manuals to the Western Australian Aero Club when it wanted to carry out combined exercises with the army, and in order to get into the air at all the would-be trainees had to pay a nominal fee of ten shillings to the club for the use of its aircraft. At that time there were no RAAF units stationed in Western Australia.

The magazine had kind words to say about Britain and Holland which it regarded as the only countries seriously attempting to establish commercial airlines by contrast to others in which they were an offshoot of the military forces, though this was rather hard on the Americans whose network of domestic and Central American routes was already under way.

It was especially unkind, since an Adelaide Aero Club pilot, W. D. G. Robertson, had just returned from the USA where he had spent fourteen months and had acted as demonstration pilot for such companies as Stinson and Fairchild. He wrote a series of articles for the magazine, including one headed "Autogiro makes rapid advance in America". Kellett Autogiro Company was at that time leading the way in these prototype whirlybirds.

Not all the predictions published were accurate. The editor was pleased to note that Christmas bookings on the Qantas routes in western Queensland were especially heavy, a number of passengers being schoolboys and schoolgirls who "accept flying as it should be accepted, not as a dangerous and questionable means of conveyance, but as the best and quickest means of travel". The editor said that such youngsters with their modern ideas were the future hope of Australian aviation. He also approved a speech by Air Marshal Sir John Salmond, RAF, during the disarmament conference. Salmond and the editor were ignored at the world's peril.

Salmond was battling against cuts in air forces. He claimed that air power must inevitably cause statesmen to hesitate before declaring war, and that if war came the mobility and striking power of air forces would rapidly bring it to an end. Britain did not have sufficient aerial might early in the war to deter German actions, and the war dragged on until supremacy of technology could be achieved.

Improvements to flying such as short wave radio and all-metal construction, new uses for aircraft such as the RAAF helping to fight bushfires, new airlines such as Holyman's service from Tasmania to the Bass Strait islands, and new appointments in the RAAF were all reported in the magazine.

The advertisements were equally informative. You could learn to fly at the Lasco School of Flying, Melbourne airport, on no deposit and easy terms; buy a reconditioned Gypsy Moth for an unquoted price which included flying instruction; apply for a job as Chief Instructor and Manager of the Victorian branch of the Australian Aero Club.

There were some pathetic little ads from flyers seeking work, too. Flying Officer G. Jenkins, late RAAF, RCAF, with 2,100 hours' flying time in all types of aircraft, urgently required a position and would go anywhere.

Another "do anything, go anywhere" ad was inserted by Cyril Freeman of Tasmania, who described himself as a "Young man who wants job among 'planes, wishes to become pilot. Studying I.C.S. course aero fitters and riggers."

Whatever happened to Cyril Freeman? Did he finally achieve his ambition—even in that Depression year when so many families were living on "sustenance"? Similar, perhaps more sophisticated, advertisements still appear in magazines such as *Aviation Week* (USA), *Flight* (UK), and *Aircraft*.

Aboard the "Giant mail planes of New England Airways" you could "save hours and enjoy the marvellous coastal scenery" by flying from Sydney to Brisbane for £7, to Lismore for £6, and to Coffs Harbour for £5. A mere £2 would take you from Brisbane to Lismore. The flight from Sydney to Brisbane took five-and-a-half hours.

From Commercial Aviation Co., Parafield, South Australia, you could buy a DH Gypsy Moth "like new" for £500, a brand-new Avro Avian for £625, a DH9 equipped to carry three passengers for £400, and a five-passenger Fokker Universal Monoplane for £2,200. But in the days when a new six-cylinder Chrysler car cost only £250 these prices were still beyond the reach of most people.

Aircraft did not hesitate to engage in dogfights with the Commonwealth Government. In November 1930 the Royal Dutch Airlines (KLM) and Royal Dutch Indian Airways (KNILM) offered through their representative in Australia, A. Bakker, to run an England–Australia airmail service. The Commonwealth bureaucrats did not even have the courtesy to acknowledge this offer, but in June 1932 it was repeated.

The Dutch suggested that they would inaugurate the airmail service and run it until such time as Imperial Airways, or any other British or Australian airline, was prepared to take it over. The advantage to KLM and KNILM would be that of having extra mail and passenger revenue, and the Australians of course would benefit by regular rapid communication with Europe.

Still the bureaucrats did not deign to give an official reply, and on 1 April 1933 *Aircraft* thundered that "the issue has been confused by the publication of 'statements by officials' of the Defence Department in the daily press. These statements have given the impression that the Dutch companies are endeavouring to compete with Australian concerns, whereas the reverse is the case."

The editor pointed out that Mr Bakker was not asking for a subsidy, a fact which the anonymous officials were turning into a vice instead of a virtue by saying that "the government could not exercise any control over unsubsidised services".

"By delaying the Dutch at every turn, the [Defence] Department is laying itself open to the suspicion that there is some reason which it cannot make public for its attitude," the editor wrote.

And he compared the "efficiency and speed" of Dutch airlines with that of Imperial Airways, saying that KLM showed almost 100 per cent adherence to schedule whereas Imperial Airways had been unable to maintain a regular six or seven days London–Karachi service. "Delays in reaching the destinations have been both frequent and considerable," the editor said.

He wondered whether the British and Australians were unwilling to let the Dutch "give it a go" in case they set such a high standard of service that "a British company following it will be forced to provide the latest and most expensive equipment, or be subjected to awkward comparisons".

Presumably the Dutch gave up in despair because KLM never ran the first airmail service. It was inaugurated by Qantas Empire Airways Ltd in 1934.

Air pageants and "mock dogfights" were popular Aero Club enterprises and the newspapers were still as enthusiastic about flying as they ever were. The Queensland *Daily Mail* gave a series of flying scholarships, but by contrast the Commonwealth Government issued a statement to say that it would no longer pay the cost of searching for missing aviators. Though, "if the exigencies of the service permit", it would allow RAAF aircraft to be used on such searches on condition that the cost of operations was paid.

In that period the Aero Clubs were becoming increasingly important and were playing an ever more vigorous role in the development of Australian aviation. The Aero Clubs have a long history, dating back to 25 October 1914 when a group of army officers at Point Cook decided to form an Australian Aero Club to be affiliated to the Royal Aero Club of Great Britain. Major H. A. Petre, who had travelled from England to inaugurate the AFC school for the Australian Government, was the first president of the Club. The first secretary was Captain T. W. White, who later became the Honourable Sir Thomas White, KBE, DFC, VD.

The Australian Aero Club had scarcely been formed when most of its members went overseas to the various fighting fronts of the first World War, but when the surviving airmen returned a number of them devoted themselves to forming State branches of the Club. With the assistance of civilian flyers or would-be flyers, they formed individual sections of the Club in Victoria, New South Wales, Queensland, and South Australia.

These State branches operated in fairly loose contact until March 1926, when the Australian Aero Club Federal Council was formed. To begin with it consisted only of representatives of the Victorian and New South Wales Clubs, but South Australia joined in 1927, Queensland and Tasmania in 1928, and Western Australia in 1929. In that year a new constitution was adopted and the name was changed to Associated Aero Clubs of Australia. In 1948 the name was changed again, to the Aero Club Federation of Australia, and since 1960, when the prefix was granted by the Queen, it has been the Royal Federation of Aero Clubs of Australia.

The Federation, in its various forms, has been a powerful force in the popularising and development of Australian aviation. Initially the emphasis was perhaps on the "popularising" side of aviation affairs, but that was an important task in the days when the average citizen was somewhat suspicious of aircraft and airmen.

The Aero Clubs played their part by organising aerial pageants, air races, and receptions for pioneer aviators when they landed in Australia at the end of their historic flights. All these events were enormously popular and well attended. Young men and women were able to see that their dream of flying could actually become reality. They could inspect aircraft, talk to flyers, join the local branch of the Aero Club, and at last experience that unforgettable moment when they took off on the first training flight.

By the 1930s the Aero Clubs began to realise that they could play a bigger and more important part than simply that of glamorising aviation. Many of the members were men who had flown in the first World War: men between thirty and forty, with sharply-etched memories of the gigantic effort which the Allies had been forced to make to win that conflict. Now, it was starting all over again. There was a tendency to look upon Hitler and Mussolini as figures of fun, but many of the ex-officers had an ominous feeling that it was all beginning again. And, next time, there would be real war in the air. Men and machines would be needed urgently, but so far as anyone could see the Australian Government was doing little, if anything, by way of preparation.

But there was at least one way in which the Aero Clubs could help: by training young men to fly. A man who felt confident at the controls of a DH Moth could take the step to flying military aircraft without great difficulty, whereas it would take a long time to train suitable pilots if the RAAF had to start from scratch.

The Aero Clubs were willing to do their part, but the problem as always was money. Aircraft had to be maintained and serviced, instructors had to be paid, the airfields had to be kept in proper condition. However much a young man yearned to fly, the chances were that in those Depression years of the 1930s he would not be able to afford the lessons. So the Aero Clubs approached the Government with a simple solution. Subsidise the cost of lessons, and the Clubs would do the rest.

The scheme got off the ground in a fairly small way, but as soon as the second World War broke out the Aero Clubs were ready to provide immediate training and other facilities. They concentrated on basic flying training for cadet flyers so that the RAAF could devote itself to intensive instructor training.

Naturally the Clubs concerned themselves with many other aspects of aviation, and these included the selection of flyers to receive the annual Oswald Watt Gold Medal. This award was instituted in 1921, under the terms of the will of Lieutenant-Colonel Walter Oswald Watt. Watt was a member of a wealthy New South Wales pastoral family, and he travelled widely in Europe. He learnt to fly in 1911, and when the first World War broke out he was in England. He was so irritated by what he saw as the slow response of the British authorities to the emergency that he hurried over to France and joined the Aviation Militaire section of the French Foreign Legion in 1914. In 1916 he transferred to the AFC with the rank of captain, and became the flight commander of 1 Squadron. Before the end of the war he was promoted to lieutenant-colonel in command of the Australian Training Wing in England. In 1921 he dived into a Sydney harbourside pool, struck his head on submerged rock, and drowned before he could be rescued.

A Gypsy Moth and her crew pause to refuel on an outback airstrip while competing in the 1939 trans-Australia East-West Air Race

His will left £500 to the Australian Aero Club to be used for the annual presentation of a gold medal to "the Australian, who in Australia or overseas, puts up the most brilliant performance in the air during the year". The executors have agreed upon two slight amendments to the bequest, so that the medal may now be awarded to an Australian who has made "the most notable contribution to aviation" during the year.

The Aero Club has been exacting in its bestowal of the award and in a number of years has not been able to find a suitable recipient. When it was given to Australia's great veteran flyer, Sir Norman Brearley, in 1973, it was only the thirty-fourth occasion since 1921. The award for 1974 was bestowed on Sir Lawrence J. Wackett, DFC, AFC, in well-earned recognition of his many years of service to Australian aviation.

The list of recipients of the Oswald Watt Gold Medal is a roll of honour of Australian aviation. The first was F. S. Briggs for flights from Melbourne to Perth and Brisbane in 1921. Others include I. E. McIntyre, for his seaplane flight around Australia in 1924 and his flight to the Solomon Islands in 1926 with Group Captain R. Williams (later Air Marshal Sir Richard Williams); Dr C. C. Fenton for his Flying Doctor services in the Northern Territory in 1937; and Captain Edgar Percival who in 1936 set three records in one year: England–Morocco and back in one day, the speed record in the King's Cup Race, and the Folkestone Speed Trophy.

Hinkler and Kingsford Smith, of course, were certain winners. The medal was awarded three times

Another variant of the great Moth family: a Fox Moth used on one of Australia's domestic airways during the 1930s. Passengers sat in the cabin above the mainplane and the pilot's cockpit was situated to their rear (*RAAF Museum*)

to Hinkler, in 1928, 1931, and 1932, and four times to Smith, in 1929, 1930, 1933, and 1934.

Since the second World War the medal has been awarded to seventeen Australians, including Sir Hudson Fysh of Qantas who, like Sir Norman Brearley, was recognised "For a most notable contribution to aviation in Australia". Others have been Martin Warner, for his British Empire Altitude Record in a Gull IV Glider on 30 December 1950, when he climbed to 23,500 feet; Squadron Leader P. G. Fisher for his England–Australia speed record of twenty-three hours, five minutes in a Canberra bomber on 12 May 1952; and Squadron Leader D. W. Leckie who led the RAAF Flight on the Australian Antarctic Expedition in 1958.

The Australian Aero Clubs were thoroughly established between the wars, and after the second World War they really came into their own. In the uncertain period of the 1950s and 1960s when the so-called "cold war" was bubbling and seething in many parts of the world, the Aero Clubs were given the responsibility of training National Servicemen, Air Training Corps Cadets, General Reservists, University Air Squadron Cadets, and some Citizen Air Force Cadets. When this scheme ended the Aero Clubs had given more than 40,000 hours of actual flying-time instruction to say nothing of all the auxiliary activities.

The training of civilian flyers was not pushed aside by this military involvement, and it gradually increased in importance. Many of Australia's commercial flyers received their flight training from the Aero Clubs, and the RFACA introduced a unique low-cost Airline Training Scheme.

Many of the Aero Clubs, of which there are now fifty-nine with more than 12,000 members, are located in areas which are not served by commercial operators. These have given opportunities for career training and helped to develop Australia's pool of trained pilots, as well as giving active help in such emergencies as bushfires, floods, and accidents. The voluntary efforts of the Aero Clubs and their members have established airfields in places which otherwise would have no aerial facilities.

Since 1948 the RFACA has been the Australian representative of the world's oldest international aviation organisation: the Fédération Aéronautique Internationale. In this capacity it has been responsible for assisting and controlling sporting aviation in all its manifestations from gliding to building model aircraft and including parachuting, ballooning, and the amateur construction of aircraft. Under the general control and guidance of the RFACA, specific responsibility for such activities as these is taken by seven organisations: the Aerostats Balloon Society of Australia, the Australian Aerobatic Club, the Australian Parachute Federation, the Australian Women Pilots' Association, the Gliding Federation of Australia, the Model Aeronautical Association of Australia, and the Ultra Light Aircraft Association of Australia. Each of them has acknowledged the important part which the RFACA has played in helping and guiding them.

Nowadays the RFACA is in close control of all aviation record attempts, air races, and similar activities. It issues the permits, ensures that the activities are carried out in accordance with the rules and regulations of the Fédération Aéronautique Internationale Sporting Rules, and certifies and registers all national records. When appropriate, such national records are registered with the FAI as world absolute or world class records.

In 1933 the Australian Aero Clubs had not risen to their present position of recognition and authority and thus played a rather peripheral part in one of the great aviation events of the 1930s. On 1 May 1933 *Aircraft* magazine asked "Who Will Win the World's Greatest Air Race?" as the headline to an article sub-titled "Super-speed machines will compete for Big Prizes Offered for Melbourne Centenary Event". The article went on to discuss an announcement which had set the aviation world in an uproar.

Sir Macpherson Robertson the "Chocolate King" had announced that by way of celebrating the centenary of the founding of Melbourne he offered five prizes for a London–Melbourne air race to begin on 20 October 1934. The first prize was £10,000 and a gold cup; the second £1,500; and the third £500. The most meritorious flights in relation to power and size of craft competing under sealed handicap would receive prizes of £2,000 and £1,000.

The conditions for entry were a good deal tighter than those for the first England–Australia flight. Competitors must follow Jim Mollison's route from England to Australia and check in at a number of points such as Baghdad, Allahabad, Singapore, and Darwin. Any person or nation could enter for a fee of £50, which would be refunded after the race, and there was no limit to the power or type of machine or the number of crew. Replacements were permitted en route, with the exception of the person in charge of the aircraft who must be in it at the finish.

Refuelling in the air, unknown to Ross and Keith Smith, was permitted, and the aircraft could fly at night. Every aircraft must have an airworthy safety certificate, and its flyers must be licensed. Emergency rations, lifebelts, and smoke signals must be carried on the aircraft which would take off from England "on a given signal" for the race to Flemington Racecourse.

It was expected that C. W. Scott's record of eight days and twenty hours would be at least halved by the race.

Aircraft grumbled a little at the conditions, especially the "sealed handicap" which would make it impossible for flyers to know whether they were winning or losing until the end of the race, and also at the clause "the most meritorious flight in relation to power and size": the writer pointed out that this would match a Lockheed Orion, with a range of 1,000 miles at 200 m.p.h., against an Armstrong Whitworth Atalanta with a range of 400 miles at 130 m.p.h.—yet the smaller aircraft could win because of the "most meritorious" loophole.

The writer made many predictions and asked many questions. He thought that Frank Hawks would enter his new specially-built Northrop Gama *Sky Chief*, with its range of 2,500 miles and top speed of 250 m.p.h. And that this American would be joined by his compatriots, Jimmy Doolittle, who held the airspeed record of 296 m.p.h., Harley W. Lake, Roscoe Turner, J. H. Haizlip, and J. R. Wedell in competing for the prize.

Dutch and German pilots were mentioned and there was a picture of "the new German high speed commercial monoplane, the Heinkel 70, which has a top speed of 218 miles an hour and cruises at 203 miles an hour". It had an uncanny resemblance to the later Heinkel fighter-bombers.

This Lockheed P3B Orion, photographed during the 1960s, shows the enormous change in aircraft design that occurred during a mere thirty years (*RAAF Museum*)

"Provided the handicap conditions are sufficiently attractive, several local Australian pilots are keen to make the flight," the writer concluded.

This was the understatement of 1933. For example Ray Parer, then battling for a living by flying in New Guinea, heard the news by radio and jumped into action. The problem was that he did not have a suitable aircraft and could not afford one. His New

The Fairey Fox in which Ray Parer flew in the second Great Air Race, photographed in Sydney (*by courtesy of the Sydney Morning Herald*)

Guinea aircraft were all short-distance load carriers, whereas the race would demand a long-distance speedster. He wanted a Lockheed, which would cost up to £7,000, but that was out of the question.

He had almost given up hope when he saw in *Flight* an advertisement for a Fairey Fox, headed "the only British 'plane with a chance in the MacRobertson Air Race". It was offered at the unbelievable price of £200 with a spare engine, but he found that it was one of the machines originally built for 12 (Bomber) Squadron RAF in 1926 at a cost of £7,000 apiece, with 430-hp Curtiss D12 water-cooled engines giving a top speed of 156·5 m.p.h. They had been leaders in their day but had soon become obsolete.

Parer snapped it up, even though he would have to raise about another £2,000 for general expenses. His mates on the New Guinea goldfields had a whip-round which raised £1,400, and he persuaded Geoff Hemsworth, one of the pilots of Parer's Pacific Aerial Transport, to fly with him as co-pilot. They left for England on 26 June 1934.

By that time sixty-four acceptances, from Britain, New Zealand, Australia, the USA, Italy, France, Holland, Sweden, India, Denmark, Portugal, and Germany had been listed for the Speed Section and Handicap Section of the great race. The Australian entrants were Kingsford Smith in a Lockheed Altair; Jimmy Woods and Horrie Miller in a Lockheed Vega; Parer with his Fairey Fox, entered by the "New Guinea Centenary Flight Syndicate"; Warren Penny in a Vultee 6; D. T. Saville in what was at first mysteriously described just as a "low wing monoplane"; and Mr and Mrs C. J. Melrose in a DH Gypsy Moth.

The pilots from other countries included such famous names as C. W. Scott, Campbell Black, Jim and Amy Mollison (flying a DH Comet for Britain), Wiley Post, Harold Gatty, and Alan Cobham.

They flew a variety of aircraft ranging from a Comper Kite to a Cessna monoplane, from an Airspeed Courier to a British Klemm Eagle. The pilots had been financed by oil companies, aircraft manufacturers, and such organisations as the Irish Hospital Trust Ltd and New York–London–Moscow Air Liners Incorporated.

There were as many tips as on Melbourne Cup Day: Flight Lieutenant Scholfield, a former Schneider Trophy pilot, tipped Kingsford Smith to win, the Dutch pilots for second place, and Wiley Post for fourth to sixth place.

Colonel F. J. Fitzmaurice, flying a Bellanca for which he was reported to have designed "a new secret method of refuelling which will fill the largest tank in a few minutes", said that he and his co-pilot

Amy Johnson was perhaps the most famous of a group of women flyers who received great acclaim during the 1930s for the flying skills and endurance which enabled them to establish a number of new records (*Hawker Siddeley photograph*)

E. W. Bonar would win both the Speed and Handicap sections for Britain, though "Wiley Post, Parmentier and Moll, Clyde Pangborn, Gatty and Smith will contest second place".

Amy Mollison was optimistic, and said that she and Jim in their DH Comet with its top speed of 230 m.p.h. would "have the throttle open all the way".

Ray Parer was in his not unusual situation of narrowly controlled disaster. He and Hemsworth had found the Fairey Fox to be in a barely airworthy condition and were working on it, with the aid of three technicians, right up to the day on which aircraft were supposed to be flown into Mildenhall airfield where the race was to begin. In fact they and some others had to apply for an extension of time because the entry list had been closed on 1 June.

Sir Richard Fairey, head of the Fairey Aircraft Corporation, was much against them entering at all because he thought they were sure to strike trouble which would reflect on his firm. Parer just nodded and, in defiance of race regulations, went ahead with fitting two extra fuel tanks. Luckily the regulations were changed shortly before the race and he was allowed to carry the extra petrol.

Six days before take-off time, Parer went down with a dose of malaria but leapt out of bed when the race authorities said that if he did not turn up at Mildenhall he would be disqualified.

At Mildenhall they found that the Fox's radiator was leaking and they had to fit another on the day before take-off. When King George V and Queen Mary inspected the line-up on 19 October, the number of actual entrants had been whittled down to twenty-one. The others had all withdrawn for various reasons, and, instead of flying in the England–Australia race, Kingsford Smith was making the first west-to-east crossing of the Pacific in his Lockheed Altair.

Those flying for Australia were Ray Parer and Geoff Hemsworth, James (Jimmy) Woods and Horrie Miller, and Mr and Mrs C. J. Melrose. There were ten British finalists, three American, two Dutch, two from New Zealand, and one Danish team.

At the last moment the arrogant Colonel Fitzmaurice had to withdraw because his aircraft *Irish Sweep* totalled a load of 8,000 pounds, 2,600 above the load limit for the race. He said that the extra was necessary for his plans—perhaps including his secret refuelling device.

The finalists began taking off from Mildenhall soon after dawn on 20 October and the race was soon over. C. W. Scott and Campbell Black, in their DH Comet *Grosvenor House*, covered the 11,350 miles of the Speed Section route in what was then the aston-

TOP: A general view of the apron at Mildenhall airfield a few hours before the start of the Melbourne Centenary Race (*Flight International*). BOTTOM: Essendon Aerodrome in 1936 (*McGrath Collection*)

ishing time of seventy hours, fifty-four minutes, and eighteen seconds.

The Dutch pilots Parmentier and Moll won the Handicap Section, and second and third speed prizes were won by the Americans Turner and Pangborn and the British team of Cathcart-Jones and Waller.

As for poor Ray Parer, he seemed to be setting yet another record for the most accident-prone flight from England to Australia. The race had been won before he even got as far as Paris, and from then on he and Hemsworth encountered a series of major and minor problems ranging from water in the fuel to a sandstorm over the desert. They flew in to Melbourne on 13 February 1935, and Parer told the reporters, "At least we've beaten my previous record by about four months!"

While the limelight was playing on such aviation greats as Amy Johnson, Jean Batten, and Jim Mollison, a Queensland housewife was quietly racking up her own score of records. She was Mrs Lores "Harry" Bonney of Brisbane: a South African woman who in 1917 had married Harry Bonney, a leather goods manufacturer. Bonney was a cousin to Bert Hinkler, and Lores Bonney had her first taste of flying in 1927 when Hinkler took her up in his Avro biplane.

Her name had been almost forgotten until the aviation writer Terry Gwynn-Jones published an article about her in the Brisbane *Sunday Mail* of 18 November 1973, yet she was one of the most remarkable flyers of her time and was the first woman flyer to be awarded the MBE.

Her flight with Hinkler gave her an overwhelming desire to fly her own aircraft, and she took lessons at Eagle Farm. Uncertain of her husband's reactions to the idea she did not even tell him until after she had made her first solo flight—only to learn that he had known all the time. And, as a mark of his approval, he gave her an aircraft: a de Havilland Moth biplane which she called *My Little Ship*.

Over the next five years she established records which have not been equalled by any other Australian woman pilot, and by only a few men. On 5 January 1932, in *My Little Ship* with its cruising speed of a mere 80 m.p.h., she set a new Australian record by flying solo over the 1,000 miles from Brisbane to Wangaratta in one day. Then after testing herself on several long flights she decided to try a flight around Australia. She made it successfully and landed back in Brisbane on 27 September 1932—though she had some narrow escapes en route.

Mrs Bonney won the Qantas trophy for outstanding performance by any Queensland pilot in 1932, for her round-Australia flight. But it had only whetted her appetite for the real challenge: that of flying from Australia to England. That was the most difficult way to fly the 11,000 miles from Brisbane to London because it is against the prevailing winds, and Mrs Bonney says she chose it because "Everyone else was coming the other way."

She followed, in reverse, the route taken by flyers who had "come the other way": Brisbane–Darwin–Singapore–Calcutta–Persia–Europe–London. Like many of the others she had her full share of adventures and accidents but she landed at Croydon airport, London, on 21 July 1933, and was awarded the MBE for the flight. Four years later she set a new record: the first flight from Australia to South Africa.

She hoped to make one more long-distance flight, from Australia to London via the Aleutians and New York, but *My Little Ship* was destroyed in a hangar fire at Archerfield three months before the outbreak of war. She retired from flying and now lives on Queensland's Gold Coast: a great but, until recently, almost forgotten flyer of Australia's pioneer aviation era.

7

The First Great Airlift

Gold was where you found it in Papua New Guinea . . . and daredevil airmen solved all the massive problems of jungle transport

Gold was found in Papua New Guinea as early as 1899 by the Germans who were then trying to develop their share of territory, and when the first World War broke out it was being worked in fairly considerable quantities around the coasts and on some of the islands.

Some of the gold was washed from the rivers, and it was not long before some fortune-hunters thought that it would be a good idea to see where the rivers came from. In 1925 prospectors fought their way from the coast up to the source of the Waria River, then clambered through the tangled jungle on the almost perpendicular Kuper Range. They descended into the Bulolo Valley and the biggest Australasian gold strike since the Golden Mile was discovered.

The diggers, who swarmed to the Bulolo Valley from Salamaua and Morobe as soon as they heard the news, had to solve a few enormous problems.

One was that of getting into the Bulolo Valley. If you were as hard as nails, capable of driving a team of natives carrying your gear, and weren't afraid of cannibals, malaria, snakes, and the other denizens of the jungle, then this was fairly easily handled.

The second was that of keeping yourself alive in the valley. You had to carry every ounce of food and equipment in with you, and before long the miners' insanitary camps were ravaged by epidemics.

The third and most troublesome was that of getting the gold out of the ground. It was not simply a matter of pick-and-shovel work or of squatting by the river with a tin dish. The easy pickings soon disappeared, and miners found that they had to wash huge quantities of river soils to flush out the gold.

This needed heavy and complex machinery, and there was no way to get it over the ranges. For most of the time the track through the jungle was little better than a mud slide, and the local natives had more sense than to hire themselves out to carry several tons of metal.

The solution, of course, was air transport. And so, in Papua New Guinea in the early 1930s, the first of the world's great airlifts got under way.

It was, relatively, a far greater feat than the celebrated Berlin Air Lift of the 1940s. That exploit used far larger and more powerful aircraft, flying short distances over fairly straightforward terrain, and helped by a sophisticated air navigation network.

The Australian pilots who flew in and out of the Bulolo Valley had to weave their way through and over mountain peaks often hidden by thick cloud. They used aircraft which had never been designed for such heavy cargo handling. And they had to grope their way down on to airfields which had been hacked from the jungle, and which certainly did not possess such refinements as radio beacons.

The air service was begun by Guinea Gold No Liability, founded by C. J. Levien. Levien soon became tired of paying native carriers £40 apiece for the ten days' climb to the goldfields, especially since the Government would not allow the natives to carry more than fifty pounds per head and a good deal of this had to consist of their food. An airlift seemed the only solution, so in 1926 he contacted Nobby Clark, an ex-AFC pilot living at Rabaul. Clark helped him to plan the airfields and selected the sites at Lae from which the aircraft would take

TOP: HMAS *Albatross*, the seaplane carrier of the RAN, photographed in about 1930 from a Seagull III of No. 101 (Fleet Co-operation) flight RAAF (*RAAF Museum*)

CENTRE: Hoisting a Supermarine Seagull III amphibian aboard HMAS *Albatross*, which carried out the Barrier Reef survey in co-operation with aircraft of No. 101 flight during the late 1920s and the 1930s (*RAAF Museum*)

BOTTOM: One of the Seagull III amphibians on a Queensland beach during the Barrier Reef survey (*RAAF Museum*)

off, and at Wau where they would land near the goldmines.

Levien then founded Guinea Airways Limited and ordered from Sydney a DH37 capable of carrying a payload of 600 pounds. Captain E. A. ("Pard") Mustar was taken on as the company's first pilot. The aircraft was shipped from Sydney to Rabaul, where Mustar had had an airstrip cleared, and on 30 March 1927 he flew it from this airstrip to the one which had been prepared at Lae.

The first two flights from Lae to Wau had to be aborted, because Mustar could not find the airstrip. This is hardly to be wondered at. Only 800 yards long, it ran up the side of a hill with a gradient of one in twelve and aircraft had to land uphill and take off downhill. An important part of the ground crew was a gang of natives who rushed forward to chock the wheels as soon as an aircraft slowed to a stop, to prevent it from rolling backwards.

Mustar made the first successful flight in to the goldfields on 18 April 1927, and in the first six months he carried 80,000 pounds of supplies and 150 passengers. Guinea Airways Limited continued to serve Papua New Guinea until the Japanese invasion, when they moved to South Australia and the Northern Territory. They flew regular services from Adelaide to the State's country centres until they merged with what was then ANA and became South Australia Airways, now Airlines of South Australia in the Ansett group.

As soon as Australia's airmen heard about the airlift, a number of them rushed to take part, for it offered them the flying opportunities which were still hard to find in Australia. One of the flyers, of course, was Ray Parer, who was running a tiny road-transport service with a little farming on the side, on King Island in Bass Strait.

He raised enough money to buy an old DH4 from Qantas, and took her by sea to the islands. With typical "Parer luck" he wrecked her on the first take-off and landed in hospital. The venture seemed at an end until a successful prospector lent him enough to have the aircraft repaired.

Parer's bad luck with aircraft was balanced by his good luck with friends. He had the gift of making friends because he was always ready to help them, and naturally this was reciprocated. It was Australian "mateship" transplanted from the bush to the skies.

When the DH4 was repaired Parer found work for her, but despite that he could not make enough money to keep her flying and in 1930 his embryo airline was liquidated and combined with Morlae Airlines. He flew with this company with varying fortunes including the loss of a new and uninsured Fokker F3. Ray Parer's mates then set him up in business again, and he formed Pacific Aerial Transport with the help of some local independent pilots who added themselves and their aircraft to his operation.

Pacific Aerial Transport and some smaller groups challenged the dominance of Guinea Airways, but they were never to catch up. During the first nine months of operations, Guinea Airways carried 367,000 pounds of cargo between the coast and the goldfields, together with 498 passengers. To begin with these were willing to pay £33 6s 8d for the twenty-five-minute trip from Lae to Wau though eventually this fare was reduced to £5. The return flight cost only ten shillings, perhaps out of consideration for prospectors who had not made their expected lucky strike on the goldfields.

It was a hair-raising flight. Lae airfield was at sea-level, and was 600 yards long but only 100 yards wide. A short distance inland the mountains rose to heights of more than 6,000 feet so pilots had to take off into a sharp climb and gain height immediately, rising to about 7,000 feet to clear the mountains which were sometimes hidden by cloud. From this height they had to begin a rapid descent to the Wau airstrip, which was 3,200 feet above sea-level, in a precipitous valley. So the pilot had to lose about 4,000 feet of height as quickly as possible and descend almost vertically.

Guinea Airways expanded rapidly. In January 1928 they bought their first two all-metal Junkers aircraft: monoplanes equipped with Bristol Jupiter engines and with payloads of about 1,800 pounds apiece. The first was shipped to Rabaul and then flown in to Lae as a seaplane, and the second was flown from Melbourne to Lae. By the end of 1933 the airline was servicing twenty-seven airfields in a semicircle up to 300 miles from Lae and was providing easy transport of a kind undreamed of in Papua New Guinea a few years before.

In December 1929 the airline carried out one of the world's first airlifts of heavy engineering equipment. Guinea Gold did not have sufficient capital to properly work the goldfield so together with a Canadian firm, Placer Development Ltd, had formed Bulolo Gold Dredging Ltd. This concern was faced with the task of transporting two complete dredges over the mountains into the valley. They had considered the idea of building a road especially for the job, but this would have been enormously expensive. In consultation with Guinea Airways and the Junkers Aeroplane Company they planned an adaptation of Junkers G31 three-engined aircraft to carry the various parts of the dredges. The planning included the making of models of the cargo compartment of the aircraft and models of all the parts of the dredges to make sure that they could be fitted into the compartment. The aircraft carried up to 7,000 pounds on each trip, which was only just enough for the heaviest dredge section. The goldmining company bought two of the aircraft, which were flown by Guinea Airways under a management fee, and in April 1931 the operation began. Twelve months later

One of the Junkers freight aircraft used for the world's first great airlift, in Papua New Guinea during the 1930s (*RAAF Museum*)

the two aircraft had carried a total of 2,500 tons of machinery, which included a hydro-electric plant as well as the two dredges, into Bulolo. During the next two years two more dredges and many tons of supplies were flown in. By the end of 1933 Guinea Airways had flown more than 1 million miles in New Guinea, and had carried nearly 7,000 tons of cargo and 12,220 passengers. With its fleet of seven aircraft, of which none had had a serious accident, this Australian-run operation was one of the world's safest and most efficient airlines, although its name was hardly known outside New Guinea.

Ray Parer's Pacific Aerial Transport also regularly flew the twenty-five-minute service between Salamaua and Wau, carrying everything from passengers and livestock to goldmining machinery. The company prospered sufficiently to buy a Fokker F3, but with Parer at the controls this crashed on take-off from Salamaua on 18 March 1931. Parer and the passengers escaped serious injury but the aircraft was a write-off.

He replaced it with a Fokker F7A, which could carry ten passengers and lift one-and-a-quarter tons. Even so it was not big enough to carry some of the bulkier pieces of equipment and on one occasion Parer had to dismantle the fuselage, hoist a piece of machinery aboard, rebuild the aircraft round it, and take it to pieces again at Wau.

John Godwin's book *Battling Parer* tells several anecdotes of Parer's time in Papua New Guinea. One of these says that Parer always carried a bottle of beer as part of his navigation equipment. If he was uncertain about a landing ground he would drop the bottle, and if it burst on impact then the splash of white foam showed that the ground was hard enough to land on.

Despite or perhaps because of his happy-go-lucky flying habits the business prospered, and he invested in a goldmine. This too was successful and it is said that during his time in New Guinea Parer made a total of £1·5 million. But it seemed that he spent it while he made it, for he had to borrow the money to take part in the 1934 air race.

Aircraft magazine of the 1930s has many reports of flying in New Guinea. On 1 May 1933 a report said that the pilots on the Lae–Bulolo service carried out eight or nine return trips a day carrying a total of between sixty and eighty tons of passengers and freight.

Major de Havilland went to Lae with two Fox Moth biplanes to join the goldfields fleet, and on his first take-off from Lae airfield he carried a load of 350 pounds in the cabin of one of these little single-engined aircraft. This was steadily increased until after fourteen return journeys to the goldfields he was carrying 675 pounds a trip.

When de Havilland went to New Guinea, seventeen aircraft were flying on the service out of forty-three which had been on the Lae–Wau service since 1926. Nine of them were British and the others were Junkers and Fokkers. Among the British-built aircraft was a DH61 biplane which had logged more than 2,900 hours' flying on the service and had needed minimal maintenance apart from the replacement of two wings damaged during a crash landing.

An article headed "Well-known Commercial Pilot for New Guinea" reported that "Yet another of the old guard of commercial pilots and prominent flying instructors is bound for New Guinea, the 'Airmen's Eldorado'." He was Edward Wilmer Leggatt, known as "Bill", who had served in the RAF until he was shot down and taken prisoner by the Germans. He returned home to become one of Australia's earliest pilots in the easygoing days before the regulations had been framed. One of his principal jobs was that of delivering Melbourne newspapers to the Victorian country districts, using an Armstrong Whitworth. He said that this "paper round" involved some of the most dangerous flying he had ever known. The landing grounds were no better than paddocks, there were no reliable weather forecasts, and the

Beardmore engine of his "Ack-W" was liable to cut out at awkward moments.

He served in the RAAF from 1923 until 1926 and then became the first full-time Aero Club instructor in Australia with the Sydney branch. He was only a week ahead of "Pard" Mustar, who had just taken charge of the Melbourne branch of the Aero Club, but soon left it for the more adventurous life of a pioneer pilot in New Guinea.

While Guinea Airways were flying their regular commercial service some other Australian pilots were pioneering other air routes through the territory. One of them was Ian Grabowski, who in 1933 helped Mick Leahy and his brother, well-known explorers of the day, to discover a "hidden valley" of which they suspected the existence. Flying a Fox Moth, Grabowski and the Leahy brothers landed in the valley to the terror and astonishment of natives who had never before seen a white man.

The success of Guinea Airways aroused considerable interest in professional circles in Australia, not only those of aviation but also of mining. The Western Mining Corporation, then, as now, one of Australia's largest mining companies, decided that aviation could be the answer to gold prospecting in Central Australia. The company was the first to carry out a full-scale geological survey in Australia, although the Mackay Aerial Survey expedition had been carried out for scientific purposes. The company ordered two new de Havilland twin-engined DH84 Dragons specially equipped for the survey. The aircraft were modified for use in rough outback conditions, and were fitted with the "most comprehensive radio installation seen in Australia in a medium size civil aeroplane". This included direction-finding equipment as well as radiotelegraphy and radiotelephone sets. Aerial photography was no longer new, but the aircraft was fitted with the most modern air camera of the time: the massive Williamson Eagle IV.

The Government was now getting into the act, and in May 1934 announced that a complete air survey of Northern Australia would be carried out by the RAAF. *Aircraft* applauded the idea, but stated that "The RAAF is . . . simply not equipped with proper personnel or material for such an expedition . . . only established aerial survey companies can do the job. Obviously, the only and proper way is for the government to call tenders among private organisations."

The Government's thinking must have been on similar lines because it called for civil tenders although Captain C. R. Robbins, of a South African company, pointed out that "aerial surveyors do not pretend to find mines any more than they can find areas of good soil or good timber if they do not exist".

Ray Parer lands on a New Guinea beach during the 1930s
(Pacific Publications Pty Ltd)

8

Brearley Leads the Way

The dapper young pilot who survived a crash in No Man's Land established Australia's first scheduled airmail service

On a fine hot morning early in 1921 Norman Brearley opened his front door and picked up his copy of the *West Australian*. He walked with it to the breakfast table, opening the pages as he went. Before he had even reached for his teacup he saw the Commonwealth Government advertisement calling for tenders for the first airmail service—from Perth to the north-west of Western Australia. It called for a weekly service from Geraldton, 200 miles north of Perth, to Carnarvon, Onslow, Roebourne, Port Hedland, Broome, and Derby. This meant a round trip of 2,400 miles over country ranging from desert to barren hills, from an almost unknown coastline to the vast sheep and cattle stations of the interior.

The Commonwealth offered a maximum subsidy of £25,000 a year and would build and maintain suitably-equipped airfields and emergency landing grounds. Contractors had to specify the number and type of aircraft and engines to be used, which must be of British manufacture.

Brearley hastily swallowed his breakfast and hurried down to his unofficial airfield on the shores of the Swan River. The two Avros which he had imported from England stood in their hangar; his mechanic Harry Critchley was working on them. "Have a look at this, Harry," Brearley said. Together they pored over the advertisement and it seemed fairly obvious that it had been inspired by Brearley's pioneering flight from Carnarvon to Onslow and back to Perth in August 1920, and that it had been influenced by the local Member of Federal Parliament.

Brearley had made his first Perth flight in one of his two Avro 504 aircraft on 2 August 1919. It was an exhibition flight from the East Perth oval of the Western Australian Cricket Association. The display was attended by some problems because the oval was surrounded by electric light wires to floodlight the Trotting Association track which ran around the ground. The Association refused to temporarily remove the wires, and Brearley managed to break them twice during the exhibition. Otherwise it was a complete success and was the first of a number of money-making shows in Perth and various country centres. Some of these were commissioned by the Government by way of "publicity stunts" for the Peace Loan being raised at that time. He flew a few charter flights, including one which carried a journalist to meet the Prince of Wales on his visit to Australia in 1920, and one to drop welcoming leaflets on a shipload of British migrants arriving in Fremantle. Brearley charged £3 15s for a ten-minute flight, and made a good living during twelve months of barn-storming in and around Perth.

He then thought of introducing aviation to the real outback; to people who had never even seen an aircraft. He shipped an Avro to Carnarvon and sold joyrides in that part of the world. Like all his country trips, this one was attended by various adventures and misadventures attributable to the bush airstrips and generally primitive conditions. When he flew to Onslow from a cattle station north of Carnarvon, he found that Onslow was smothered by a thick red dust storm and he had to grope his way down into the wide main street of the town. His usual system was to telegraph ahead to stopping places and ask for a white sheet to be pegged down on any suitable landing

The Intelligence Officer of 450 Squadron RAAF, in Libya 1942, stands on the fuselage of a wrecked German aircraft and scans the sky through his binoculars for signs of the squadron's return from an operation. The Adjutant and some of the pilots not on duty wait in a group nearby, while the "hospital gharry" stands close to the runway in case it is needed. The 450 Squadron aircraft were flying close escort to a flight of twelve Bostons in a devastating daylight raid on Martuta airfield. (*painting by W. A. Dargie: Australian War Memorial*)

A Handley Page Halifax four-engined bomber, of an RAAF squadron based in the United Kingdom, is silhouetted against bomb blasts and searchlights during a night raid on German installations. (*painting by Dennis Adams: Australian War Memorial*)

HRH the Duke of York (later King George VI) inspecting RAAF Guard of Honour at the opening of the first federal parliament in Canberra 1927, accompanied by Air Marshal Sir Richard Williams and "Harry" Cobby, the highest scoring Australian fighter pilot of the first World War (*RAAF Museum*)

ground. On this occasion he could not even see the sheet.

At twenty-nine, Brearley had completed the longest accident-free flight then made in a light aircraft by a private flyer in Australia. He was the right man in the right place when the Civil Aviation Branch was established on 16 December 1920 under Lieutenant-Colonel H. C. Brinsmead. The Government was beginning to think seriously about making use of aircraft to communicate over the immense distances of Australia. Apart from this, there was a plan to encourage civil aviation and thus provide a training ground for the embryo RAAF saving the cost to the taxpayer.

Now the tenders for Australia's first airmail service were being called. Brearley burnt the midnight oil for several weeks in working out a proposition. It was difficult: he knew nothing about preparing a tender, his business friends knew nothing about aviation, and the whole project was breaking completely unknown ground. To most people, aviation meant one of two things: the daredevil exploits of flyers in the first World War, or a brief but unforgettable joyride in a barnstormer's flimsy little aircraft. Only such men as Brearley could foresee the future of commercial aviation. He had had it in mind as an officer in the RFC, and felt that the bombers of the first World War could be developed into passenger aircraft. He says, "In general I treated flying in a cold and scientific manner . . . To me aviation was a serious industry in its own right, with unlimited commercial and economic importance. Aviation was a business." Not many men thought like that in those days although some tiny airlines were staggering off the ground in Britain and America.

For his airline tender, Brearley had to solve problems which no one had even thought of. Most of the area to be flown over had never even been seen from the air. He had to plan the location of fuel dumps and airfields, work out how many aircraft would be needed to keep the service running regularly, and plan for the engagement of competent pilots. Even though the hiring might seem to be the least of his problems, with so many ex-pilots looking for work, he knew that the choice would have to be made with extreme care. A good many pilots of those days had been rushed into the air to fill the gaps in the RFC or the AFC and, as Brearley had seen during the Peace Loan drive in 1919 when such airmen were used for advertising stunts, many of them took unnecessary risks. There were a number of crashes—of the type which would soon ruin an airmail service.

Another problem was communications. Radio equipment was still uncommon in aircraft and in any case there were no ground-to-air radio stations. But there was a single-wire telegraph line running the 1,200 miles from Geraldton to Derby; and Brearley obtained permission for his pilots to tap this line if they made forced landings. The PMG Department co-operated to the extent of providing a portable telephone for each aircraft, and the postmasters and telegraphists of the north-west never let Brearley down.

The weather was not likely to be so co-operative. The north-west coast averages about one cyclone a year, so all the airfield buildings would have to be securely anchored. Such details as these had to be considered for the tenders which Brearley submitted. He gave the Government several options at various prices. The most comprehensive would call for payment of the full £25,000 annual subsidy, and in case this should be granted he made tentative arrangements for the Bristol Aeroplane Company in England to ship six 250-hp Puma-engined Bristol Tourer biplanes to Fremantle. These could not arrive any earlier than late October 1921, so Brearley's starting date for the tender was 5 December 1921.

On 2 August he was told that his contract had been accepted and he would be paid the full £25,000. There had been only one other tender, from a man whom Brearley had taught to fly.

Now Brearley's problems really started. The Bristol Tourers had to be paid for on acceptance, and these would be only part of the expense involved. His idea had been that of launching a public company if he gained the contract, and he had spoken to a number of people about this and thought he was sure of financial backing. But he had no sooner ordered the Bristols than some of these backers changed their minds. A tip from a friend made him visit H. V. McKay, the Melbourne manufacturer of farm machinery which was used all over Australia—including the famous Sunshine Harvesters. McKay backed him by arranging for the Government to guarantee an overdraft at the Commonwealth Bank, and the first prospectus of Western Australian Airways Ltd was issued on 29 August 1921. Financially at least, Australia's first airline was off the ground.

Brearley chose his first pilots at tests carried out at Point Cook. They were Charles Kingsford Smith, Len Taplin, R. Fawcett, H. A. Blake, and V. R. Abbott—all men who had flown over the Western Front in France. His chief engineer was Peter Hansen, the technical warrant-officer from the AFC who had helped Brearley to assemble his first Avro aircraft. Then countless final details were taken care of, including a small mirror for each pilot so that he could flash to searching aircraft in case of a forced landing.

On 4 December 1921 Australia's first official airmail aircraft took off from Perth for Geraldton. To mark the occasion, three of the Bristols took part in the flight and were greeted enthusiastically at Geraldton. Next day they continued northwards. Tragedy struck almost at once. Brearley, with two passengers and the mail bags, flew in line with the other aircraft piloted by Taplin and Fawcett. They flew steadily along the rugged coastline and crossed the Murchison River north of Geraldton, and then Brearley saw Len Taplin going down. He watched him land safely in a homestead paddock, and then found that Fawcett's aircraft had disappeared. Brearley landed, and was greeted by mounted stockmen and told that Fawcett's aircraft had crashed. Fawcett and mechanic Edward Broad were dead. They were buried in the little cemetery of Murchison House cattle station.

It was a bad beginning, but it settled Brearley's determination on one point: he would not continue the service until proper airfields and emergency airstrips had been prepared by the Commonwealth. He

Part of West Australian Airways fleet of aircraft at Perth's first official airfield at Maylands, in 1924 (*Sir Norman Brearley*)

Amy Johnson (left) received the kind of adulation later reserved for pop groups. The crowd in the above photograph mobbed Perth's Savoy Hotel in an attempt to see the "flying heroine" of the 1930s (*photos: The West Australian*)

had already been shocked by the condition of the so-called airfield at Geraldton, and if a suitable emergency field had been constructed anywhere near Murchison House then Len Taplin, who had been forced to land by a misfiring engine, could have taken advantage of it. It was thought that Fawcett had crashed because he had followed Taplin down to see how he fared on a forced landing in rough country. He had been preparing to help if possible.

Brearley started all over again by making a survey of the route, as a result of which some improvement was made to the airfields. He flew with Kingsford Smith and Peter Hansen, and they had a narrow escape when an engine fault caused a forced landing on the beach south of Broome. Eventually their Bristol had to be dismantled and hoisted aboard a coastal steamer for shipment back to Fremantle.

Brearley carried on a running fight with civil aviation authorities, over the condition of the airfields, for several years. He and his pilots became used to landing and taking off under extremely difficult conditions. The official landing ground at Onslow became a quagmire after heavy rains, so the pilots used a near-by coral marsh even though official permission to use this ground was denied.

But by the end of its first year in operation, WA Airways had made a small profit and had been commended by Colonel Brinsmead for the regularity of its service. Despite the initial hold-ups, the service flew nearly 100,000 miles in nine months, and in eighteen months it was to total 250,000 miles. Ninety-seven per cent of the service had been flown to timetable. When Qantas began operations, and experienced some teething troubles, Brinsmead asked Brearley to supply full details of his operations for the benefit of the Queenslanders. Brearley sent a lengthy report of his airline's operational procedures and these were of great help to Qantas. In Hudson Fysh's book *Qantas Rising* he wrote, "It must be recorded that Norman Brearley had established the first well-organised air transport operation in Australia in 1921, and their undertaking was superior to our own more groping endeavour. They showed the way to us all."

WA Airways ran with almost monotonous safety and regularity on the north-west service, with the original Bristol Tourers being replaced by DH50s. These new, bigger de Havillands were a help: a good deal of freight was now being carried and the combination of passengers and cargoes had made for some uncomfortable Bristol flights. Most of the flights were uneventful, but there were a few unusual incidents. In 1927 a pilot flew through such dense fog that he descended to the height of the telegraph line in order to keep his bearings, and then decided to land on one of the emergency airstrips until the fog cleared. The airstrip was typical of those which Brearley had been arguing about and consisted of soft sand. The aircraft came to a dead stop and could not take off again. Pilot Bertie Heath tapped the telegraph line and asked for help to be sent in, and a Bristol flew in with axes, shovels, and food for Bertie and his four passengers. The party managed to cut enough brush to make a track firm enough for a take-off—after two uncomfortable nights in the bush.

TOP: Inside an Australian aircraft factory in the 1930s. Note the plywood fuselage of the aircraft being assembled (*Sir Norman Brearley*)

BOTTOM: Hans Bertram's Junkers aircraft under repair after crashing in Western Australia (*The West Australian*)

Norman Brearley started the Perth Flying School at Maylands aerodrome in 1927, and with the help of a Government subsidy this ran successfully for a number of years. A private pilot could gain his licence after about thirty-two hours' flying for a cost of about £30. A number of the early graduates joined the RAAF when war broke out.

In 1928 Brearley arranged for Bert Hinkler's welcome in Perth, when he flew in there after completing the first solo flight from England to Darwin. Hinkler had written to Brearley and asked him to arrange a fenced landing ground with turnstiles, close to the city, so that an entrance fee could be charged. Some 20,000 people paid to see Hinkler land his Avian on what is now Perth Oval.

Brearley's aerial activities in this period included an air survey of 400 square miles of the Darling Range east of Perth, for the Western Australian Government. This was a new venture for Brearley, but he tackled it with the usual gusto. Once again, with his unusual and consistent drive, his easy shouldering of responsibility, he had to solve all problems from scratch. He floated a company called Air Surveys Limited, and on a trip abroad learnt all that he could about aerial photography and mapping. When the survey began, he and his brother Stan had to work out a new way of flying in order to keep straight courses over the monotonous bush country. They shared control of the plane, with Stan keeping the aircraft level and at a uniform height while Brearley stood up with his head through a hole cut in the cabin roof and did the steering. The survey company worked profitably for a few years, until work ran out, and was then wound up.

In 1928 an even more exciting opportunity arose. The Government sought tenders for a five-year mail and passenger service between Perth and Adelaide. Brearley, as the leader of the nation's largest and most successful airline, was determined to win the contract.

WA Airways had the finance, the pilots, and the experience. They had the best chance of gaining the contract even though other airlines were likely to be strong competitors. Certainly there were some new problems to be worked out, but these were not too difficult. The contract called for night flying, so it would be necessary to install a line of rotating beacons so that pilots could find their way from one light to another. In those days it would not be possible to make the flight in one hop, so a hostel for a total of twenty passengers and crew would be installed at Forrest, 400 miles east of Kalgoorlie. But something much bigger and better than DH50s would be needed for the service. It would not be difficult to obtain larger aircraft because many overseas manufacturers were by that time producing multi-engined aircraft to carry substantial loads of passengers and freight. Apart from heavier-than-air aircraft, airships were also available. These were still being used for passenger services in various parts of the world, and continued to be used until the *Hindenburg* disaster in 1937. The British gave them up after the crash of their R101 on her maiden flight. If it had not been for this disaster then she and her sister-ship, R100, would have opened an eleven-day service from England to Australia. Brearley had even been asked to help to choose the site for a mooring mast near Perth.

The aircraft Brearley chose for his main tender was the mighty de Havilland Hercules. It had been used on the Cairo to Karachi route of Imperial Airways, and with its three 450-hp Bristol Jupiter engines it could cruise at 100 m.p.h. with a payload of more than 3,000 pounds.

WA Airways won the contract and Brearley went abroad to finalise details of the equipment his airline would need. In the USA, where he saw the aircraft being built by the Americans, he inspected the Sperry rotating beacons he planned to use. It was on this trip that he rejected the General Electric airfield lighting system which had already been approved by the Government. When he reached England he bought instead a huge landing light made by the lighthouse engineers Chance Brothers. In England he also visited the de Havilland factory where he arranged for some modifications to be made to the four Hercules he had ordered for the new service.

On 26 May 1929 the first official passenger and airmail service took off from Perth for Adelaide. The single fare was £18. Two of the new Hercules flew together to Kalgoorlie, with Norman Brearley and Bertie Heath as pilot and co-pilot of one and Stan Brearley and Eric Chater in the other. It was a proud day not only for WA Airways but for Australia. A special stamp had been designed for the airmail carried by the new service, and there had been immense publicity. Brearley and his colleagues, with the constant support of Brinsmead, had gone through a meticulous programme of flight training in the huge aircraft and had attended to innumerable details designed to make every part of the service run smoothly.

The two Hercules had a great welcome at Kalgoorlie, and Stan Brearley and Chater stayed there to give demonstration flights while Norman flew on to Forrest. This was a real landmark in Australian aviation. The huge Nullarbor desert over which they were flying had until quite recently been an almost impassable barrier between eastern and western Australia. The only communication had been by sea until the Transcontinental Railway was opened in 1917. Now, only twelve years after the first train had puffed across the Nullarbor, a plane-load of passengers was droning steadily above it at 100 m.p.h.

The new service received a rapturous reception in Adelaide, and on 30 May 1929 the *Advertiser* carried a lengthy article headlined, "BY AIR FROM THE WEST: Arrival of Giant Triple-Engined Mail Planes: Flight of 1,450 miles in 14 hours: Adelaide to Perth Service Starts on Sunday".

The reporter wrote: "Seventy-five feet across the wings and 19 feet high, the Hercules de Havilland biplanes, which are to carry mails on the Adelaide to Perth service, made a great impression on the spectators at Parafield yesterday afternoon. The first—City of Perth—arrived from the westward at the moment she was expected, and after sweeping across the aerodrome considerably less than 1,000 feet from the ground the giant machine paid a visit to the city before landing a few minutes after 1 p.m. . . .

"The throbbing roar of her three engines spoke of herculean power, the breadth of wing and beauty of line gave her grace, and responsiveness in manoeuvre was evidence of airworthiness and skilful handling. When the machine passed over the aerodrome to visit the city, Captain H. C. Miller went aloft in a Moth to welcome the air liner. Both returned in a few minutes, and while the Moth tumbled and indulged in lively aerobatics, the giant came to earth—losing height and gliding in against the breeze, and at last meeting the ground almost imperceptibly. When the City of Perth taxied into position near the barriers, a round of cheers was heard from the cabin of the machine. The propellers having stopped, the passengers emerged, smiling and happy, having participated in an event of national importance, the successful trial flight of an inter-State mail machine.

"The second machine—City of Adelaide—arrived at 3.40 p.m. Each of the machines weighs 9,935 lb., and 15,660 lb. when fully laden.

"'Now for a smoke. I haven't had one since we left Ceduna.' That was the remark of one of the men passengers on the City of Perth, as soon as he set foot on the ground. He had no other complaint—if that could be called one—the rest of his remarks being highly complimentary to the City of Perth's air qualities. Apart from the discomfort smoking might cause when the windows of the cabin were drawn shut, the risk of fire could not be entertained in mid-air. Hence the sign—No smoking.

TOP: One of the DH Hercules airliners used to inaugurate the Perth–Adelaide passenger services (*Sir Norman Brearley*)

BOTTOM LEFT: Norman Brearley congratulates one of his key pilots, Eric Chater, in front of a Hercules (*The West Australian*)

CENTRE RIGHT: Maylands airfield in 1929, with Hercules airliners in front of West Australian Airways new hangar (*Sir Norman Brearley*)

BOTTOM RIGHT: Two of West Australian Airways DH Hercules trimotor airliners. They were the first of this type to be fitted with tail wheels and enclosed cockpits (*Sir Norman Brearley*)

"All the passengers appeared to be wearing wadding in their ears to lessen the effects of the noise from the engines, but they said conversation could be maintained by raising the voice. Supplies of wadding are contained in hygienic receptacles above the heads of passengers. Some played bridge during the flight. One of the woman passengers was enthusiastic about the experience, and said she would not have missed it for anything. All spoke highly of the skilful handling of the machine. Mr. H. Grenville was of the opinion that the comfort of the City of Perth was superior to that of the air liners used in Europe. The former inter-State cricketer, J. W. Rymill, described the trip in one word, 'Great!' He had been on a visit to his brother in Perth . . .

"Major Norman Brearley said the City of Perth had no difficulty in maintaining the speed schedule of 100 miles an hour. On most parts of the journey the plane had averaged 110 miles an hour. The first stage of the flight, from Perth to Kalgoorlie, a distance of 350 miles, was covered in three and a half hours. The second stage, from Kalgoorlie to Forrest was 400 miles in length. This span was covered in three hours and 35 minutes. Ceduna was reached after a flight of three hours and 10 minutes, 340 miles being covered. At Ceduna 92 passengers were carried for trial flights. They were all impressed with the machine, and with the value of the service which was being inaugurated. The last stage of 355 miles was accomplished in three and a half hours . . .

"Everything on the trip ran like clockwork, said Major Brearley. The passengers had expressed their views of the flight in very definite terms. They considered that Australia had a wonderful bridge across the continent.

"Questioned concerning flying conditions between Perth and Adelaide, Major Brearley said they were about as near to perfect as it was possible to have them. He was familiar with the route, having flown across from Perth 18 months ago. As far as climatic conditions, ease of navigation, and topographical features were concerned, almost everything was in favor of the air line. Normally the plane could carry 14 passengers, but by the use of sling seats (especially for joy-riding), this number could be increased to 21, in addition to the two pilots. The planes had a cruising range of 525 miles, which meant that there was plenty of petrol in reserve, even on the longest stage of the flight. Each plane carried 300 gallons of petrol, which was consumed at the rate of 55 gallons an hour when the three engines were in operation, and the plane was travelling at 100 miles an hour. Each of the four planes purchased by West Australian Airways, Ltd., cost £15,000. No mails were carried on this trip. It was an unofficial journey."

On 1 June 1929 the Governor of South Australia, Sir Alexander Hore-Ruthven, christened the *City of Adelaide* and there was a celebratory air pageant at Parafield. The first Adelaide–Perth airmail service took off from Parafield at 9.30 a.m. on Sunday 2 June 1929 and should have arrived in Perth at 1 p.m. on 3 June, but the aircraft was delayed by a sandstorm and the mail missed the London-bound mail steamer at Fremantle. It was the only failure that WA Airways was to experience during seven years of flying the Perth–Adelaide–Perth airmail and passenger service.

The Depression which began in 1929 soon became worse, and in 1930 it seemed that the Government would discontinue subsidies to airlines and have the mail carried by the RAAF. Brearley made a special trip to Melbourne to protest, and convinced Prime Minister Scullin that the scheme was impracticable. The service continued, though the Hercules aircraft were replaced by Vickers Viastra twin-engined monoplanes. The disadvantage of the Hercules had been their maximum cruising speed of 100 m.p.h. which was frequently cut down by strong head winds, whereas the Viastras could cruise at 140 m.p.h.

Two of the Hercules were sold to Imperial Airways, and in 1934 a third was sold to Captain E. J. Stephens who was one of those flying freight from Lae to Wau for the Bulolo goldfields. The aircraft seemed ideal

TOP: The scene at Maylands when the new service was inaugurated (*Sir Norman Brearley*)

BOTTOM: This model Hercules was suspended and illuminated outside the Perth GPO to celebrate the first Perth–Adelaide airmail service (*Engineering Division, PMG's Department, Western Australia*)

for this service, but Stephens needed some instructions so Brearley agreed to fly up to New Guinea with him. It was a fairly eventful trip. Tropical rains forced them to spend four days at Somerset on the tip of Cape York Peninsula, and they had to live on fish, rock oysters, and damper. Then they took off for Port Moresby across the Torres Strait, which was beyond the normal range of the Hercules. So they had stacked extra four-gallon tins of petrol in the cabin and had to pour the petrol from the tins into a large fuel drum with a hand-pump attached so that it could be pumped into the top wing tank of the aircraft. The cabin was full of petrol fumes even with all windows open, but at last they landed safely on the little uphill airfield at Wau.

In the same year WA Airways lost the north-west airmail contract, for the service which they had pioneered and handled successfully for thirteen years, to MacRobertson Miller Airlines. It was a blow, especially during those Depression years in which the Government had imposed new contract terms on the Perth–Adelaide service, paying only for mail actually carried instead of a fixed annual subsidy. WA Airways already had dropped the passenger fare from £18 to £12, which was the same as the first-class rail fare, and to make the service more efficient and economical they bought a smaller aircraft, the de Havilland Dragon-Rapide twin-engined biplane which carried eight passengers at 130 m.p.h. But even this could only delay what Brearley, in his book *Australian Aviator*, calls the "controlled spin" caused by the Depression years. On 30 June 1936 Australia's first airline was sold to a new company known as Australian National Airways. Brearley retired briefly, soon returning to uniform. He eventually became Commanding Officer of 4 Service Flying Training School at Geraldton during the second World War. His next appointment was Commanding Officer at Tocumwal, an RAAF training centre. During his term here, this man who had done so much for Australian civil aviation experienced what he has described as "the proudest moment of his life": the presentation of wings to his son, Pilot Officer M. Brearley.

Group Captain Norman Brearley, when commander of Tocumwal RAAF training centre in 1944, presents pilot's wings to his son Maurice (*RAAF Museum*)

9

Across the Bass Strait

The seafaring Holyman brothers founded an airline to link the Island State with the mainland, but it claimed one of their lives

While Norman Brearley was making Australia's first airline a trimly efficient service, several other companies entered the field. The only one still remaining is Qantas, which began as a small domestic carrier and has now become one of the world's great international airlines, no longer running any domestic flights.

Qantas, Ansett Airlines of Australia, and Trans-Australia Airlines make up the current trio of giants. They have a magnificent reputation and because they have become such household names it is easy to forget that Australia has many other flourishing air services, large and small, with a huge backing of supporting industries.

And this is nothing new. Soon after Brearley's pilots began their first scheduled flights, the Larkin Aircraft Supply Company won the Adelaide–Sydney airmail contract. Harry Butler tried to begin a service between Adelaide and the Yorke Peninsula and was the first man in the southern hemisphere to carry mail over water. Horrie Miller made a premature attempt to open a flying-boat service over the same route. He later went on to make such a name for himself with various flying exploits that the founder of Guinea Airways asked him to pioneer the service with his DH9. But Horrie's father had been an unwise speculator in goldmining enterprises, and Horrie did not think his DH9 would suit the arduous conditions. He decided against the offer and "Pard" Mustar was chosen instead.

H. J. Larkin, whom Horrie Miller describes as "probably the most pushing and prominent in the Australian aviation field", soon struck trouble on the Adelaide–Sydney service. The railways objected to his competition so he was obliged to start and finish flights at Cootamundra instead of Sydney, and his ex-RAF DH9 aircraft were too small. They could carry only two passengers jammed into the open cockpit.

He decided therefore to build his own aircraft, and opened a factory in Melbourne. By the 1930s he had become more involved in aircraft production than in flying.

The story of Kingsford Smith and Ulm's airline has been told, but they were not the only service in eastern Australia. The other, G. A. Robinson's New England Airways, was a direct descendant of the horse-and-buggy days.

Robinson came from a transport family which progressed from horse-drawn transport to motor trucks, and was, in fact, one of the first to realise the potential of the motor truck. He built up a profitable network of motor deliveries but soon began to see that an even faster type of transport was available.

Like Brearley, he had many problems to solve. Unlike Brearley, he was not a flyer and so did not have experience to aid him. But he bought a small single-engined cabin monoplane and inaugurated New England Airways on 17 July 1931.

The service ran only twice-weekly from Sydney to Brisbane and return, but the demand grew so steadily that Robinson invested in three tri-motor Avro 10 aircraft and on 1 September 1932 began a daily service up the coast to Brisbane and back.

With Wing Commander L. J. Wackett as his Aeronautical Engineer-in-Charge, and Keith Virtue

One of the great men of Australian aviation: Sir Lawrence James Wackett. Among his numerous achievements he was General Manager of the Commonwealth Aircraft Corporation from 1936 to 1961 (*Commonwealth Aircraft Corporation*)

and Tommy Young as his chief pilots, the airline rapidly built up an enviable reputation for safety and service. By 31 August 1933 it had flown a total of 5,663 hours, covering 587,500 miles, and had carried 10,995 passengers.

"Wing Commander L. J. Wackett has stated on more than one occasion that he has never seen a more loyal, enthusiastic, or more efficient team of men than those employed by New England Airways," *Aircraft* reported on 1 December 1933.

Robinson had big plans for his airline and in 1934 went on a world trip to examine the latest trends. His transport and airline companies gave him considerable financial substance and in England he talked to such men as Lord Sempill, Lord Southborough, W. M. Hughes, and the managing director of the British Pacific Trust. He returned to Australia to say that these men would be members of the board of a big new airline, incorporating New England Airways, which would fly as far north as Townsville and at least as far south as Melbourne.

He had flown on the rapidly-growing American airlines and had "travelled across the North American continent in a 12-sleeper plane, and had slept in perfect comfort", and obviously visualised such aircraft being used for the proposed new airline.

He spoke rather vaguely of using multi-engined aircraft which would fly the through service from Brisbane to Melbourne in six hours, with every comfort for the passenger including radio music. Obviously these were to be either American machines, because "the British aircraft which he had inspected were not suitable for the services contemplated", or aircraft built in Australia.

The Commonwealth Government, in that Depression period, had forbidden the import of American aircraft, so Robinson may have planned to build them under licence in Australia. "The company would, no doubt, enter upon construction here on a scale not hitherto seen in Australia," he said.

The new company, under the name of Airlines of Australia Ltd, was formed and operated until 1942, but it never fulfilled Robinson's hopes. He was just a little too early.

Among others who had soared into prominence were the Holyman brothers. Like Robinson they came from a long-established "transport family" but in their case the transport had been by sea and not by land.

Their grandfather, Captain William Holyman, settled in Tasmania in 1854 when he was twenty, and worked on coastal vessels until he had saved enough for a down-payment on the eighteen-ton ketch *Cousins*, in 1861. That was the start of a shipping company which still runs ships out of Tasmania. Some of its earliest vessels were commanded by Captain Holyman's sons James, William, and Thomas.

Naturally Ivan and Victor Holyman, of the third generation, also went to sea. They obtained their master mariner's certificates, but when the first World War broke out Victor took to the air instead of the sea and became Captain V. C. Holyman, RFC.

During the 1920s Tasmania became the home of a flourishing branch of the Aero Club. It was based at Launceston, under Captain Joe Francis who had taught Victor Holyman to fly. Both Victor and Ivan were members of the club, and in March 1933 the Victor Holyman Cup was presented to the winner of its thirty-six-mile Aerial Derby: part of an Air Pageant attended by three Bristol Bulldogs and three Westland Wapitis of the RAAF. The pageant drew huge crowds, and the isolated Tasmanians were perhaps more air-conscious than the people of any other Australian State.

"A regular air service [from Launceston to Hobart] is bound to be established in the near future," Joe Francis said after the pageant.

In June 1932 L. McK. Johnson began a regular service from Launceston over the 120 miles to Whitemark on Flinders Island. In three months his Desoutter monoplane made fifty-six trips on schedule, carrying a total of eighty-five passengers plus mail and freight. To people in a hurry to get to or from the island, the service was considerably more attractive than the Holyman steamer which called only once every ten days.

The Holymans who had pioneered regular shipping services were smart enough to see that the future lay in the air—and smart enough not to begin a fare-cutting war with Johnson. Instead, Victor relinquished command of the *Laranah*, then running between Launceston and Melbourne, and formed Tasmanian Aerial Services Limited. Johnson was brought into this company, which continued his service to Flinders Island, and with a DH83 Fox Moth imported by Victor, flew regularly to King Island too. The aircraft rudders were painted in the same design as the shipping company's house flag: a white star on a red ground.

The flight from Launceston to Flinders Island took seventy-five minutes and cost fifty shillings. The longer flight, to King Island, cost £4 2s. Special charter flights cost £10 to Flinders Island, £15 to King Island. Several medical rescue flights were made.

Pilots who flew the service had to be top-notch men who knew every inch of the route and had almost a sixth sense for the rapidly-changing weather conditions. They were flying above a rugged, wind-swept coastline and over long stretches of open water, without radio aids or reliable weather forecasts. The Holymans, from a family which had accumulated a huge store of experience gained through decades of sailing the Tasmanian waters, were ideal for the task, but even this experience did not protect them all the time. One spell of bad weather caused Victor to force-land at Woolnorth Point in north-west Tasmania, and the company's mechanic had to go there to replace an oleo leg broken off against a rock.

But despite such minor problems the airline forged ahead. The Hart Aircraft Service of Melbourne, which flew a once-a-week service to Launceston with their Avro 10, *Tasman*, decided that they would try for a slice of the Flinders Island trade even though "landing conditions were not the best", and before long were making regular visits. In Tasmania, the Windhover owned by Matthews Aviation Co. was making sightseeing flights for tourists, and a steady stream of pilots was being trained by Joe Francis at the Aero Club. The Windhover, an amphibian, was also used on the Melbourne–Launceston service, and made record time by covering the 300 miles in two hours, twenty minutes.

Tasmania was certainly an air-minded State. As early as 1922 a Vickers flying-boat had been employed by the Government to carry out aerial mapping work, and in the same year H. J. King made an aerial photographic survey of Launceston for the City Council. The total cost to the council, including twenty hours' flying, was £80.

In the winter of 1932 the Holymans ordered a DH84 Dragon twin-engined cabin biplane, described as "an airliner in miniature", to be delivered from England. Major H. de Havilland, in Sydney, took delivery of the aircraft, assembled and tested her, and flew her to Melbourne where the Holymans took over.

The DH84 was the most modern commercial aircraft to be delivered in Australia. WA Airways had also ordered them for the east–west run. Among her equipment was a specially-designed short wave radiotelephone, then a comparative novelty. Named *Miss Launceston*, she began service on 4 September 1934, carrying passengers who paid £5 for the three-hour flight from Melbourne to Launceston.

The Commonwealth Government was now talking of subsidising a regular service to Tasmania, and as a first step landed a party of navy and air force technicians to erect an experimental radio station. There was talk of using flying-boats on the route, since flying-boats were still regarded as a safer proposition over the ocean. They had a chance of floating in case of a forced landing, and could use harbours as landing fields without the need for costly constructions ashore.

Tenders were called for a Melbourne–King Island–Wynyard–Launceston–Hobart service, which caused Victor Holyman to comment that the Government would have to do something about building airfields in the south of Tasmania. The Commonwealth, perhaps unintentionally, riposted by enforcing section 12 of the Air Navigation Regulations against the Holyman service. This forbade aircraft, except sea-planes or flying-boats, to fly for more than fifty miles across open water, thus preventing the *Miss Launceston* from flying the fifty-five miles from King Island to the mainland. The service was continued via Flinders Island, which came within the distance, and the Government kindly contributed £250 towards a new airfield at Pats River on the island.

Three firms competed for the Government contract, all on about equal terms. Matthews Aviation had the Windhover amphibian and a Saunders-Roe flying-boat, *Cutty Sark*; Hart Aircraft Service had an Avro 10 which it had bought from Smith and Ulm when the first ANA collapsed; and the Holymans had their new Dragon. The Avro 10 was an old stager on the Melbourne–Launceston route, and was the one which Jim Mollison was flying on the day the *Southern Cloud* disappeared. On that day the weather was so bad that "the passengers declined to enter the aircraft" so Mollison took off alone, and was an hour late at Launceston.

The Holymans won the contract simply because their proposition was better than the others. Their shipping connections had enabled them to obtain a guarantee of financial backing from the Melbourne shipping company of Huddart Parker, and the Union

Australian National Airways DH86 airliner *Lepena* photographed at Mascot in 1936. Victor Holyman and eleven others vanished in an aircraft of this type off Wilsons Promontory, Victoria, in October 1934 (*McGrath Collection*)

Steamship Company of New Zealand. With this backing they were able to state that they would buy two new de Havilland Commonwealth type airliners, the DH86, for the service. The DH86 had four engines and with its top speed of 170 m.p.h. was one of the world's fastest airliners, and had proved so popular that de Havilland's could hardly keep up with the orders.

The award of the contract to the Holymans caused a certain amount of bitterness, because Hart and Matthews had been maintaining the Bass Strait service before the Holymans took to the air in the commercial sense. Both companies withdrew from what they termed the "experimental services", and *Aircraft* regretted the passing of "the rather noisy old Avro and the slow Windhover". The writer thought that passengers in newer aircraft would "miss half the thrill of crossing the Bass Strait", but perhaps for the majority of them this was a thrill which they would gladly forgo.

Tasmanian air-mindedness continued to flourish. There was such a rush of would-be flyers, men and women, that Joe Francis and his two part-time assistants at the Aero Club could not handle them all. A Melbourne–Hobart Centenary Air Race, financed by Sir Macpherson Robertson who was giving the prize money for the London–Australia race, was arranged. *Aircraft* said that the Flinders Island route would be best for this race, because it had emergency airstrips about thirty miles apart and "intermediate rocks to which an airman could cling while awaiting rescue".

Aircraft seemed to take rather a poor view of the Holymans as they prepared to run their new service under its new name of Holyman's Airways Pty Ltd. Victor Holyman announced that he would be employing "the five best procurable pilots in Australia" for the Bass Strait service. *Aircraft* sniffed, "With the numerous opportunities now offering for good pilots it appears that Holyman's Airways will have to pay through the neck for the five best in Australia. Anyhow, who is going to decide who are the five best?"

On 1 October 1934 Victor Holyman flew the first DH86 on the opening flight of the new service. Like Norman Brearley's, it was launched with a tragedy. On 19 October the aircraft in which he was flying with a co-pilot and ten passengers was sighted off Wilsons Promontory but was never seen again, and no trace of it has ever been found.

Both seamen and airmen have long since learnt to fear the treacherous weather and waters of the Bass Strait, and even the DH86 was still only a rather flimsy aircraft with a plywood fuselage. Its four engines would not have saved it from a massive storm of the type which the area can brew up almost out of a clear sky. Exactly a year later another Holyman DH86 was lost off Flinders Island, with five deaths, but by that time Ivan Holyman had taken over the management and was determined upon success.

He looked for a more suitable aircraft for the short but hazardous flight. The Government had lifted its embargo on the import of American aircraft, so he decided on the Douglas DC2. With its successor the DC3, the "biscuit bomber" as its C47 variant was named by the troops to whom it dropped supplies in

TOP: Birth of an airline. The DH Fox Moth *Miss Currie* with which Ivan and Victor Holyman began a Launceston–Flinders Island service in 1932 (*McGrath Collection*)

CENTRE: Australian National Airways Douglas DC2 airliner refuelling at Mascot in 1937. Named the *Loongana*, she and her sister *Bungana* set new standards for airline travel in Australia (*McGrath Collection*)

BOTTOM: This Douglas DC2, one of the first metal fuselage aircraft in Australia, flew for the RAAF late in the 1930s (*RAAF Museum*)

New Guinea, this aircraft was to revolutionise air transport in Australia as in many other countries. "This machine is considered one of the best aircraft in the world," *Aircraft* wrote of it in 1934, and of all the aircraft of that period it bore the closest resemblance to those of today. Metal-skinned and capable of carrying eighteen passengers, the twin-engined DC2 could cruise at 180 m.p.h. From 1933 to 1945 thousands of DC2 and DC3 aircraft were built for civil and military service and they gained a fantastic reputation for low-maintenance reliability, flying ease and comfort, and virtual indestructibility unless they were shot down or the pilot ran into the side of a mountain.

One of the DC2 aircraft bought by Ivan Holyman, and named the *Bungana*, made her first Bass Strait crossing in July 1936. By September 1946 she had logged 29,000 hours' flying time and covered 4·75 million miles, and seemed fit for a million more. But the time had come for the "old girl" to go into retirement.

While the Holymans were establishing their service, another shipping company had gone into the aviation business. This was the Adelaide Steamship Company, which founded Adelaide Airways Ltd. This began by flying small passenger aircraft from Adelaide to such South Australian country centres as Mount Gambier. Shipping companies were becoming very conscious of both the potential and the threat of aviation, especially when, like the Adelaide Steamship Company, they carried passengers from one port to another along the Australian coast. And overseas shipping lines trading to Australia began to feel the same way. They thought that passengers might decide to leave ships in Fremantle and fly to Sydney, cutting down the travel time by several days and also cutting the ships' profits.

One of the companies was the Orient Line, which had had friendly relations with the Adelaide Steamship Company for more than half a century. Norman Brearley's problems in reducing costs on his Perth–Adelaide service, and his loss of the north-west mail service, were being fairly freely discussed in *Aircraft*, so the news soon got around through the Adelaide Steamship Company and Ivan Holyman's connections in both shipping and aviation.

Someone must have suggested: "Why not a really big airline, financed by shipping companies? It could combine Holyman's outfit with Adelaide Airways, take over West Australian Airways, and then we could look for other routes—maybe buy out existing airlines. That way, we can lick 'em *and* join 'em."

Five shipping companies joined in this project, and in April 1936 Sir Alan Anderson, a director of the Orient Line, called on Norman Brearley in Perth. Yes, Brearley said, he would sell his airline—at his own price. He thought it was the ideal solution to his problems, especially those of re-equipping with expensive new aircraft like the DC2s which cost £35,000 apiece.

Sir Alan sailed on to Melbourne and talked it over with the others. Brearley gave permission for a director of the Adelaide company to inspect his accounts and they sent a Mr A. D. Hill to Perth. He gave a favourable report and the shipping companies, with Ivan Holyman who represented both shipping and aviation, made their decision. They telephoned Adelaide, and Hill rushed to find Brearley who, waiting to hear the outcome, was also in Adelaide. He was playing golf, and Hill caught up with him on the eleventh tee. In *Australian Aviator*, Brearley says:

"'What do you think it is?' he [Hill] burst out.

"'It's an easy par four hole, but I want to try for a birdie,' I said, aware that in a moment I might learn that I was a retired flyer at forty-five years of age.

"'Great Scott, man, it's not golf, it's your airline! Your price has been accepted!'"

Thus Australian National Airways was born. With Ivan Holyman as managing director it was registered on 1 July 1936 with a capital of £500,000 provided by the shipping companies. Adelaide Airways Limited and Holyman's Airways Pty Ltd were incorporated into the new company.

With the money, thrust, and travel expertise provided by Holyman and his backers, the new ANA got off to a flying start. Its next move was to buy the shares which British Pacific Trust Ltd owned in G. A. Robinson's Airlines of Australia Ltd, once New England Airways, which was completely incorporated into ANA in July 1942. From then on the airline founded by the son and grandson of shipmasters forged steadily ahead, and the rest of its story is linked closely with that of the other Australian airlines.

This Ford trimotor monoplane was among the passenger aircraft used in the mid-1930s (*RAAF Museum*)

10

Qantas Takes Off

The "Flying Fysh" and his friends opened up the outback with their Queensland and Northern Territory Air Services, and then expanded into the British Empire

Hudson Fysh, like Ivan and Victor Holyman, was a Tasmanian. And like the Holymans he came from a substantial and respected Launceston family. The main difference was that the Holymans were brought up with a nautical background whereas that of young Hudson tended towards legislation and religion. He was named after J. Hudson Taylor, a missionary to China.

He was born in 1895 and in due course sent across Bass Strait to be schooled at Victoria's Geelong Grammar. He was a shy and retiring lad who fared poorly at academics but loved the outdoors.

For eighteen-year-old Hudson Fysh the outbreak of the first World War meant adventure, as it did to thousands of other young Australians. He, like them, had been trained under the "Kitchener scheme" which put schoolboys into cadet uniforms and then passed them into the reserve battalions of the Australian Army. Fysh had been trained with the 26th Light Horse Regiment in Tasmania and then with the 70th (Ballarat) Infantry Battalion in Victoria, so he was in khaki a few days after 4 August 1914.

Fysh did not plan to fight the war on foot, so lost no time in wangling his way back into the Light Horse with the aid of a horse given to him by his uncle, Willie Grubb. When the troopships sailed with the 1st AIF, Fysh and his horse were aboard. They were landed in Egypt and marched to the huge training camp at Heliopolis, near Cairo. Unknown to him, a young man named P. J. "Ginty" McGinness was among the 20,000 Australians of all arms who were assembling at Heliopolis.

For gruelling months they were trained to hair-trigger fitness, expecting to be sent to fight the Turks in the desert, until the day when they found themselves boarding troopships again. This time it was at Alexandria, and they were bound for a place that most of them had never heard of: Gallipoli. And the Light Horse were to fight as infantrymen.

Fysh and McGinness, still unknown to each other, survived the fighting at Courtney's Hill, Pope's Hill, Shrapnel Hill, and the other Gethsemanes of that fearful and useless campaign. In late 1915 they were back in Egypt again. McGinness, wearing the ribbon of the Distinguished Conduct Medal which he won on Gallipoli, transferred to 1 Squadron, Australian Flying Corps.

Fysh was back with his horse, and with the 3rd Light Horse Regiment crossed the Suez Canal and took part in the great battle at Romani where British and Australian troops gave the Turks a bloody nose. Fysh had given little thought to flying so far—except to curse the German airmen who bombed the horse-lines in the desert.

He was promoted to corporal, then called to the major's tent and told that he would be commissioned as lieutenant, to replace Lieutenant Ross Smith who had transferred to the AFC.

The idea of flying instead of riding across the hot, dusty desert began to appeal to young Hudson Fysh, and when his Light Horse troop was stationed next to 1 Squadron's airfield he decided to transfer. He became an observer under Major Richard Williams who was to become commander of the RAF/AFC group in the Middle East, and who was to go on to make such a mark on Australian aviation.

Qantas "foundation members" at a re-union in 1939. From left: Sir Hudson Fysh, Sir Fergus McMaster, Arthur Baird, and P. J. "Ginty" McGinness (*Qantas Airways photograph*)

Fysh met McGinness, admired his flying skill and dashing competence, and flew with him as observer-gunner. They made a good team with McGinness at the controls and Fysh at the twin Lewis guns on their swivelled track.

They travelled home together in the *Port Sydney*, and parted when Fysh disembarked at Hobart with the rest of the Hobart contingent. The parting was short, because McGinness persuaded Sir Samuel McCaughey to back him in an entry in the 1919 England–Australia race. He sent for Fysh, who sold his new car at a profit and enlisted their old engineer Flight Sergeant Arthur Baird, who gave up his new garage business in order to join in the fun.

But they had hardly reached the mainland when Sir Samuel died, and the executors of his will could not carry out his promise to finance an entry in the air race. But Fysh and McGinness had no time to mope. Major-General James Legge, Chief of the General Staff, asked them to plot an air route from Darwin to Longreach in Queensland.

The survey had to be done, by car, over land which had rarely been travelled by white men. Those who had done it had gone by horse or on foot.

The two young men jumped at the chance, and the job provided them with months of excitement and a steadily-growing conviction. Now that an air route had been surveyed, and airfields established for competitors in the air race, the huge country should not be abandoned to the slow, difficult old methods of transport which were all it had ever known. An aircraft could go as far in a day as a horse and buggy in a month.

Such ideas were effervescing in their minds when Fysh watched the Vickers Vimy drift in across the flats near Fannie Bay Gaol, and inspired the shy young man to make his first public speech. "I am quite safe in saying that within five years we will have a regular [air] service between England and Australia," he said. "This will automatically make Darwin the first port of call in Australia on the great aerial highway."

In his book *Qantas Rising*, Fysh commented: "But twelve years were to pass before, in 1931, I was back in Darwin as one of those piloting the first England–Australia Experimental Air Mails; and it was fifteen years, in 1934, before the service I predicted actually started."

There was a lot of flying to be done before then. The story of how McGinness secured finance has been told in an earlier chapter, and by 21 May 1921 they had a capital of £3,300 of which McGinness had subscribed £1,200, Fysh £600, and the graziers Fergus McMaster and Ainslie Templeton £500 and £1,000 respectively.

Fysh and McGinness hurried to Sydney to meet Arthur Baird and order two Avro 504K biplanes, with Sunbeam Dyak engines instead of the old Gnome rotaries, from Arthur Love at Mascot. Only one was delivered, which whittled their capital down by £1,500, so the technically innocent airmen ordered the one and only Avro triplane ever to be delivered to Australia. Probably this was lucky for Australians, because the "tripe" was a cursed machine, although it was supposed to be the most advanced British aircraft of the day and it was the only real triplane design "since the Armistice".

It was to cost Fysh and McGinness even more than the £2,798 which they paid out to Love, and was a constant source of trouble.

But with their Avro 504K and a war-surplus BE2e which they ferried north for the grazier Charlie Knight who changed his mind and sold it to them for £450, they were in business. Both machines made it safely to western Queensland and the first board meeting of the new company was held in the Winton Club, Winton, on 10 February 1921. One of the board's decisions was that the company's headquarters should be established in Longreach, Queensland, and they remained there until the move to Brisbane was made in 1930.

Longreach was hot. Day burst on the plains and simmered steadily until evening. Old Model T Fords coughed down the red dust of the almost empty streets, past weatherboard houses topped with corrugated iron. When ice was available the iceman's rickety dray carried it from one house to another to replace the melted chunk in the iceboxes. Most people relied on Coolgardie safes, cooled with water seeping through hessian. Housewives suffered over woodstoves in lean-to kitchens of galvanised iron, their irons heating alongside the pot which was boiling

Rocket and cannon-firing Bristol Beaufighters of 455 Squadron RAAF attacking a German convoy off Mandal, on the Norwegian coast, on 15 July 1944. (*painting by R. Honisett: Australian War Memorial*)

A fine portrait of an RAAF officer of the second World War, painted by C. Dudley Wood. He was Wing Commander J. M. Whyte, Commanding Officer of 2 Squadron flying Lockheed Hudson general reconnaissance bombers from Hughes airstrip, south of Darwin. (*Australian War Memorial*)

Alexander Eaglerock (Curtiss OX5) VH-UGK over Sydney in the early 1930s. Probably more Australians had their first taste of flying in this aircraft than any other in Australia to that time, with the possible exception of the *Southern Cross* (*McGrath Collection*)

the salt beef. Peppercorn trees hung limply in the still air.

That was life in an outback town, and no one really expected that it could be much different. But outback folk were quick to realise that, at last, their isolation was ended. The aircraft, which were roaring, unpredictable contraptions, were their key to the future. With Fysh and McGinness as pilots, and Arthur Baird keeping them in the air, the little airline flourished.

In 1922, with two Armstrong Whitworth FK8 aircraft, the "big Ack-W", of which one was bought from Horrie Miller for £400, they began a regular service over the 577 miles from Charleville to Cloncurry. They covered the distance in less time than it would take a man to ride thirty miles on horseback. From Longreach to Cloncurry took four-and-a-half hours, and when they carried eighty-one-year-old Alexander Kennedy as passenger he recalled that when he was thirty-four it had taken him eight months to battle over the same distance with a bullock wagon.

In 1925 they were flying on to Camooweal as part of their master-plan to provide a regular service between all the main centres of western Queensland plus an eventual link from Charleville to Brisbane and across the Barkly Tableland to Darwin. The Brisbane link did not come until 1929, when they also had a service from Cloncurry to Normanton. The Brisbane link gave them a total route mileage of 1,484.

By 1926 they had decided to do away with the middleman and build their own aircraft, and under licence from de Havilland they built their first DH50—as did Brearley's WA Airways. Their teething troubles were over, especially now that Colonel Brinsmead had put them in touch with Norman Brearley who gave them the benefit of all that he had learnt about running an airline.

Arthur Baird, as engineer, was kept busy. McGinness had three crashes; another pilot, Lester Brain, had four; and Baird himself had three. Fysh had only one, when he put the Armstrong Whitworth FK8, G-AUCF, into the ground during take-off from the tiny airstrip at Jericho. But the pilots walked away from all the crashes despite the frantic pace which all of them, especially Hudson Fysh, pursued in order to keep the service flying. Fysh's willingness to work and the driving personality which was emerging from his previous shyness caused him to be named managing director of Qantas, though this rank did not tempt him to avoid his full share of the flying.

The idea of a flying doctor service had been mooted as early as 1917, when a Victorian medical student, Clifford Peel, suggested to the Reverend John Flynn, who was superintendent of the Australian Inland Mission, that aerial ambulances could solve the painful problem of carrying sick or injured patients through the rugged outback conditions. Ten years later Flynn talked it over with Hudson Fysh, and the first branch of the AIM Aerial Medical Service was established at Cloncurry in 1928 with Qantas providing a DH50 aircraft. The fuller story of the Flying Doctor Service will be told later.

Qantas now had a competitor: a branch of Jim-

Larkin's Lasco Airline. Larkin was the terror of the airline business, and it seems that the most civil thing anyone said about him was that he was "tempestuous". Apart from building his own aircraft down in Melbourne, he made frequent raids upon any route which seemed to offer an opportunity. While Qantas was trying to obtain the contract to run airmail from Charleville to Brisbane, Larkin brought his monster DH61 aircraft into Queensland. This was a real touch of the future: an eager-looking aircraft with a sleek fuselage behind the giant radial engine and four-bladed air-screw. To convince the Government that they could do the job, Qantas began a daily unsubsidised service between Brisbane and Toowoomba with a DH50A, but it was so unprofitable that they had to close it in less than a year. The Brisbane–Cloncurry airmail contract was secured in 1929, but Larkin managed to grab the Camooweal–Daly Waters section of the route.

Also in the year 1929 what seems to have been a feud between Larkin and the civil aviation authorities exploded into the headlines, when he accused M. L. Shepherd, the Secretary for Defence in 1928, of touting for bribes. Shepherd was suspended from duty until the charge could be investigated by a board under the chairmanship of H. P. Brown, Director-General of Postal Services. The inquiry opened in Melbourne on 7 February 1929, and was fully reported in the Australian newspapers. *Smith's Weekly* devoted a full page, complete with photographs and sketches of the protagonists, to a review of the inquiry, which held Larkin's charges to be unfounded.

Larkin appears to have been the *enfant terrible* of Australian aviation between the wars. It seems that he had no hesitation in making enemies, but this was mainly because he was so urgently intent on establishing a viable Australian aircraft manufacturing industry and on taking part in the establishment of Australian airlines. To a man of such thrusting impatience the delaying tactics of bureaucrats were a real burr under the saddle. His Larkin Aircraft Supply Co. was indeed a serious attempt to produce aircraft wholly Australian designed and built from local materials, and the Lascot and Lasconder, designed by W. S. Shackleton, were strong and simple machines. Larkin's company also produced thirty-two DH Moths, under licence, for the RAAF, but eventually failed financially.

The *Smith's Weekly* reporter wrote, "No one studying the physiognomy of young Mr Larkin, Managing Director of the Larkin Aircraft Supply Co., would doubt that he was an intrepid fighting man in air warfare. Eagle-shaped head, close-cropped and parted neatly in the middle, but not quite, eyes deep set and a little close together, a hard, straight mouth, and a way of glaring hard at the person he is addressing, he is just the man who would fiercely believe that enemy aircraft existed for the purpose of being 'crashed'."

Colonel Brinsmead seized the opportunity to speak his mind about Larkin, and in the *Westralian* of 8 February 1929 he is reported as saying that "for many years past Larkin has been making my life a misery".

According to Larkin, Shepherd made a thinly-veiled request for a bribe when he sat next to Larkin at a dinner being given for Bert Hinkler. Larkin had used this meeting to express his complaints against Brinsmead and the civil aviation authorities in general, which appears to have been his favourite topic of conversation.

Shepherd's reaction to this, according to Larkin, was to explain that papers were presented to the appropriate Commonwealth Ministers in two different piles. One of these would be signed without

The DH84 Dragon was another de Havilland aircraft to prove its worth during many different flying activities in Australia. This one, wearing the RAAF roundel, was photographed in about 1941 (*RAAF Museum*)

Building DH Dragons. Wing and fuselage sub-assembly at Mascot, NSW, in 1940 (*Hawker Siddeley photograph*)

reading: the other would be read before signing. Anyone who wanted to make sure that his request got through without difficulty ensured that it was placed in the first pile and, Larkin alleged, Shepherd asked, "Why don't you pay?"

Shepherd's story was quite different. He said that he listened to Larkin's complaints, told him to put them in writing so that they could be investigated, and when this was done had a report prepared by the "finance officer of the central administration". He said that he had five or six conversations with Larkin about Colonel Brinsmead, in which Larkin "once referred to Brinsmead as a crook and said that he did not get a fair deal from him. I invited Larkin to put his charges in writing and substantiate them."

The whole burden of Larkin's complaints seems to have been his belief that WA Airways and Qantas were receiving a "better deal" from Brinsmead than he was. According to *Smith's Weekly*, it all began in July 1921 when Larkin gained the airmail contracts from Sydney to Brisbane and Adelaide. His tender of four shillings for every mile flown was accepted by the Government, and he was supposed to start the services in 1921. He was unable to start flying because he could not obtain aircraft in time, and was granted an extension of time until April 1922.

But in March 1922 he "pointed out that there had been more delay in getting in the share capital, and he indicated an intention to default on the contract unless the Government came to light with a whopping big subsidy."

"Massy-Greene was Minister at the time and said he was not going to have a gun pointed at him by Larkin to get a subsidy, and the demand was refused. But in order to get the service going he agreed to grant Larkin an overdraft of £15,000 in the Commonwealth Bank on condition that the company helped itself to the extent of [issuing] 25,000 shares.

The DH89 Dragon Rapide, immediate successor to the Dragon, was another classic aircraft which gave many years of service in war and peace. Among its other activities it was used by the Royal Flying Doctor Service for outback medical work. The RAAF aircraft above was photographed during the second World War (*RAAF Museum*)

"The next delay was when the Crown Solicitor found that Larkin's subsidiary company had been illegally constituted and was ineligible to make contracts with the Government.

"Starting of the service was further postponed until September. The machines had not come and Larkin applied, without success, for the loan of Air Force machines from the Government . . . Then he applied for leave to start without the approved machines provided for in the contract. He proposed to use a Sopwith Antelope (which had been bought second-hand and cheap) with a Sopwith Wallaby (a rebuilt machine which had crashed), and a second-hand DH4.

"Brinsmead advised strongly against this by pointing out that civil aviation must, above all, be safe. To start with these old machines, with the engines badly worn and practically no spare parts, would be heading straight for disaster."

Apparently this early opposition by Brinsmead rankled so much in Larkin's mind that he came to believe Brinsmead was "a crook": a feeling which extended to all the civil aviation authorities and led at last to his accusation against Shepherd.

When Qantas won the Brisbane–Cloncurry airmail contract, the company decided to make the long-discussed move to Brisbane. As part of the new operation they opened a flying school at Brisbane's Eagle Farm airfield. Lester Brain was chief instructor. Later he was joined by Russell Tapp and C. W. A. Scott, who served as a Qantas pilot before he went on to win the England–Australia air race of 1934. "Scott was too brilliant to be stable," Fysh said. Scott's father was Master of the King's Music, and Charles, besides being a poet, pilot, and heavyweight boxing champion of the RAF, was also a musician. This balance of intellect and physical skill made the lasses of Longreach adore him. He was fired by Fysh when he crashed a DH50. Fysh later changed his mind and rehired him.

The flying school, using four DH Cirrus Moths, was a success. Its aircraft were used for charter flights as well as instruction. In *Qantas Rising* Fysh tells the story of Russell Tapp taking a newspaper man to photograph a new liner coming into Moreton Bay. As soon as they took off, the photographer started to stand on the front seat of the Moth. "Siddown," Tapp called through his Gosport tube, the communication device from one cockpit to the other.

"There's a snake in the cockpit," was the woebegone reply, and the photographer took a swing at the reptile with his camera which was lost over the side during the mêlée. Tapp hurried back to Eagle Farm, but at first the snake could not be found. It was eventually located coiled tightly inside the metal guard covering the front throttle of the Moth.

Qantas built the first private hangar at Eagle Farm, and opened another flying school at Longreach. This venture was killed off by drought and turbulent flying conditions which forever discouraged some of the pupils from flying. But Queenslanders generally were becoming enthusiastic flyers. In 1932 the Queensland Aero Club flew nearly 2,000 hours and trained twenty-four pilots, and in 1933 there were nearly 400 applicants for a flying scholarship donated by the Brisbane *Daily Mail*. Station owners, such as O. T. Baker of Bowenville, were buying private aircraft, and Ron Adair was using his Hawk Moth to carry prospectors to the outback.

But Hudson Fysh and his colleagues now had bigger things in mind than their Queensland services. By 1930 the idea of a regular airmail service between Australia and England had been discussed countless times although the Federal Government was largely noncommittal especially after the Depression took hold of Australia. It seemed that if the airmail contract went to anyone it would be to Imperial Airways. Many Australians still looked to Britain for a lead in this area, and Imperial Airways was one of the very few overseas air services flying in those days. At last, in 1930 Imperial suggested an experimental trip. They were to fly the mail as far as Darwin, where it would be picked up by Qantas and taken to Brisbane, then carried by an aircraft of Kingsford Smith's ANA. The airliner left Croydon airport, London, on 4 April 1931, but crash-landed at Koepang on Timor. Kingsford Smith and G. U. Allen immediately flew the *Southern Cross* to Koepang and brought the mail to Darwin. Qantas then took it over and the rest of the service was carried out as scheduled.

The whole trip had taken twenty-four days to Brisbane: hardly a record-breaking attempt because sea mail took only about a month from London to Sydney. But the return flight, with Smith flying the mail from Darwin to Sittwe (then Akyab) in Burma, was more successful. A second outward flight, again with Smith and his *Southern Cross* participating, shaved even more off the time.

With their usual eager desire to be first in the field, and disregard of possible problems and difficulties, Smith and Ulm then applied to the Australian Government for a permit to operate a permanent airmail service between Australia and India. By way of establishing their claim they organised an unsubsidised return Christmas mail flight from Australia to London. The public responded wildly and offered 52,000 articles for the Christmas airmail. "Scotty" Allen took off with the load from Melbourne on 20 November 1931; Horace Brinsmead was his only passenger.

The Avro X tri-motored machine performed well as far as Alor Setar, on the Malay Peninsula. There, the centre engine cut out on take-off and Allen had to abort his trip. As soon as Smith received the news, he took off in the *Southern Cross* from Australia, picked up the mail in Alor Setar, and carried on to London.

Brinsmead had not waited for Smith. He was impatient to get to London and discuss the continuing question of how the England–Australia airmail service should be organised. He took the KLM service, but the aircraft crashed on take-off from Don Muang airport, Bangkok. Five passengers were killed but Brinsmead, though badly hurt, survived. He was eventually brought home to Australia, but he never recovered from his injuries and died in 1934. His civil aviation post was taken over by Captain Edgar Johnston, but the loss to Australian aviation at that crucial stage was inestimable.

Brinsmead's accident completely changed the balance of power in the still unresolved problem of the overseas mail service. Brinsmead had worked well with Norman Brearley, of WA Airways, whereas Brearley and Johnston were mutually antagonistic. They had argued too often while Johnston was still Brinsmead's second-in-command. However, it may have been pure coincidence that when Johnston became Controller of Civil Aviation Brearley lost his Perth–Wyndham airmail service to the newly-formed MacRobertson Miller Airlines. Brearley had been keen to participate in the overseas airmail service and had taken the first steps by linking up with a Dutch ship which called at Wyndham and carried the mail on to Indonesia, whence KLM flew it on to Europe.

Brearley respected KLM as a highly efficient service, whereas he had reservations about Imperial Airways. Ulm agreed with him, but Hudson Fysh preferred the British airline. A series of conferences between the three airline chiefs made no progress towards a firm understanding, and it became obvious that Fysh would not agree to the formation of a new company which incorporated his own Qantas, Brearley's WAA, and Smith and Ulm's ANA. Later he wrote, "This we were anxious to avoid because we knew Qantas and its way of life would vanish if it came about."

He may have been right, because there was no doubt that Ulm and Brearley were drawing closer together. They continued to do so even after it became obvious that they would not attract Fysh away from his own plans. When Qantas and Imperial Airways made their first agreement, early in 1933, Ulm and Brearley began to concentrate on domestic air services. On 1 March 1934, in an article headed "W.A. Airways Prepare Elaborate Plans for Operation of Mail Services". *Aircraft* magazine wrote of a new company, to be called Commonwealth Airways Ltd, to be formed "by a group of Australian interests, including W.A. Airways Ltd, and a number of eastern States concerns represented by Mr C. T. P. Ulm".

The proposed new company did not eventuate. Brearley sold his airline and withdrew from aviation, and Ulm met his death in the air.

The Qantas "way of life" continued successfully. Its first million flying miles had been completed in 1930 and the company had racked up a number of "firsts". In 1923 Prime Minister Bruce and his wife hired one of the company's DH50s for a taxi trip to west Queensland: the first such flight by an Australian Prime Minister. In March 1927 Qantas built the first private hangar in Brisbane, and opened the Brisbane Flying School. Later this was handed over to the Aero Club. The first DH Moths in Queensland were imported by Qantas in 1927. In May 1928 the company flew the first Flying Doctor services for the Australian Inland Mission. In the same month it opened Australia's first regular daily air service, between Brisbane and Toowoomba.

Now Qantas was to be first in another direction: that of flying Australia's first regular overseas airmail service. The lengthy argument as to how the service should be operated had been resolved in January 1933, when after a three months' investigation the Inter-Departmental Committee on Air Communications submitted its report to the Australian Government. The committee's report contained many suggestions for the reorganisation of internal air services, and then discussed the competitive offers made by Imperial Airways and KLM. Imperial Airways' bid was preferred. Even though a link with Imperial Airways would call for a Government subsidy while KLM would operate without such a subsidy, the committee pointed out the "disabilities that would follow acceptance of the Dutch proposals". These included very high airmail postage, which at 2s 6d per half-ounce would be beyond the reach of many people in those days; national prejudice against the use of a foreign service; adverse effect on the development of Australian civil aviation; and the fact that a Dutch company and not the Australian Government would benefit if the service flourished.

The committee suggested a once-weekly, mails-only service, using multi-engined landplanes fitted with approved flotation gear and direction-finding wireless, to carry mail to Singapore where it would be transferred to Imperial Airways. But "Passengers should not be carried in landplanes over the Timor Sea."

The service should enter Australia at Darwin instead of Wyndham, and the 2,325 air miles from Darwin to Singapore should be flown via Timor, Bima, Surabaja, and Djakarta, which eliminated "all water crossings of any magnitude except the Timor Sea". It was estimated that the service would carry mail over this section of the route in seven days, thus creating a seventeen-day service to London at a cost to the public of 1s 6d per half-ounce of mail.

Qantas owned five of these Catalina flying-boats and used them on the Perth–Ceylon service, known as the Double Sunrise Service, from 1943 onwards. They were also used on Pacific Islands services. The "Cat" played many parts during her long life and was widely used by the Allied air forces during the second World War (*Qantas Airways photograph*)

After that, things began to move. Imperial Airways "declined to comment" on the report, but registered a company in Australia. A. Bakker of KLM addressed the Air Convention held in Melbourne in March and gave a lengthy exposition of how much his company's proposals would benefit Australia. *Aircraft* magazine tended to take his side, especially because KLM offered a faster service and had a better record for keeping to schedule than had Imperial Airways. The magazine pressed for an early decision, but the picture changed when Imperial Airways began talking about sending one of their aircraft to Australia on a trial flight. It became obvious that if the British company scented any chance of success then all the Australian commercial airlines would be pushed aside. In no time Imperial Airways would be running a service to Darwin and then perhaps to Sydney, and the Australians would be left only with the crumbs from the imperial table.

The redoubtable Larkin Aircraft Supply Company entered the lists by offering to run an unsubsidised service from Darwin to Koepang, where the mails would be transferred to KLM aircraft, but Larkin's move does not seem to have been taken very seriously. It was, after all, only four years since his attack on civil aviation authorities.

On 18 January a new company was formed: Qantas Empire Airways Ltd. It was owned in equal parts by Qantas and Imperial Airways, with forty-nine per cent apiece, and with two per cent of the shares held by an "umpire": the Australian, Sir George Julius. His function was to ensure that the new airline operated in the best interests of all concerned, but Fysh called him "the umpire who never had to blow his whistle".

Hudson Fysh's subservient link with Imperial Airways did not please every Australian. Many, like Brearley, disliked the idea of being tied to the British and felt uneasily that Imperial Airways would use the connection as a means of entering Australian aviation. A newspaper cartoon of the day shows a well-satisfied John Bull diner looking down at the skeleton of a fish (Fysh) on his plate. The caption is "The Imperial Meal".

The new company, with a capital of £200,000, planned to fly its first airmail service from Australia to Singapore on 10 December 1934, though the Government had not yet accepted any of the tenders which had been submitted. The Qantas Empire Airways tender specified that the company would buy five de Havilland DH86 aircraft, specifically designed by de Havilland's for the service, to fly the Brisbane–Singapore route. The aircraft had four Gypsy VI engines and could carry ten passengers and a crew of two. They were then the fastest British airliners, with a top speed of 170 m.p.h. and a cruising speed of 145 m.p.h. Official tests had shown that the aircraft could function satisfactorily even if two engines cut out.

The tenders were opened on 19 April 1934, and the Prime Minister announced that Qantas Empire Airways was the successful tenderer for the airmail contract, with a price of 3s 1d per mile during the first year, decreasing to two shillings a mile during the fifth year; a total of £318,426 during the five-year period. Fysh said that elaborate precautions had been taken to safeguard passengers, and that only five would be carried on the sea crossing from Darwin to Timor. The aircraft would be fitted with radio and radio direction-finders and the Government was to provide radio and weather-forecasting stations at Timor and Darwin, together with a Darwin-based twenty-three-knot patrol boat which would be in constant radio touch with aircraft making the 520-mile flight over the Timor Sea. Each aircraft carried a rubber lifeboat capable of carrying eight people. The anticipated trip time from Brisbane to Singapore was five days, with overnight stops at Longreach, Daly Waters, Koepang, and Surabaja, and intermediate stops for meals and aircraft servicing. The company believed that the time would eventually be cut to four days.

The company also submitted an alternative tender, offering to include a link between Cootamundra in New South Wales and Charleville in Queensland for a total cost of £339,486, but this sector went to C. A. Butler who flew the mails up from New South Wales to connect with the overseas service. The Western Australian mails were to be flown on a Perth–Katherine link, and this contract was won by MacRobertson Miller Airlines, recently formed by Horrie Miller with the backing of Sir Macpherson Robertson. The Tasmanian mails were brought across from Hobart, and fed into the various connecting links, by Holyman's Airways Pty Ltd which won the tender for a service flying six days a week between Melbourne, Launceston, and Hobart.

The starting date of 10 December 1934 had been set because it was the anniversary of the Ross and Keith Smith landing in Darwin in 1919. Lester Brain went to England to bring home the first DH86 and landed successfully, but the second aircraft, piloted by Captain Prendergast, of Imperial Airways, crashed at Bairnsdale, near Longreach. It was the first of a number of unexplained DH86 crashes.

It seemed that the problem would force postponement of the 10 December starting date, but Imperial Airways juggled their rosters and equipment and announced that they would be able to use one of their Armstrong Whitworth Atalanta biplanes to fly a Singapore–Darwin connection. Qantas could link Darwin to Brisbane with its smaller DH61 and DH50J single-engined biplanes.

Actually there were three starts to the service: from Brisbane on 10 December, from London's Croydon airport on 8 December, and from Singapore on 16 December. Qantas, which rarely missed a publicity

Aircraft of the transition period when designers and builders were producing a variety of machines to meet the increasing demands of military and civil aviation. FROM TOP: A Wackett/Tugan "Wackett" of the RAAF; a Short Empire flying-boat during the second World War; and an Airtruck freighter (*All photographs from RAAF Museum*)

S25 Mark IV Sandringham flying-boats were used by Qantas from 1950 to 1955 on services from Australia to Papua New Guinea, the Pacific islands, Fiji, New Hebrides, and Lord Howe Island (*Qantas Airways photograph*)

opportunity, made the most of the Brisbane start from Archerfield airport. The official take-off ceremonies were attended by the Governor-General; Prime Minister Joseph Lyons; the Queensland Governor, Sir Leslie Wilson; Minister for Defence, Sir Archdale Parkhill; and the Director-General of Postal Services, H. P. Brown. Fergus McMaster, one of the founders of Qantas and chairman of Qantas Empire Airways, missed the show. His aircraft was grounded at Longreach by a dust storm.

Two aircraft were needed for the first flight to Darwin. Captain Lester Brain flew *Diana* and Captain Tapp flew the *Hippomenes*, starting out with a total airmail load of 1,267 pounds, 4·75 ounces. They picked up mail from the feeder services en route, and delivered a staggering total of 55,967 items to the Imperial Airways machine *Arethusa* at Darwin.

Australia's first regular overseas airmail service commenced operations fairly smoothly, despite minor mechanical and human problems, but the Qantas executives were anxious to take full charge of their own Brisbane–Singapore sector. The fault in the DH86 aircraft was believed to be that of weak fin posts, and when two more machines were delivered the Qantas engineers soon made the necessary modifications. On 25 February 1935 Captain Brain and First Officer F. W. Stevens flew the first complete Qantas overseas operation in the DH86 Royal Mail aircraft *Melbourne* from Singapore to Darwin. There were no passengers aboard, and the civil aviation authorities kept a nervous eye on the operations until they were sure that they were being flown safely and successfully. They were satisfied by 17 April 1935, when Major H. Philips and Lady Mountbatten were the first fare-paying passengers.

Qantas had been through some shaky times; both in its early years when it seemed quite often that the infant airline would never make the grade, and more recently during the Depression era, when Hudson Fysh and other executives took salary cuts. But the fiscal year of 1934–35 showed an encouraging upturn: almost £5,000 profit. Obviously the overseas operations were here to stay, and Lester Brain was placed in charge of operations with Baird as chief of works and engineering. C. O. Turner, an Australian accountant who had been working with Imperial Airways, was appointed QEA accountant and eventually became the company's chief executive.

Hudson Fysh kept an eagle eye on every detail of operations. Qantas was becoming an Australian institution, and like all institutions it attracted the probing of the Press and the prying of the critics. When the second DH86 of Holyman's Airways was lost, in October 1935, the engineering witch hunt was on again and Qantas had to ride out the storm. Perhaps it was the gospel of efficiency preached by Hudson Fysh which kept passengers flying safely in Qantas DH86s.

Captain "Scotty" Allen was the big mystery of those early years. The petrol consumption figures which he turned in were far below those of other pilots, and eventually Fysh sent a memo to Brain. It said, "I think the continuance of phenomenally low petrol consumptions obtained by Captain Allen requires careful investigation. The figures show results far below what the engine makers say is possible."

The mystery was solved when Allen explained how he played with the DH86 throttles all the way from Singapore to Brisbane, constantly easing them back in a kind of game which he played to keep the engines running at full pitch on minimum petrol consumption. It was only one of the tricks which QEA pilots taught themselves as they kept the service running.

By comparison with the aircraft then being turned

out by such firms as Douglas in America and Fokker in Germany the DH86 looked antiquated even before it was delivered, and Hudson Fysh realised that the aircraft would not see out the five-year airmail contract without seeming hopelessly outmoded. The choice of replacement aircraft was difficult. Even though American, British, and European airlines were regularly flying considerable distances with land-based aircraft, the thought of that 520 miles of sea between Darwin and Timor still seemed like a terrifying barrier. And of course Imperial Airways aircraft flew regularly over wide stretches of sea. So both QEA and Imperial Airways favoured flying-boats. It was a type of aircraft which had progressed immeasurably and had many advantages, of which perhaps the greatest of all was that it was literally a boat, able to survive with its passengers in quite a heavy sea, whereas a land-based aircraft usually sank rapidly.

Imperial Airways was making the change to flying-boats on a number of services when QEA was established. The Short flying-boats of the three-engined Calcutta and four-engined Scipio and Scylla classes soon came into regular use, and the Scylla was described as having "the most luxurious saloons ever prepared in an aeroplane". It could carry thirty-nine passengers and a total payload of four tons: weights which designers doubted could be supported by the undercarriages of that period. The flying-boats were built for comfort and safety rather than speed, and those who knew them in their heyday speak nostalgically of their luxury and elegance. Certainly they were not record-breakers, but those who flew round the British Empire before and just after the second World War enjoyed their stately, leisurely progress as they droned across the world, with nights spent at comfortable hotels while the flying-boats lay moored in a lake or seaport.

The decision to use Short flying-boats for QEA was not difficult to make. The problem was that of setting the creaking machinery of Government in motion to accept the new idea. After many misgivings it had approved the use of land-based aircraft for the overseas service; now, after only three years, it was being asked to make a fresh decision and to provide new facilities. A seaplane base had to be established, and it seemed that the argument as to what were the correct facilities would never end.

Finally the decision was made to establish the base at Rose Bay, Sydney. It was a logical step, and to take it even further the planners decided that the QEA flying-boats should fly all the way to England though with a change of pilots, to Imperial Airways men, at Singapore. They would touch down at the Imperial Airways flying-boat base near Southampton.

A QEA office was established in London, and on 10 June 1938 the first two Short S23 Empire-class flying-boats were flown into Rose Bay by Lester Brain and Arthur Baird. The first Sydney–Southampton service, flown by the *Cooee* under Captain P. W. Lynch-Blosse, took off on 5 July 1938. There was little more than a year of peacetime flying to run.

11

Reg Ansett and Horrie Miller

Two very different personalities had similar dreams of efficient air transport. They succeeded separately, and then one was taken over by the other

An illuminating review of Australian aviation in 1939, published by the British aviation magazine *Flight* on 9 November, gives a clear picture of the huge steps which had been taken in the twenty years since pilots returned from the first World War had been eking out a living by giving joyrides or barnstorming in Peace Loan publicity stunts paid for by the Government. In 1919 Australia had the barest minimum of civil aviation facilities, and those were largely funded by enthusiastic amateurs. Australian production of civil aircraft had only just begun, at the Australian Aircraft & Engineering Co. Ltd run by H. E. Broadsmith and the Avro men he had brought out from England. There were no scheduled airmail or air passenger services.

In 1939 the box on the front page of the *Flight* article proclaimed: "Ultra-short wave radio beacons in use on inter-capital routes; the Melbourne to Perth route and vice versa covered by Douglas in one day; all airlines subsidised to enable them to run at a reasonable profit by temporary contracts, awarded for six months from last July; air services co-ordinated and rationalised. All-metal American aircraft on every main route; Commonwealth Aircraft Corporation in production with 'Wirraways'; fifty Lockheed Hudsons ordered by the R.A.A.F.; and Mr. Clapp organising the government scheme to manufacture Bristol Beauforts. These developments in Australian aviation have all occurred within the last twelve months."

A great part of this progress had been made because of the impulse given by Joseph Lyons, Prime Minister until his sudden death in April 1939. Lyons and other Australians could see the certainty of war, and he announced a vastly expanded programme of defence spending. This programme was backed up by improvements in internal communications, and in a country like Australia with its huge distances it was logical to take advantage of all the improvements in aircraft and their auxiliary services. The Civil Aviation Board was finally upgraded from an offshoot of the Defence Department to a full Ministry, and in 1939 H. V. Thorby was appointed the first Minister for Civil Aviation.

The establishment of the Department of Civil Aviation was inspired partly because of the recommendations of yet another inter-departmental committee—consisting of members of the Defence Department, Post Office, Commonwealth Railways, and the Civil Aviation Board—and partly because it was obvious that the whole pattern of Australian civil aviation was becoming confused. Some airlines were subsidised and some were not; the airport and navigational facilities lagged behind those required for a developing country; and the Civil Aviation Board was subject to constant criticism from those who had to operate the nation's airlines and other aerial services.

The situation was brought into tragic perspective by the crash of the ANA airliner *Kyeema* in October 1938. The Douglas aircraft tried to find Essendon airport in low cloud, but instead flew into Mount Dandenong. Fourteen passengers, including Charles Hawker of South Australia who many people believed a likely contender for the post of Prime Minister, died together with the crew of four. The

Top: Tiger Moth fuselages being treated in the "Dope Shop" at Mascot. The women doing the work show that this is probably a wartime photograph, but it could have been taken at any time from 1935 onwards (*Hawker Siddeley photograph*)

Bottom: A DH Moth Minor monoplane in the line up preceding an RAAF aerial display in 1937 (*RAAF Museum*)

The DH Flamingo was one of the beautiful "new breed" of aircraft designed and built during the late 1930s (*RAAF Museum*)

inquiry showed that the crash had been caused by pilot error and that the pilot had not used his radio direction-finder to discover his position. The Accidents Committee report, which went to the Cabinet, criticised various aspects of the Civil Aviation Board. Joseph Lyons acted immediately by separating civil aviation from defence and creating the department.

By that time ANA had expanded considerably. With their subsidiary, Airlines of Australia, they ran four Douglas DC3s and three DC2s, and flew from Melbourne to Perth in thirteen hours. They operated services from Townsville in the north to Hobart in the south, but the direct Sydney–Adelaide route had been lost to Guinea Airways. This Australian company based in Papua New Guinea had by the late 1930s become the largest freight-carrying airline in the world and one of the most profitable. It had been seeking an opportunity to step on to the mainland and seized the chance which came when Adelaide Airways Ltd, a tiny company operated for about two years by the Adelaide Steamship Company and Coast Steamships Ltd, was absorbed by the Holyman interests as part of the formation of the second Australian National Airways. Adelaide Airways had been running small passenger aircraft to such outlying points as Port Lincoln, Cowell, and Ceduna in South Australia, and when the airline was taken over the Civil Aviation Board appointed Guinea Airways to carry on those runs. This was a useful lever for the Papua New Guinea company, and it soon was running a service between Adelaide and Darwin as well as the Adelaide–Sydney flight. This it had snapped up while ANA were awaiting delivery of the aircraft to use on that run.

But the wheel of fortune turns, and the time came many years later when Guinea Airways was absorbed by ANA, after the latter company had been taken over by Reg Ansett. Now it is one of the almost forgotten airlines of Australia.

ANA had made an earlier stab at Guinea Airways. An article in *The Aeroplane* of 28 September 1938, headed "Nineteen Years of Australian Aviation", describes how ANA discussed amalgamation with Guinea Airways and continues, "Then all was off and Guinea Airways stalked through the dead heart of Australia to Darwin. Two years of unsubsidised running reduced the bulge in their bank balance to a hole. But the experiment was justified, for although they lost the Sydney–Adelaide route they will receive £40,000 annually from the Government for the Central Australian route.

"Then came Ansett—one of the really bright young men in Australian aviation. He made a phenomenally successful mark for himself in road transport, was wiped off the road for all practical purposes by the Victorian Government in a wild frenzy to save the railways, and quietly resolved that he would take to the freedom of the skies. He did.

"He maintained a splendid service between Melbourne and Hamilton before deciding to spread his wings in competition with Australian National Airways (to Sydney) and Guinea Airways (to Adelaide). He, too, has got his reward, by becoming the subsidised service between Sydney and Adelaide, from which Guinea Airways withdrew after receiving the Centralian contract. Thus airlines are made."

Numerous small airlines were operating in that period. They included Adastra Airways, which operated a daily service between Sydney and Bega; North Queensland Airways; and Mandated Airlines, run by W. R. Carpenter Ltd. Using DH86 airliners, this last company flew regularly over the 2,500 miles from Sydney to Rabaul by way of north Queensland and Port Moresby, including the 425-mile sea crossing from Cooktown to Port Moresby.

By late 1938 Australian domestic airlines were flying 10 million miles a year. There was talk of forming a new company comprising Qantas, Imperial Airways, and Union Airways of New Zealand, which would first institute a regular trans-Tasman service and, in 1940, extend it from New Zealand to Vancouver. This would have completed the round-world chain of air services linked together by various Commonwealth airlines, but war came before it could be implemented, and when the war was over the whole pattern of aircraft and aerial services had changed beyond recognition.

Australia in the immediate pre-war years had become thoroughly air-minded and it seemed that the sky was the limit for all the competitive airlines which, with or without Government subsidies, were striving to take over a solid share of the available routes and even to extend them. Most of them vanished from the domestic airline scene during the war, and out of them all only two principal figures survived: Reg Ansett and Horrie Miller. Of those two, Ansett was to become the household word while Horrie Miller retired gracefully to his cattle property in the north-west of Western Australia. His name, now linked to Ansett's, lives on with MacRobertson Miller Airlines, an Ansett subsidiary.

Both Horrie Miller and Reginald Myles Ansett are self-made men. Ansett went to work at fourteen, as a knitting-machine mechanic; Miller's first job was as assistant to a bread carter. He lost that one when he forgot to close the door of the cart, and the bread was strewn from one end to the other of St Kilda Road, Melbourne. His next job was car cleaner and rouse-about for a car hire service, and this began a love affair with machinery which lasted for the rest of his life.

For Ansett, the prospect of a life spent in repairing knitting machines was not good enough. He decided to "go north, young man", to see whether the wide open spaces had something better to offer in the way of opportunity. At nineteen he was aboard ship en route to Darwin. "I met with a remarkable man about to form a survey party," he recalls. "It was a good way to learn the Territory, so I signed on. I was the axeman, later the mechanic."

"The tallest, toughest timber in the world always grows on the survey line," he said ruefully. "I learned axemanship and I'm still proud of it. Drop a tree on a survey line and the stump stays there. Each stump is a memorial to the survey party that made that line. I used to leave them smooth as a billiard table."

Ansett thought that the Northern Territory would offer a good opportunity for growing such crops as peanuts. When he had learnt something about the country, and built up his stake a little, he planned to take up land and become a farmer. Destiny planned otherwise. "The Depression hit. Survey work was abandoned. There was no future in the Territory then, but I learned the problems of areas which were unserved by transport."

Another of the "greats" of Australian aviation: H. C. Miller, known to everyone as "Horrie." This photograph was taken in the 1960s

Transport: that was the key to the future. Australian transport services were bad. Once a man got a few miles from the cities, he might as well have been back in pioneer days. Certainly the railways had been running from about the 1880s onwards, but Australian railways then had a poor reputation: a history of "muddle and mismanagement" as *The Aeroplane* magazine said in September 1938.

Fast transport had long intrigued Ansett. He obtained his pilot's licence, No. 419, in 1929, just before leaving for the Territory. He was not sure what he was going to do with it, and certainly he could not afford to buy an aircraft, but like many other young Australians he had been inspired to take flying lessons by the trans-Pacific flight of Ulm and Kingsford Smith. These two Australians had shown that, given sufficient ambition and daring, the future of aviation was unlimited. Unlimited, exciting . . . and uncertain.

Ansett's first transport company took to the roads. From Hamilton, Victoria, he started a regular car service to Melbourne. Few people then owned private cars, and the idea of a quick road trip to the city, instead of grinding along in the outmoded rolling stock of Victorian Railways, had instant appeal. Ansett's success alarmed the railways. He was actually offering a better, quicker service than they could—and he was extending his operations to other towns, and beginning to carry freight as well as passengers. Slowly the bureaucratic juggernaut began to turn, intent on crushing this intruder.

But even the railways could not stop a man taking to the air—though by 1938 the Railways Commissioners of both Victoria and New South Wales were urging the institution of State airline services to be controlled, naturally, by the railways. Fortunately for the future of Australian aviation, the State Governments were not impressed.

On 16 February 1935, with the help of local graziers, Ansett bought his first commercial aircraft and began a service from Hamilton to Melbourne. There was some brief competition from Adelaide Airways, which opened a service from Adelaide to Melbourne via Mount Gambier and Hamilton, but Ansett held his own and continued to expand. He had a stroke of bad luck on 28 February 1939, when his Lockheed 10 Electra, *Ansirius*, was destroyed by fire on the ground. But it did not even break his stride.

The second World War dealt a body blow to many of the smaller Australian airline operators. If they were not forced to fold up during the war they did not survive the fierce competition immediately after it. Ansett won through both of these periods. His airline served the more densely populated areas of Australia, so there was more business than he could handle during the war, and when the world lurched suddenly from war to peace after the A-bombs were dropped he was well established to carry on.

So was Horrie Miller, now contentedly flying on the other side of the continent. Apart from the fact that they had both had to make their own way in the world, the only similarity between Miller and Ansett seems to be their involvement in flying. They have no physical resemblance. Miller is sandy and solidly built, like his Scottish father who left him in charge of a Victorian dairy farm while he went off to try for a fortune in Western Australian goldmining shares. The fortune did not eventuate and young Horrie had to make his own living. Ansett is dark, lean, seemingly as poised and eager as the racehorses he loves. Yet there is an underlying similarity between the two: a determination, which manifests itself in different ways.

In Ansett, it was a determination to succeed. In Miller, it was a determination to learn more and more about flying. Flying, to Ansett, was one of the routes towards success. To Miller, it was a way of life in itself. Perhaps this is why the first took over the airline founded by the second—though Horrie Miller has always been more aggressive and purposeful than his seemingly shy and self-effacing personality would show. He must have been aggressive: his record is a saga of achievement and his name stands high in Australian aviation.

Miller's first ambition was to break into the motor business, and his father apprenticed him to a heavy engineering company making boilers and farm machinery. Young Horrie found that working in a smoke-filled factory was a poor substitute for his own idea of the "motor business": driving through the sunny countryside as free as a breeze. The idea of flying seemed even better, and he became an avid reader of every report of those infant days of Australian aviation before the first World War. He began making model aircraft, culminating in a four-foot model which he flew at night to avoid the jeers of the local larrikins. One of these night flights attracted the interest of a local car distributor, who offered young Horrie a job.

At last he was in the world he had yearned for, and it brought him into contact with Harry Hawker, Harry Busteed, and Harry Kauper: three young Australians who were to make their own names in aviation. Miller lost his job with the car firm and launched out on his own. He started a taxi service, and among his first jobs was that of chauffeuring Major H. A. Petre and Captain Eric Harrison around

Top: Horrie Miller was the mechanic who flew with Harry Hawker in this 1914 Sopwith aircraft, the first to have windows in its fuselage

Horrie Miller tows the aircraft he designed and built with Bob Cousins, en route to Point Cook for a test flight

the countryside near Melbourne as the two men sought a suitable site for Australia's first military flying school.

Miller's father died in 1913, and Horrie used his tiny legacy to buy a ticket to England, following in the wake of Hawker, Busteed, and Kauper. He sailed with his friend Bob Cousins, and because of their contacts with Hawker and Kauper, by then working in the Sopwith aircraft factory, they were taken on at Sopwith's.

Now Miller was in the world of aviation with a vengeance. He says that the film *Those Magnificent Men in their Flying Machines* might have been based on his own knowledge of 1914 aviation in England, because he became familiar with the daredevils from many countries who flew their rickety crates off Brooklands aerodrome—and sometimes came to grief in the near-by sewage farm.

August 1914 put an end to all that; Miller tried hard to enlist in the RFC or any other of the British services that would have him, but he was only one of hundreds of thousands of young men clamouring to have a crack at the Germans and he became discouraged by the slow-moving confusion of the British authorities. When he heard that the AFC was being formed in Australia he took the first ship home, and spent a good deal of the voyage in drawing up thirty-six sheets of plans of his own aircraft design. He found that he was just too late to enlist in the first contingent of the AFC, so with Bob Cousins he set to work to build his own aircraft. At last it was finished: a beautifully poised machine standing in the stable in which they had built it. Miller was in the cockpit, running the first engine tests, when the machine jumped the chocks and crashed into the end wall of the stable. Cousins was badly injured, but recovered in hospital. Miller had now been accepted for the AFC and towed the repaired fuselage of his little aircraft to Point Cook, where its graceful lines showed up the clumsy BE2s lined up outside the hangars.

Miller passed his flight training course in England, was commissioned into the AFC, and took off for the Western Front. "Having to become a fighter pilot and a commissioned officer was a roundabout way of achieving my. ambition to fly. I had no desire to outdo others in shooting enemy planes out of the sky," he wrote in his book *Early Birds*. He crossed to France with B Flight, 3 Squadron AFC, in August 1917, and spent the war in France until he was invalided back to England as one of the countless victims of the Spanish influenza epidemic that raged in the second half of 1918. "Having a strong instinct for personal survival, I did what was required of me without taking any unnecessary risks or finding anything glorious—except the sheer relief of coming out alive—in sending other airmen to their deaths," he wrote.

Miller returned to Australia as one of the flyers who wanted to turn the skills learnt in wartime to a peacetime profession. To further this ambition he ordered aircraft from War Disposals in England, and filled in time at Point Cook testing aircraft to be used in the permanent RAAF which was to be established in 1921. His first venture into civilian flying was as a member of a small "barnstorming" company, using a Sopwith Dove to give joyrides at a guinea a flight. There was plenty of excitement, especially in keeping the crowds of eager kids away from the machine, and a need for skilful flying as they manoeuvred the Dove in and out of a five-acre airstrip bounded by high-tension electricity cables. This venture was extended into a barnstorming tour around South Australia. It was accompanied by all the minor adventures customary to flyers in those days, in which a landing in a wheat paddock to clear an engine fault was all part of the day's work.

At last Miller's own aircraft arrived. It was one of the Armstrong Whitworths known as "big Ack-Ws", with Daimler engines. The first barnstorming flight ended in disaster, with a sudden gale which wrecked

TOP: Obviously prepared for a ducking, Horrie Miller stands on a Curtiss flying boat off Glenelg, South Australia, early in the 1920s (*H. C. Miller*)

BOTTOM: The Fokker Universal aircraft flown by Horrie Miller from Adelaide early in the 1930s (*H. C. Miller*)

This painting by Harold Freedman has the official title of "Aircraftwoman, W.A.A.A.F." Many Australian women served in a great variety of capacities with the RAAF during the second World War. (*Australian War Memorial*)

On 16 May 1943, the Mohne and Eder dams which supplied a large area of Germany with water and hydroelectric power were attacked and breached by Lancasters of 617 Squadron RAF in Operation Chastise. Wing Commander Guy Gibson, leading the operation, subsequently was awarded the Victoria Cross. Australians in 617 Squadron included Flight Lieutenant J. F. Leggo (squadron navigation officer); Flight Lieutenant R. C. Hay (bombing leader); Pilot F. M. Spafford (Gibson's bomb aimer); Flight Lieutenant H. B. Martin (pilot of the third attacking Lancaster); and Flight Lieutenant D. J. Shannon and Pilot Officer L. G. Knight (pilots of the Lancasters which attacked the Eder Dam). *(painting by R. S. Sherwood: Australian War Memorial)*

the machine on the ground. His accountant and public relations man, Art Kennedy, also an ex-AFC pilot, helped to repair the machine, and they took off again, criss-crossing South Australia, New South Wales, and Victoria from the edge of the Nullarbor to the Queensland border. Miller says that Kennedy's financial acumen saved the little venture from the failure which ended sixteen other "barnstorming" outfits in 1920, but the partnership broke up when Kennedy married.

Miller sold his Ack-W to the newly-formed Qantas airline, and turned to flying-boats. He and a partner managed to obtain two Curtiss flying-boats on loan for an experimental service across Spencer Gulf and St Vincent Gulf, in South Australia, with the idea of carrying passengers between the little gulf ports and Adelaide. Their main problem was weather, and Miller seemed to spend most of his time saving the frail flying-boats from being wrecked at their moorings. It was almost a relief when they were refused a permit for the service.

Miller's next venture was the purchase of a Bristol fighter which lay in its case in the huge deserted AFC hangar at Richmond, New South Wales. He worked many lonely hours in assembling the aircraft and at last got her off the ground. She was a beautiful machine but now that he had her he was not sure how to use her. This problem was solved when he met Hudson Fysh in Melbourne. Fysh had ordered a Vickers Vulcan biplane from England, but when it arrived at Longreach it proved to be quite unsuitable for Qantas requirements. The machine was known as "The Flying Pig" because of its huge barrel-shaped fuselage. Luxuriously fitted and upholstered for passenger service, it was grossly underpowered. On its delivery flight to Queensland it had had to bump along under the clouds, and the reporters which it carried were sick all over the sumptuous upholstery. At Longreach, the Qantas base, test flights showed that it could not achieve a reasonable altitude in the hot air of a Queensland summer.

Fysh told this story to Miller, and Miller told Fysh about his own aircraft. Finally Miller said, "Well, Huddy, you're stuck with an air service and no aircraft and I'm stuck with an aircraft and no service."

Miller sold the Bristol to Fysh for £1,000, delivered it to Longreach, and returned to the south. He ordered a DH9 from England, and filled in the waiting time by flying an Avro for a Brisbane firm. Their idea was that the Avro could be used for joyriding and for taxi services to outlying towns and stations, and though Miller had some misgivings because he knew the Avro was underpowered he took on the job. Sometimes the engine was so sluggish that he had to land again soon after take-off, and on one such occasion the trailing edge of the wing de-horned one of a herd of browsing cows.

He persuaded the owners of the aircraft to fit her

This Vickers Vulcan, christened the "Flying Pig," made a brief appearance in Australia between the wars (*H. C. Miller*)

with a new Curtiss engine, which improved her performance so much that they entered her in the 1924 Aerial Derby and Light Aircraft Competition run by the Aero Club of New South Wales. Some famous Australian airmen entered this event, including Wackett with his Warbler and Edgar Percival in an Australian-built Avro B4. Percival, who later went to England and greater fame, not only as a record-breaking pilot but also as designer and builder of the Percival Gull which was one of the fastest and most elegant light aircraft of the 1930s, won the Light Aircraft Competition with ease. Miller won the Aerial Derby speed and handicap sections.

Miller was not content to go on flying for other people. He wanted to go into business for himself, and he could not get the idea of flying-boats out of his mind. So, as well as the DH9, he ordered one of his Curtiss flying-boats to be shipped from Adelaide to Brisbane, where he reassembled her and floated her in the Brisbane River. She proved ideal for joyriding at the holiday resort of Southport on what is now known as the Gold Coast.

When the DH9 arrived, Miller felt that it was the first aircraft he had ever owned which gave a really good performance. He felt confident that it would launch him into a successful flying business. Instead, he rejoined the RAAF, under the persuasion of the Defence Department which was trying to form a cadre of experienced flyers at Point Cook. "I was not able to understand then, nor am I now, why at this moment in my career, a time I had awaited with some impatience and worked for so consistently, I should have decided to accept this job," he wrote in *Early Birds*.

He stuck it for just under a year, then resigned and flew off in his DH9 on another barnstorming tour.

Sir MacPherson Robertson, the "Chocolate King" whose interest in Australian aviation led him to help Horrie Miller in the foundation of MacRobertson Miller Airways

By giving joyrides and hiring out for charter services he made a fair living. Businessmen and pastoralists were beginning to realise that air travel was the best way to cover distance without wasting time, and one of Miller's passengers chartered him to fly from Melbourne to Adelaide in a hurry. When they reached Adelaide, he told Miller that he was hoping to form a goldmining company to exploit the gold discoveries in New Guinea. He asked Miller to bring his DH9 to New Guinea, and said that it was a wonderful new field for flyers. Miller decided against it and lost an opportunity to "find [himself] on the ground floor of Guinea Airways". As recounted earlier, the job went to "Pard" Mustar.

Miller met most of the aviation greats of the period: H. J. Larkin; Arthur Butler, "that volatile and most enthusiastic of air service pioneers"; and others. Butler worked first for Broadsmith in the Avro works at Sydney, then as engineer for Larkin. He learnt to fly with the Larkin organisation, but when Miller met him in Hay, New South Wales, he had not yet been taught how to recover from a spinning nose dive. Miller taught him, in "rather a hair raising experience for both of us".

Butler went to England in 1931 and tried to interest British aircraft manufacturers in an all-metal light aircraft which he had designed: the Butler Bat. He failed in this but bought a Comper Swift aircraft in which to fly home to Australia—and broke the solo record by two hours. He returned to England in 1934 to buy aircraft for his new Butler Air Transport service, but could not even get to see the managing director of Armstrong Siddeley. So he bought two twin-engined eight-seater de Havillands instead.

Miller was tiring of joyrides and charter services and sought for a permanent service of his own. He thought that Broken Hill might give him the opportunity, even though Larkin's service from Melbourne to Broken Hill was unsuccessful. Broken Hill folk have traditionally traded with Adelaide even though their city is in New South Wales, because the "Hill" is closer to Adelaide than to Sydney.

The venture was a moderate success. Based in Broken Hill, Miller provided an outback air service which included aerial photography, emergency medical flights, charter flights for prospectors and others, and the eternal joyrides which still brought in a comfortable share of the proceeds. But the one-aircraft business almost ended abruptly when the DH9 was partly wrecked on the ground, when Miller had to land in a paddock near Port Pirie during a dust storm. While waiting for new wings and a new propeller to be delivered, Miller realised that it was time to seek substantial backing and establish a regular air service. He went to see his friend David Robertson, in Melbourne, who had often told Miller to consult him if he needed finance. Robertson introduced him to his brother, MacPherson Robertson whose MacRobertson brand of chocolates was known throughout Australia. The deal was made very quickly. Robertson would back Miller with £5,000 to buy a DH61 passenger aircraft—on condition that the aircraft be named *Old Gold*. This was the best-known brand of MacRobertson chocolate.

The new company began operating in 1927 but was not formally registered until May 1928, as MacRobertson Miller Aviation Company Ltd.

The first flight of the new company with its new aircraft, capable of carrying eight passengers, was from Melbourne to Adelaide—with a full load of MacRobertson chocolate. At that time the Adelaide airfield was still at Albert Park, and Miller was the first to land on the new airfield established at Parafield, in 1928. One of Larkin's frequent arguments with the Civil Aviation Board centred around the new airfield, and he established a hangar in a paddock across the road from the landing ground.

The new airfield was opened with an aerial pageant, with participants including Flight Lieutenant McIntyre who had made the first round-Australia flight. Soon after the opening McIntyre flick-rolled a Moth too close to the ground and became the first fatality at Parafield.

MacRobertson Miller Airlines grew fast. Eventu-

Top: The Wackett Widgeon II amphibian, photographed at the RAAF Experimental Station, Randwick, in about 1930 (*RAAF Museum*)

Centre: The Wackett Warrigal Mark II, designed by Lawrence Wackett when he was the commander of the RAAF Experimental Station (*RAAF Museum*)

Bottom: The Wackett Gannet was the only aircraft designed and built by the RAAF to see service outside the Australian mainland. She was flown during the 1930s (*RAAF Museum*)

TOP: A Westland Wapiti Mark II makes a message pick-up during army co-operation exercises in the 1930s (*RAAF Museum*)

CENTRE: RAAF Short Sunderland flying-boat comes in to land after a coastal patrol (*RAAF Museum*)

BOTTOM: One of the earliest of the famous Cessna series of light aircraft: a Cessna C34 photographed during the 1930s. The VH-UYG registration shows that she was flying under Australian ownership (*RAAF Museum*)

FACING PAGE: The first Empire Airmail being loaded into an MMA DH84 Dragon at Perth, for carriage to Daly Waters and transhipment into a Qantas aircraft (*H. C. Miller*)

ally its Adelaide-based fleet consisted of the old DH9, the DH61, a Dutch Fokker monoplane with a Wright Whirlwind engine, imported from America just before the Civil Aviation Board banned the import of aircraft from the USA, and two DH Moths. The company built two hangars at Parafield and eventually flew services to Mount Gambier, Renmark, Kimba, Whyalla, and Kangaroo Island, apart from the charter flights which Miller still enjoyed because they always promised some unusual challenge or adventure. One of these flights carried an Adelaide specialist to a sick woman at Alice Springs, in the Northern Territory, and when the specialist had examined her he decided that she must be taken to Adelaide immediately. Miller was flying a Fokker Universal monoplane, and when he and the specialist and the patient were aboard he could barely coax the aircraft off the ground. It was a night take-off, and Miller suddenly realised that the aircraft would not clear the wall of the MacDonnell Ranges near Alice Springs. He made a skidding flat turn and slowly gained enough height, but the mercy mission was in vain and the patient died in Adelaide on the following day.

It is sometimes forgotten that MMA began in Adelaide, because "Mickey Mouse Airways", as it is affectionately known, has become such an integral part of Western Australian life. The westward move was made after MMA was the successful tenderer for the Perth–Katherine link on the overseas airmail route in 1934. Until then Miller and his machines were regarded as South Australians, and he played an enthusiastic part in developing aviation in that State. MMA opened a flying school in Mount Gambier, and young pastoralists of the South-East, who wanted to use aircraft in the same way that their fathers had used horses, learnt to fly at the school and then bought their own machines. The Aero Club at Parafield had plenty of members, though when Leslie Birks, scion of a well-known Adelaide family, was killed in a

MMA entered a new era with their purchase of Friendship turboprop aircraft. The Friendship in this photograph inaugurated the first regular air service to Groote Eylandt in the Gulf of Carpentaria (*MacRobertson Miller Airways*)

Horrie Miller with his Wackett VH-AIY in Broome, photographed in about 1960 (*Sydney Morning Herald*)

crash there was a public outcry against "encouraging young men to throw away their lives".

During this period Miller tried a new kind of flying: unpowered flight. Australia's earliest flyers had taken to the air in gliders, and even now that the drone of an aircraft engine had become familiar there were plenty of people who found a greater thrill in the silent soaring above the countryside, so much like a bird that the birds ignored them. Miller enjoyed the few short flights that he made but did not take up gliding seriously, though there were plenty who did. One of Australia's leading clubs was the Melbourne Gliding Club. Its flights, however, were of very short duration. A flight of three minutes, three seconds won the club's Garrett trophy for A. Borgeest in 1934. Gliding enthusiasts pointed out the value of gliding as initial training for flyers because it taught them "pure flying" that depended on their sensitivity to the controls, and the Government at last responded to their propaganda with a subsidy of £600. This was in the era when German pilots, taught to fly in gliding clubs while Germany was still forbidden to build military aircraft, were beginning to take to the air in the powered aircraft designed by Messerschmitt. Hitler had unilaterally revoked the Treaty of Versailles, and its effects on aviation development would soon tell.

"Melbourne Gliding Club Notes and News" and reports from the Geelong and Tasmanian Gliding Clubs featured fairly regularly in *Aircraft*, but the Government subsidy was gained by a letter to the Minister for Defence from Mr Pratt of Aircraft Manufacturing & Supply Co. Ltd of Geelong. Pratt founded his firm in 1920, when he imported four aircraft from overseas, and his principal interest was in gliding. He established the first Australian soaring record at Tower Hill, Victoria, and designed and built the Cadet sailplane which consistently set new soaring records. Pratt's firm ran into difficulties during the Depression and he had to stop building powered aircraft, but he continued to build gliders. Part of his letter was a protest against "a movement on foot by the Aero Club to obtain a monopoly of gliding training, as it has of the training of power pilots".

Though Horrie Miller spent little time on gliding he had not forgotten his early enthusiasm for model aircraft, and became a member of the South Australian Model Aircraft Club. Rather ironically, the club headquarters was the hangar which Larkin had built when he attempted to set up his own airfield on the other side of the main road from Parafield.

The move to Western Australia opened up completely new fields for MMA. The "Mickey Mouse" aircraft began to fly the routes left open when Norman Brearley sold his airline to ANA, and soon became familiar sights in the seaport towns of the north-west coast and then inland. Horrie Miller moved to Perth and then began to spend much of his time in Broome.

Captain Cyril Kleinig is another extremely well known figure in the world of Australian aviation. He joined Horrie Miller in Adelaide in 1930 and piloted MMA aircraft for many years before transferring to the administrative staff. He is now the General Manager of the airline, based in Perth (*MacRobertson Miller Airways*)

His daughter Robin has recorded in her book *Flying Nurse* that he has not lost any of his old delight in flying for the sheer fun of it. He bought an old Wackett trainer which he kept at Broome, and sometimes took his daughters up for a wildly exciting display of aerobatics. He was the despair of the local airfield controller because he never lodged a flight plan when the mood seized him to take off on a fun flight.

While Horrie worked in Broome, Cyril Kleinig and his associates looked after the head office in Perth. The airline thrived and prospered, but administrative problems increased. In 1953 a man adept in solving such problems became the company chairman. He was Mr R. T. (now Sir Reginald) Rushton. Eventually Sir Reginald Ansett acquired MMA as one of his subsidiary companies, but it is still administered by Cyril Kleinig under the chairmanship of Sir Reginald Rushton, with its overall policy directed by Sir Reginald Ansett.

This happened after another war had vanished into history, and during the period when Reg Ansett was expanding his transport empire faster than ever before. It cannot be said that Miller, holder of Australian pilot licence No. 25, and Ansett who holds No. 419, ever actually came to grips, but Ansett is that most irresistible type of human being: a businessman with a dream and the determination to make it work. His dream was quite straightforward: to run Australian air transport as it should be run—one huge and efficient airline serving as many Australians as possible. The second ANA was to be absorbed by this ambition, and then came most of the other smaller airlines which still served Australia after the second World War.

12 Mantle of Safety

John Flynn had visions of a time when isolated outback folk could call for immediate assistance when they were sick or injured. With the Royal Flying Doctor Service of Australia he made his dream come true

One of the best known of all Australian aviation services must be the Flying Doctor. Books, articles, and documentary films have made it known throughout the world, and its principle has been accepted in such places as East Africa which has the same problem as Australia: small communities scattered over vast distances. The romantic and adventurous nature of the service has given rise to a good many myths, which is why the Royal Flying Doctor Service commissioned a factual history, published in 1961. Since then it has had many changes but so far as the patients are concerned they have all been for the better. Other books which tell some of the story of the RFDS are *Flying Nurse* by Robin Miller; *Flying Doctor Calling* by Ernestine Hill; *Flynn's Flying Doctors* by Harry Hudson; and *Flynn of the Inland* by Ion Idriess.

No one is quite certain who conceived the idea of the service. It is certain, however, that John Flynn was the inspirer and moving spirit who literally got it off the ground.

He was born at Moliagul, in north-west Victoria, in 1880, of English, Irish, and Scottish blood: a powerful mixture which has churned restlessly in many other men before and since. His father was a schoolteacher; his mother died when he was born, and he was brought up by his grandmother in Sydney until, at five years of age, he joined his father, brother, and sister in Melbourne.

Originally he had planned to become a teacher like his father, and had actually become a relieving teacher, at the age of eighteen, in Victorian country towns. Then he received his "call to the Ministry" and began theological studies. On a date which is still uncertain, probably in 1903, he went as a Home Missionary to the little town of Beech Forest. There he "found endless interest among the pioneers of one of the most heavily timbered areas of Australia". His interest in pioneers was to stay with him all his life.

It may be the reason why he was keen to enter the mission field. He conducted a Shearers' Mission in Hamilton, Victoria, and volunteered for a mission to Korea. He was far from the conventional picture of a missionary—which in any case is usually completely inaccurate—and like most missionaries was well able to take care of himself. He became a skilled bushman and in 1910 published *The Bushman's Companion—a Handful of Hints for Outbackers.*

He passed his final theological examinations and was ordained in 1911, the year in which he went to Beltana in South Australia. By that time he had seen that his life's work lay among the outback people of his own country, and he took up their cause with ardour. He published a quarterly journal called *The Outback Battler*, and later founded *The Inlander*.

In 1912 he made his first great trip through the Northern Territory, for the Presbyterian Church of Australia's Board of Home Missions. His report on the outback communities was considered by the Presbyterian Church while it was planning the Australian Inland Mission to serve people who, in those days, lived in almost complete isolation. They needed ministers to "bring them the word of God", and doctors to bring succour when they were sick or injured. A man or woman on a station hundreds of miles from the nearest large town lived in constant

Curtiss Kittyhawks of 3 Squadron RAAF taking off in a dust storm at Antalet, in the Western Desert area of the North Africa campaigns, in 1942. (*painting by Frank Norton: Australian War Memorial*)

Curtiss Kittyhawks, bearing the squadron code letters OK of 450 Squadron RAAF, and North American P51 Mustangs with the CV markings of 2 Squadron RAAF, attacked Venice on 21 March 1945. 3 Squadron Mustangs operated in an anti-flak role, while 450 Squadron's Kittyhawks, together with 112 Squadron RAF and 5 Squadron South African Air Force, attacked shipping and harbour installations. (*painting by Alan Moore: Australian War Memorial*)

D OK

TOP: Test of the Airtourer fuselage torsional and tailplane bending stiffness by DCA representative Max Weston, in an old garage in Williamstown, Victoria. Harry Millicer is in singlet and Bill Cooper wears no shirt because the temperature was over 110 degrees. Max Weston is on Millicer's left. (*photo: H. V. Millicer*) BOTTOM: The Airtourer prototype in front of the Williamstown building rented for a dollar a week. The adults, from left, are Ellis Walker, manager and owner of Perfectus A/S Co.; Trevor Sloane, treasurer of the Airtourer group; Henry Millicer; and designer Bill Cooper. See text for full story. (*photo: H. V. Millicer*)

peril. In case of serious illness or accident, the only way to obtain medical attention was to take the patient to a doctor. Too often, he or she died on the bush track.

The matter of spiritual care could be solved by travelling ministers who made their way around the outback communities, and at least a measure of medical attention was provided by the institution of tiny hospitals, usually staffed by nursing sisters, in outback centres which sometimes could hardly be dignified by the name of towns.

The AIM hospitals were a major innovation, and gave outback folk the comfort of knowing that at least there was some kind of medical assistance not too far away—maybe only a hundred miles or so—but they were only a first step forward. They did not solve the problem of the sudden deadly accident: the man gored by a bull, the child bitten by a snake, the fall from a horse. Nor could they help the illnesses which seemed to come from nowhere and which most bushmen simply called "fever". Sometimes the patient recovered, sometimes he or she died, sometimes the illness lingered on.

The basic idea of a Flying Doctor Service was put into words by Clifford Peel, a medical student who must have had a keen interest in flying because he chose to join the AFC instead of completing his studies and becoming a medical doctor. He went overseas in November 1917 and while en route to the Western Front he wrote a letter which was published in *The Inlander* in October 1918—after Peel had been shot down on 19 September and killed in action. The letter gives technical details about the handling of aircraft and the coverage which could be obtained by flying from Oodnadatta, Cloncurry, and Katherine, and shows that a great deal of thought had been given to the idea.

But it does not mention the name of the originator of the idea of speeding airborne doctors to succour outback folk. Flynn himself, in later years, says that Clifford Peel and H. V. McKay shared in his "dearest dream", and it seems that Peel may have been the first to think of it. The important thing is that the idea *was* germinated—perhaps in one of those discussions in which ideas seem to come to life among a group. Whoever thought of it, the credit for the Flying Doctor Service must belong to John Flynn because he was the man who changed it from a dream into reality.

John Flynn and his sister, who used the pen-name of "Cousin Charlotte", would have made good public relations officers. She had already operated fund-raising schemes, including a "mile of three-penny bits", for the mission services. In public meetings, church and Sunday School talks, the publishing of pamphlets, and articles in church newspapers, the Australian Inland Mission slowly spread the idea of a Flying Doctor Service. The

TOP: A camel-drawn wagon brings a patient to the Flying Doctor in the 1930s. BOTTOM: A Fox Moth aerial ambulance of the same period (*Royal Flying Doctor Service*)

national Press took it up, and donations began to flow in. The first came from C. Alma Baker, an Englishman from the Malay States, who came to Australia during the first World War to raise money to buy fighting aircraft. In June 1919 he wrote to John Flynn that "The brave pioneers living in these practically uninhabited parts [of Australia] are now entirely cut off from doctors and nurses . . . none but the people of the 'Never-Never' can appreciate the great blessing and boon which an aerial medical service will be to them . . . on my return to Sydney, at the end of July, I will have much pleasure in giving you £100 towards your medical service."

The opening of Norman Brearley's service to the north-west of Western Australia, followed by the first Qantas services in Queensland, began to make people believe that the idea of airborne medical services was not so far-fetched after all. On 24 February 1922 the Melbourne *Herald* published an article headed "Flying Doctor Service—Help for the Bushman: Aeroplanes would save Lives". The article said that "by the end of the year it is felt that the North-West and Queensland mail services will have discovered practically all the difficulties of flying in the interior, and landing grounds will have been mapped out and prepared. The time will then have arrived for experiments with Flying Doctors."

By that time John Flynn had convinced the

Australian Inland Mission that the idea was feasible, and politicians were beginning to take an interest. The report to the General Assembly of the Presbyterian Church, in 1920, stated "After long advocacy of a medical service for isolated regions we have something to report: members of the legislature in Western Australia have taken up the idea and have induced their Cabinet to seek co-operation by the Federal Government in an experiment. Much may be done for the Bush merely by spreading ideas beyond our own organization."

In 1922, in his report to the General Assembly, John Flynn wrote "The Minister of Home and Territories has already stated in public that the aeroplane will solve the medical problems of the Territory. The Prime Minister has already promised to try out wireless telephones in the same connection. If wireless can be introduced, the problem of Church services is automatically solved."

The physical and spiritual were inseparable in Flynn's dream of the future. The "wireless telephones" which would enable isolated settlers to call for help could also be used for broadcasting religious services to the widespread communities.

But radio, like aviation, was still in its infancy. The optimistic article in the Melbourne *Herald* had given no real thought to the countless problems of outback flying in the frail short-range aircraft of that era, and there was no easy way to provide "wireless telephones" for the outback folk. The equipment could have been made available: the difficulty was that of supplying electric power to operate the transceivers. Very few of the outback communities had electric power of any kind. A few had wind-driven generators which charged banks of twelve-volt batteries, used only for lighting, but most of them still relied on kerosene lamps. And there was no point in sending radio sets to people without electric power.

As always, Flynn had faith that such problems would be solved. "Eventually—Why Not Now?" was the heading to a map published in *The Inlander* in 1920. It showed Flying Doctor "bases" located at Burketown, Cloncurry, Longreach, Charleville, Bourke, Broken Hill, Port Augusta, Marree, Oodnadatta, Kalgoorlie, Mount Magnet, Carnarvon, Port Hedland, Wyndham, Derby, and Katherine. Forty years later, in 1960, RFDS bases were located at nine of those towns: a tribute to Flynn's prophetic vision as well as to his capacity to get things done.

In 1922 Flynn wrote an article, "More About Flying Doctors—Are They In Sight?" and quoted facts and opinions given by Hudson Fysh.

Flying Doctors were indeed "in sight". Aviation was to end the era in which, for example, young Jimmy Darcy was brought in to Hall's Creek after he was seriously injured when mustering cattle. There was no doctor in the tiny bush town but there was a telegraph station. F. W. Tuckett, the postmaster, tried Wyndham first. It is 250 miles from Hall's Creek, but there was some hope that the doctor might make the trip or at least telegraph instructions. The telegraph clicked an answer: the doctor was away. Tuckett tried Derby, 400 miles distant, and received the same answer. He had some first-aid training but did not feel capable of treating young Darcy's injuries, so almost in desperation he telegraphed his old St John Ambulance lecturer, Dr Holland, hundreds of miles to the south in Perth. Holland was at home, and Tuckett took down his instructions for an emergency operation then carried it out on Darcy.

In his 1922 article, Flynn told his readers that Flying Doctors had already begun to serve outback folk, though in an unorganised manner and purely in response to occasional requests. The first recorded Flying Doctor flight was made by Kingsford Smith when he was flying for Norman Brearley. He flew

A Dragon Rapide aerial ambulance taking off from a lonely Northern Territory homestead, with patient on board, in the early 1950s (*Royal Flying Doctor Service*)

Dr Trethowan of Geraldton to attend a young Carnarvon patient. Other early flights included one in July 1925 to a station 100 miles out of Hay, that involved an alteration of the aircraft to fit in a stretcher. Another was from Melbourne to Deniliquin. According to the book *Joysticks and Fiddlesticks* by F. S. Briggs and S. H. Harris, "the doctor performed the operation, and was back in Melbourne in time for his customary round of golf at the Royal Melbourne course in the afternoon".

The most substantial breakthrough came in the form of a bequest from H. V. McKay, the man who had backed Norman Brearley's dream of establishing Australia's first airline. He was a great friend of Flynn, and in his will left £2,000 towards "an Aerial Medical Service experiment". The next step forward was helped by another of John Flynn's friends: E. T. Fisk, managing director of Amalgamated Wireless A/Asia Ltd. In 1927 AWA gave the Australian Inland Mission free use of its patent rights on condition that radios manufactured by the AIM were sold at cost price to lonely settlers and mission stations.

Hudson Fysh took a keen interest in the project from the very beginning. In June 1922 he wrote an article for *The Inlander*, in which he advocated the use of a small two-seater aircraft which could use the smallest possible airstrip. In the same year a definite Aerial Medical Service Fund was launched. The British Medical Association was approached for advice and help, and Flynn found another friend in Dr J. W. Dunbar Hooper. He convinced the BMA Council that Flynn's project was practical, feasible, and desirable.

John Flynn must have been a firm believer in the saying, "The Lord helps those who help themselves." The problem of simple, cheap radio communication had to be solved, and he set himself to the task. His aim was a small receiving and transmitting set with a range of 200 or 300 miles, costing no more than £50 and weighing only about sixty pounds. Every amateur and professional radioman whom he consulted scoffed politely at the idea. They said that the techniques simply had not been developed to build such a set.

So Flynn determined to find out how it could be done. He learnt something about radio theory and manufacture, and solved the problem of a power plant. By that time most outback communities owned some kind of motor vehicle, so the power should be supplied by a generator run by a pulley and belt attached to a back wheel. With the help of aviation pioneer Harry Kauper, of Adelaide, D. G. Wyles, and George Towns of Sydney he built and operated the outfit. On a trial trip into the outback, voice messages were sent over 300 miles and Morse signals sent to Adelaide from as far away as Innamincka.

But Flynn was not satisfied. The radio outfit might have been just satisfactory, but it was not as simple as the one which he hoped for: a radio which could be operated quickly and easily in an emergency. He and his helpers discarded the idea of telephony and decided to build a radiotelegraph outfit powered by Edison primary cells like those used by the railways. They do not need electrical recharging so could be used on outback stations without electric power.

Harry Kauper enlisted the aid of a man whose name has become inseparable from the radio history of the Flying Doctor Service: Alf Traeger of Adelaide. He and Flynn made an outback trip with three new sets: one main set to be stationed at Alice Springs, and two field sets stationed at Hermannsburg and Arltunga about seventy miles to the west and east respectively. After many trials and errors, communication was established between the three stations. Flynn and Traeger made another outback trip, to Cloncurry, in 1927 and set up an experimental main transmitter which could be heard in Sydney.

Then Alf Traeger had an inspiration. Instead of the awkward, low-powered "wet cells", he would power the sets with a generator that relied on nothing but human energy—foot power. He designed a pedal generator on which the operator sat rather as though he was riding a bicycle. The chain drive transmitted the foot power through a series of gears that turned the generator fast enough to give the essential electric current.

He built five pedal generators, and in 1928 he and Flynn made a car trip north to Cloncurry. They showed the radio set and the pedal generators to every outback community en route, and encouraged people to learn Morse Code so that they would be ready to take part in the network of radio communication. Flynn and Traeger had no doubt that it would be established—and very soon.

By the end of 1929 the first permanent Flying Doctor radio stations had been installed: a main station at Cloncurry and eight outpost stations ranging from Birdsville in south-west Queensland up to Mornington Island in the Gulf country. Their primary purpose, of course, was to summon medical help, but outback folk soon found that this wonderful new communications network put them in permanent touch with the outside world—and with each other. A few urgent telegrams had been sent during the early wireless experiments; as soon as the network became an established fact the Post Office allowed each radio outpost to become an official telegraph station.

Until June 1929 the Aerial Medical Service Fund had spent what now seems to be the tiny amount of £2,257 18s 9d on radio equipment, wages, and expenses. This does not take into account the enormous amount of time, thought, and willing assistance given by many people whose names have not been recorded. Enthusiastic workers had to experiment and improvise because radio manufacturers had no interest in building sets for the outback, so Alf Traeger's business flourished instead. The driving force was John Flynn. Because of his faith and determination the rough-and-ready field experiments of the 1920s have developed into a complete national network of radio communications, and a very important financial asset to the RFDS. The network is used for sending thousands of private and business telegrams each year, and the RFDS receives a proportion of the cost of each of these. This money, of course, helps the RFDS to continue and expand its work.

The network has overcome the problems of isolation in other ways. For specific periods each day the outpost stations are allowed to chatter and gossip with each other, passing on local news and simply talking about this and that in the way which is essential to human contentment. In emergencies such as floods and bushfires it is invaluable. And another outback service, the Correspondence School of the Air which brings children into "classrooms" stretching over thousands of square miles, has been founded on the network.

The radio service, essential though it was, was only the foundation of the Aerial Medical Service. McKay's bequest of £2,000 had the provision that a further sum of £4,000 had to be raised and that the working arrangements for the service should be approved by Civil Aviation. The General Assembly of the Presbyterian Church approved continuation of the "medical experiment" provided that the Board gave its approval and that a further sum of £5,000 could be raised. The Wool Brokers' Association gave £1,000 as did Civil Aviation, and the Commonwealth Government agreed to give Qantas a subsidy of a shilling a mile, on mileage flown up to 20,000 miles. John Flynn and his associate, the Reverend J. A. Barber, guaranteed the remainder. In Flynn's words, they went "bushranging for money" among their many friends and business contacts.

Flynn, Barber, and Dr Simpson, who had been appointed medical adviser to the Aerial Medical Service, decided that Cloncurry would be the best place for the first Flying Doctor base. It was in an area served by a reasonable telephone and telegraph network, which would serve for communication until all the radio problems had been overcome; it was reasonably good flying country without too many rugged areas; Qantas knew the area and had been flying over it regularly; and there were various small outlying hospitals which the Flying Doctor could visit.

Barber and Simpson made an inspection tour in June 1927, and when they were at Cloncurry Simpson heard that a message had been received from Mount Isa. A miner had broken his pelvis. Would Qantas send an aircraft to bring him to Cloncurry hospital? Simpson went one better. He flew to Mount Isa with the Qantas DH50, treated the patient, and brought him back to Cloncurry. It was the first recorded flight of the new Aerial Medical Service, made before the radio network was established.

Simpson and Barber recommended that Cloncurry was the ideal centre for the first base. It had the well-established Qantas airfield, a public hospital with a resident doctor and full nursing staff, and it was the rail and communications centre for a huge area.

An Aerial Medical Service sub-committee of the Australian Inland Mission approved the idea, and in September 1927 recommended a trial period of one year with a Flying Doctor centred on Cloncurry, a contract with Qantas, the continuation of radio experiments, the formation of a local advisory committee—and the raising of a further £7,000 to cover expenses. Flynn and all his hard-working assistants had to go out "bushranging" with even greater determination.

The first advertisement for a Flying Doctor appeared in the *Medical Journal of Australia* late in 1927. It specified that the doctor must attend urgent cases on the spot, and if necessary take them by air to hospital; make periodic tours of districts without medical services; and be available for consultation with isolated local doctors. The salary was £1,000 plus travelling expenses and a £2,000 life insurance policy, but no right of private practice.

Out of twenty-two applicants the medical members of the sub-committee chose Dr K. St Vincent Welch. He was perhaps a better man than they could have expected for such a pioneering task, because he was a well-known surgeon.

The contract with Qantas provided that a DH50 aircraft, capable of carrying four passengers, should be available for the service, with a provision that Qantas should be allowed to run the aircraft to Normanton with the mail. The aircraft was built in the Qantas Longreach workshops, and on the suggestion of Mrs H. V. McKay was named *Victory*.

Dr Welch began work on 15 May 1928. His first project was a series of exploratory flights, to explain the system to people living in the area and especially to urge station folk to build airstrips. At Vanrook Station, they knocked over hundreds of white-ant nests to clear a landing ground.

Welch's first flight was to the Bush Nursing centre at Julia Creek, and his next was to Camooweal, 250 miles west of Cloncurry. The aircraft had a good following wind and made the trip in two hours; the horse-drawn passenger coach usually took seven days.

Obviously Welch was the ideal man to launch the service. Now and again he was frustrated by the hard-and-fast Qantas policy of economy and avoidance of risk—which over more than half a century has paid off in the proud boast of never having lost a passenger. But when A. H. Affleck, pilot of the DH50, decided not to take off because of a dust storm, Welch had to travel 250 miles by car to bring a patient back to Cloncurry. There is no doubt that a pilot has to be the final arbiter when it comes to flight decisions, and eventually RFDS policy settled into that of a conference between doctor and pilot, with the urgency of a case weighed against flying difficulties and aimed at a solution acceptable to both parties.

The people of west Queensland appreciated the new scheme and Welch flew 20,000 miles on fifty flights. He treated 255 patients and held fifty consultations with other doctors. But some of the local doctors did not co-operate as much as had been hoped. They seemed to see the Aerial Medical Service as a threat to their own practices, and the doctor at Julia Creek made his attitude clear. Even the Cloncurry hospital staff was a little conservative. Welch had been appointed Honorary Medical Officer of the hospital, but resigned when he was not called into consultation during a typhoid epidemic.

"Life here is pretty dreadful but there is no real regret in my mind about the venture," Welch wrote to his friend Dr Dunbar Hooper on 9 October 1928. One of his problems was that he simply did not have enough to do. With no right of private practice, there were long periods of boredom in the outback town.

But despite this and some other difficulties it was clear that the Aerial Medical Service could be a success, though it nearly came to grief during the Depression. The Queensland Government withdrew an annual grant of £1,000 which began in 1929, the Commonwealth Government subsidy to Qantas ended in 1933, and the Aerial Medical Service Advisory Committee was faced with raising a minimum of £3,500 a year to continue the service. They decided that it must be closed, and asked Dr Simpson to tell the people of Cloncurry. When he arrived he found that Dr A. Vickers, the Flying Doctor from 1931 to 1934, was organising local support and was willing to take a ten per cent cut in salary. The service was saved—but the next threat came from the Minister for Health in the Queensland Government. Maybe he saw political capital in the Flying Doctor scheme, because he proposed to establish three aerial ambulance bases under the control of the Queensland Ambulance Brigade. The Australian Inland Mission was politely uncooperative, and refused to hand over to any new organisation which would be controlled by a Government.

However it was becoming obvious that if the seed sown in Queensland was to grow into a national organisation covering the entire Australian outback then some kind of overall management system would have to be established. There were three ways in which this could be done. The Commonwealth or State Governments could take over; local organisations, assisted by the Australian Inland Mission, could run services based on specific centres and covering the surrounding areas; or a national "private" system could be arranged, assisted by the AIM if necessary.

Flynn, who had got the whole idea off the ground, now seemed content to leave its further development to others. On 6 May 1933 he wrote to Dr A. Vickers, "It is good to have all sorts of half-formed ideas which await the turn of events. My own plan has always been to float in a sea of half-formed ideas, waiting for mystical commands—which have sounded from time to time—to bring selected samples to the sandy shore, where they may be transmuted into realities."

The idea of a national "private" organisation was the one which found most supporters. Government control would to some extent have been directly opposed to the original inspiration of the service. All those concerned with it in the early years believed in practical Christianity: that those who have should help those who have not. City folk, secure in their possession of medical, dental, and hospital services, could be practical Christians by helping the outback folk who were isolated and largely without services of any kind, even companionship. And, of course, the outback people would be shown how to help themselves. In countless cases they have done so and are among the most ardent supporters, financially and otherwise, of the RFDS.

During the mid-1930s, a series of meetings and conferences hammered out the basic form which a national Flying Doctor Service should take. Businessmen, the Red Cross, the Aero Clubs, the Country Women's Association, the Australian Inland Mission, the H. V. McKay Charitable Trust, and Civil Aviation were among those who took part.

Between them they established a set of basic principles. It was agreed that the Cloncurry operation had proved the potential success of a national Aerial Medical Service and that several more bases should be established "in areas at present beyond the area of operation of the Cloncurry base". Each of the State systems should be autonomous, but should co-operate and be affiliated under one name. Every outback habitation should be linked to a base by radio or telephone. Special equipment should be maintained to help search parties seeking people lost in the outback. Appropriate local benefit funds should be established, and "interested citizens" and Governments should be asked to help.

Dr Welch's feelings about the boredom of outback life for a doctor who had nothing to do but wait for an emergency had not been forgotten, and the suggestion was made "That isolated doctors should be furnished with facilities to devote part of their enforced leisure to medical research; and that provision should be made for those who have faithfully completed terms of service, to engage in adequate post-graduate studies before resuming practice elsewhere."

The final point of the draft constitution covered the

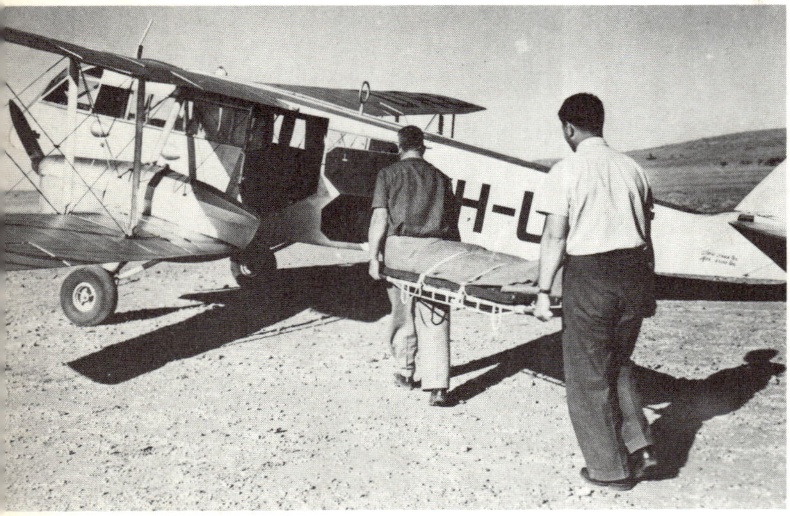

essential need for all medical, ambulance, and nursing services to be effectively co-ordinated within each isolated district.

The enthusiasm of the citizens and organisations who were prepared to work towards such ends was not matched by all of the doctors then in private practice in isolated areas. Dr Simpson wrote to John Flynn on 28 November 1933, "I feel a bit upset about the Broken Hill doctors. They are definitely antagonistic [but] I do not think their opinions count for very much." He believed that if public opinion favoured a Flying Doctor base at Broken Hill, the doctors would accept it favourably.

The basic points worked out at the various private and public meetings and conferences, especially the Citizens' Conference of 30 October 1933, became the structural form of the Royal Flying Doctor Service today. The Service operates in seven separate Sections, not divided according to State boundaries. Each Section is responsible for running itself effectively, and all are affiliated to the national organisation which is only concerned with overall policy and takes no part in Section affairs. The Sections based within Queensland, Victoria, and South Australia all fly into areas of the Northern Territory, and the NSW Section regularly flies into Queensland and South Australia. In Western Australia the Eastern Goldfields Section operates from Kalgoorlie and the Western Australian from bases at Carnarvon, Port Hedland, and Meekatharra. Derby and Wyndham are administered by the Victorian Section.

The Royal Flying Doctor Service of Australia is a truly remarkable organisation. Perhaps it functions so effectively because a large number of people are working together in an area unaffected by the profit motive which shadows so many business operations or by the bureaucratic delays and jealousies which seem inseparable from Government-controlled operations.

The organisation which was still known as the Australian Aerial Medical Service was operating in Queensland, New South Wales, Victoria, South Australia, and Western Australia before the second World War. The name was changed to the Flying Doctor Service of Australia in 1942 and to the Royal Flying Doctor Service in 1955, when the Royal Charter was given to the service after Queen Elizabeth II and the Duke of Edinburgh visited the Broken Hill base in 1954. A Tasmanian base was opened in 1960, and the first School of the Air, using the RFDS radio network, began on 20 September 1950 at Alice Springs. Long before that, a Sunday School of the Air had been operated over the Queensland network.

While the methods of what was to become the RFDS were being worked out in the cities, various outback flyers were operating just such a service in an informal way. Horrie Miller was among those who flew a number of rescue missions, and when

TOP: E. J. Connellan, of Connellan Airways, Alice Springs, prepares for a Flying Doctor mission. CENTRE: Doctor and pilot carry a stretcher patient to a Dragon Rapide aerial ambulance in the 1950s. BOTTOM: A Connellan Airways aerial ambulance over the Northern Territory in the same era (*Royal Flying Doctor Service photographs*)

E. J. Connellan established his air service from Alice Springs to Wyndham in July 1939 he soon became involved in medical flights. Both MacRobertson Miller Airlines and Connellan Airways were regular supporters of the FDS in Western Australia and Central Australia.

One of the most remarkable men of the 1930s was a true "Flying Doctor" although he was not connected with the FDS. He was Dr Clyde Fenton. In his book *Flying Doctor* he says that he was passionately interested in flying from early youth. He learnt to fly with the Melbourne Aero Club, but like most young men of that era had no hope of owning an aircraft and could rarely afford to hire one. When he qualified as a doctor he went to England and joined the RAF which seemed to be one way of combining his dual interests. The RAF did not hold him for long. Soon after he joined he acted as defender at a court martial which tried one of Fenton's fellow officers for drunkenness at a civilian dance. Fenton used the officer's own story as evidence for the defence, but other witnesses showed that this story was untrue. After the court martial Fenton told the officer "what I thought of him in good Australian phraseology, more picturesque than complimentary". The officer complained to the commanding officer and Fenton was reprimanded for "conduct prejudicial to the maintenance of good order and discipline". In disgust, he resigned from the RAF and returned to Australia.

In 1928 he visited Darwin, and agreed to help the local medical officer, Dr Cook, who was toiling under a huge case-load. He stayed five months and saw the immense problems of serving a practice that stretched over thousands of square miles, with no way of carrying patients except by car or buggy. Even this became impossible in the wet season.

Aircraft seemed the obvious answer, and he tried to interest John Flynn in the idea of a doctor actually flying his own machine to outback communities. According to Fenton, Flynn's reply was "One man, one job," and "You are twenty years ahead of your time."

Fenton was one of the applicants for the first position of Aerial Medical Service doctor at Cloncurry, but was unsuccessful. He tried hard to find backing for his own scheme to fly himself to patients, but no one was interested in those Depression years. Then Dr Cook "used his persuasive pen on officialdom in Canberra", and in March 1934 Fenton was appointed medical officer at Katherine, with a mileage allowance if he used an aircraft for transport. He found a second-hand DH Gypsy Moth that was available for £500 and went to see John Flynn again. Flynn was adamant. "One man, one job," he said again. In Flynn's picture of the Aerial Medical Service, properly qualified men should fly the aircraft, and properly qualified men should fly with them to administer medical care. This principle is still adhered to by the Royal Flying Doctor Service. The service believes it is better for the doctor to be freed from the strain of a possibly hazardous flight,

MMA has had a long association with the Flying Doctor services. This photograph shows a patient being unloaded from an MMA aircraft, under charter to the Flying Doctor, at Derby in 1950 (*MacRobertson Miller Airways*)

Top: A new DH Drover aircraft purchased by the Royal Flying Doctor Service of New South Wales parades on the tarmac during dedication ceremonies at Mascot (*Hawker Aircraft Limited*)

Bottom: The interior of a Drover aerial ambulance showing that there was ample room for the stretcher patient at left and seated passengers at right, and plenty of space for the doctor to administer treatment while in flight (*Royal Flying Doctor Service*)

so that he can concentrate entirely on his patient. There have been some exceptions to this rule, especially in emergencies.

Fenton persisted. He obtained a loan of £250, and his mother lent him £250 even though she had a horror of flying. He bought the aircraft, took a quick refresher course in flying, and took off for the outback.

Perhaps it is not surprising that Fenton did not appeal to the somewhat austere John Flynn, who had been nurtured in the Presbyterian tradition of temperance and moderation. Fenton was an extrovert: a daredevil flyer and one who enjoyed the good things of life. His reminiscences are sprinkled with accounts of parties and he shows that he had a true Australian appreciation for beer—even when he had to pay five shillings for each unchilled bottle of it at Camooweal. Beer and sandwiches for an in-flight snack appealed to Clyde Fenton.

He was a flyer of the old breed and would no doubt have made a name for himself if he had been old enough to fly in the first World War. To him, flying was a way of life and all the risks were part of the fun —and he liked plenty of fun on the ground. Some of his antics backfired. Together with other Northern Territory identities, he was awarded the Coronation Medal of King George VI, and at the accompanying ceremony he pinned on two Coronation Medals awarded to other people plus Dr Cook's Jubilee Medal. The Administrator's wife came bearing down on him "like an avenging angel" and ripped them off.

But the gusto with which he enjoyed life did not prevent him from being an excellent flyer and a good doctor. One of his most publicised exploits, which brought a storm of criticism as well as of applause, happened in mid-1934. When Civil Aviation heard about it they threatened to cancel his flying licence.

The Northern Territory had asked Fenton to fly Dr W. G. Woolnough, a Commonwealth geologist, to carry out some surveying work at Ord River. Fenton flew him there safely and was waiting to return when he received a message asking him to go to the help of a woman who had been gored by a bull at Wave Hill Station, 130 miles away. It was late in the afternoon, and even though Fenton had never made a night flight he decided to make the attempt. Woolnough agreed to fly with him.

Fenton telegraphed the manager of Wave Hill Station to light fires to mark a landing ground, but found later that the telegram was not received in time. He took off from Mistake Creek with Dr Woolnough, and flew through bright moonlight which made it easy to pick out landmarks. He saw what he believed to be Wattie Creek, and flew along it until he thought that he had made a mistake. Woolnough agreed that it might be the Victoria River instead, and so Fenton believed he had drifted southwards. He continued flying for another thirty minutes, but began to worry when he saw no sign of buildings along the river. He decided to descend for a closer look, and tried to decrease engine power to let the aircraft glide down. There was no response, and when he moved the throttle in both directions the engine still droned steadily on. He realised that it was disconnected and that he had no control over the engine.

Top: Inside an aerial ambulance in flight, with the nurse at left, doctor and pilot in front, and the patient lying on stretcher at right. Bottom: A DH Beagle aerial ambulance of the NSW section in flight (*Royal Flying Doctor Service photographs*)

He knew that he had to land as soon as possible, and looked desperately for Wave Hill. At last he saw white buildings on the riverbank, but knew that they were not Wave Hill. When he checked the map he realised that he was over Pigeon Hole, fifty miles north of Wave Hill, and that the waterway had been Wattie Creek after all. He had flown north of his destination.

The only way he could land was by shutting off the ignition and hoping to glide in to a controlled landing. He snapped off the switch, but as the ground came nearer he saw that it was patterned with fences and that a mob of horses was galloping to and fro beneath him. When he switched on the engine again it came to life with such a roaring jerk and vibration that he thought it would break loose, but he had no alternative tactics to those of gaining height, switching off for a glide down to look at the ground, and switching on again, to the accompaniment of another roaring vibration, when he could not see a safe landing place.

He decided to make for Victoria River Downs Station, about twenty miles closer than Wave Hill. He followed the road from Pigeon Hole until he saw what looked like a fine stretch of level grass. Fenton brought the aircraft down for what should have been an easy landing, but the grass was five feet high and covered the rough, hard-baked soil of a dried swamp. The aircraft pancaked heavily but did not seem to be damaged, so Fenton and his passenger climbed out and camped beneath it. Fenton had an uneasy night, "brooding on the things that might (and should) have happened to us".

In the morning Fenton repaired the throttle and they tried to take off, but he could not coax the aircraft off the ground especially since he had to manoeuvre through anthills, stumps, and other obstacles. They agreed that he should make a solo attempt, and Woolnough watched as the aircraft at last left the ground.

As soon as Fenton rose above a belt of timber he saw Victoria River Downs Station straight ahead, and he turned to drop a note to Woolnough. His passenger did not see the first note, so he wrote another on a scrap of paper and flew close to the ground, holding up the paper so that Woolnough could see it. He dropped the note, turned his head to see Woolnough chasing it, and almost immediately hit the ground. In his concentration on the note he had flown too low.

The aircraft turned on its back and was a complete wreck, but Fenton was miraculously unhurt. He struggled clear and they plodded to Victoria River Downs, and eventually made their way to Katherine and then to Darwin.

Fenton was now without an aircraft, so he telegraphed Qantas to ask whether they had one for sale. Qantas had one to offer, but the deal hung fire for some weeks until Fenton received a message saying that a woman at Beetaloo, thirty miles from Newcastle Waters, was dangerously ill in childbirth. The aircraft had been flown to Darwin but it still belonged to Qantas and Fenton could not fly it without permission.

He asked J. A. Waters, Acting Administrator of the Northern Territory, to contact Canberra immediately. Canberra responded with unusual speed and gave immediate approval for purchase of the aircraft. Fenton took off at once, flying straight across country with the most scanty route directions and by perfect navigation he landed at Beetaloo Station in the dark. He was able to save the woman's life.

On his way back to Darwin he landed at Newcastle Waters and found another woman in need of medical attention. He treated her, flew on to his home base, and received an urgent call to Pine Creek, where a third woman was dangerously ill. Next came a man at Katherine who had fractured both sides of his jaw, and Fenton flew him to Darwin and helped in the operation.

Such missions as these earned the high respect of Northern Territory folk. In an article on 1 September 1934, headed "Valuable Service Rendered By Flying Doctor", *Aircraft* magazine lauded Fenton's efforts and said that "All through the northern end of the Territory people are praising the surgical skill and flying ability of Dr Fenton. They applaud the devoted and unselfish work he is doing with a full realisation of the immense risk he runs.

"Dr Fenton has been accused of being a dare-devil for having undertaken a night flight over dangerous country. It is our wish to see that fair play is meted out to him, and that airmen, at least, shall recognise that this flight is no escapade, but was undertaken, seriously counting the cost, at the urgent call of an inescapable duty . . . Dr Fenton was a severe loser by the wreck of his machine, and deserves sympathy, not criticism, for his heroic attempt."

Fenton escaped official action over the Wave Hill episode, though Civil Aviation's threat to take action resulted in a feud which continued while he flew in the north. But he was too light-hearted and daring to avoid censure for very long. On one night flight back to Darwin he spotted a film being screened at the open-air cinema, and swooped so low that the aircraft roared past the screen. Tom Harris, proprietor of the cinema, took it in good part and demanded that Fenton and his passenger should pay the price of two dress-circle seats. But the police and other citizens were outraged, and when Fenton went to Katherine he found a Civil Aviation message telling him that his pilot's licence was suspended until further notice and that the certificate of airworthiness of his aircraft was suspended until navigation lights had been fitted.

He replied to this telegram, sent by a Civil Aviation official in Darwin, with a telegram consisting of "a single pungent word".

After that, Fenton had to obtain special permission to make every urgent medical flight, and as these came thick and fast a constant stream of telegrams flowed to and from Civil Aviation. The Press had a wonderful time with the story, and at last the matter was allowed to fade into obscurity. There is little doubt that Civil Aviation, like Government departments everywhere, often took itself far too seriously.

Fenton's experiences were similar to those of many Flying Doctors who were to follow him: urgent calls to isolated places to set broken bones, diagnose and treat sudden painful illnesses, arrange for chronic sufferers to be taken to hospital, and in fact do the combined work of a general practitioner, surgeon, and casualty ward officer over thousands of square miles. He had several crashes and forced landings and escaped serious injury in all of them, and by way of a busman's holiday flew to China and back. This, of course, involved a double crossing of the dreaded Timor Sea.

In 1937 the Commonwealth Government provided him with a DH Fox Moth fitted as an ambulance machine, and as a well-known Northern Territory identity he continued his work into the 1940s.

Fenton was never a member of the Royal Flying Doctor Service or any of its antecedents, but was a Government-employed medical officer who used his own flying skills and, at first, his own aircraft to perform what would otherwise have been an almost impossible task. It he had not flown, he would have been physically incapable of reaching the patients who needed his help.

In 1963 this writer discovered what may be described as a link between Dr Fenton and the RFDS, albeit a very tenuous one. In about 1929 Fenton obtained a DH53 Hummingbird, serial number 118. She was a tiny, dainty, and rare machine, really quite unsuitable for the rugged conditions of the north, and one of only a few built by de Havilland's. In fact she could carry only two gallons of petrol.

Fenton made an unsuccessful attempt to take off in her from Wyndham airstrip, and when he stepped out of the damaged aircraft he decided she was not for him and that he would have to abandon her. He sold her to two locals, and for six years the tiny monoplane stood in a shed near the airstrip, gathering red dust on her fabric wings and graceful body. She had not been registered with Civil Aviation and did not even have registration markings.

In 1935 Dr Ralph J. Cato arrived in Wyndham as resident medical officer. He too was a trained flyer, but when he made a number of flights for the Aerial Medical Service he was not allowed to pilot the aircraft. Flynn's dictum of "one man, one job" meant that Cato was flown in MacRobertson Miller aircraft under a contract with the Aerial Medical Service.

But once a pilot, always a pilot, and Cato pricked up his ears when he heard about the Hummingbird. He inspected her, thought that he could repair her, bought her, and sent to de Havilland's for Hummingbird plans. Slowly he rebuilt her, literally learning all about the aircraft from the inside out. Not only did he learn all the limitations of the aircraft and her structure, but he was able to fit her with a new engine, a little 40-hp Aeronca, and to install a new petrol tank which increased her capacity to seventeen gallons. When she was ready for the air he christened her *Icarus*.

When his period of service expired he decided to fly the *Icarus* home to Perth. It was a fearsomely long trip for such a tiny aircraft: down the coastline where mangroves clung like fungus to a log, over the broad empty riverbeds that wound through the red, pocked land, past mountain ranges rising stark and red and shimmering in the heat. The heat was the worst of it; *Icarus* was too tiny to give any protection to Cato as he sat in the exposed cockpit.

He reached Onslow safely, then flew on to Carnarvon. The engine began to stutter when he was seventeen miles en route, and he had to bring her down for a forced landing in the desert. He found the fault and took off again, after the only major problem on a flight of 1,500 miles. At other places he used bicycle and tractor parts to keep *Icarus* flying.

He landed at Maylands airfield, Perth, on 30 May 1937, after a flying feat which astonished the world of Australian aviation. His only navigational equipment had been a schoolboy's pocket compass.

It aroused the interest of bureaucracy, which did not applaud Cato's achievement. The Civil Aviation Board refused to grant a certificate of airworthiness to the *Icarus* because of her "dubious history" and unauthorised modifications. Cato wheeled her into a Maylands hangar and she lay there for years, until he gave her to the Perth Technical College for ground instruction. Later she was moved to the College's Fremantle annexe, to gather dust in another shed.

I had heard rumours about this rare and long-forgotten Hummingbird, and like many other aviation enthusiasts attempted to track her down. Only three DH53s had been imported to Australia, and only one was known to survive in Britain, in the famous Shuttleworth Trust.

Eventually I tracked down the *Icarus*, unaware of her history and that I had attended Air Training Corps meetings with the son of her former owner. The Education Department, then her official owner, had forgotten it had the aircraft. At last an official said that I could have her, provided that she was taken away from the Fremantle annexe.

She was in a dark and rustic outbuilding in that stone-walled establishment. The gloom could not conceal the fact that she had been roughly treated: not so much by age as by rough handling. Obviously the maintenance staff had regarded her as just so much rubbish, and at some time, during the summer burning-off of long dry grass, she had been wheeled outside and burnt. All that remained were the engine mountings and a shapely wooden strut: useless remnants, but to be cherished as a reminder of Dr Cato's feat.

If John Flynn and Clifford Peel had not been first in the field with their idea of an aerial medical service, then it is possible that the Commonwealth Government would have followed up their provision of a flying ambulance to Dr Fenton by instituting some more widespread medical aviation service. In that case the Royal Flying Doctor Service would not exist in its present form.

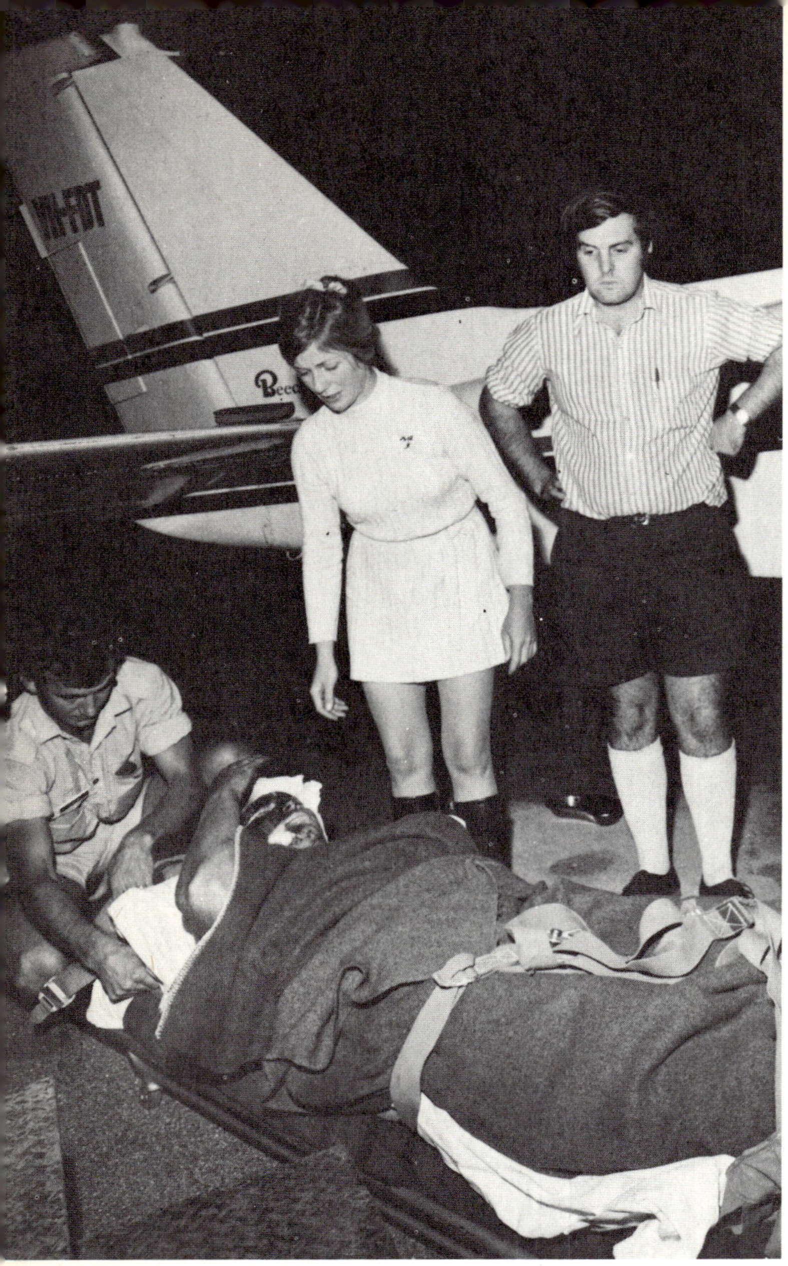

The records of the RFDS contain many stories of hazardous flights, but it must always be emphasised that the service has no admiration for daredevil flyers. The first concern is the patient's safety; the hazards are those of bad weather encountered while taking medical aid to a sufferer or flying him back to hospital, or those of having to land under difficult or unusual conditions. The flying safety record of the RFDS, like those of all modern Australian aviation services, is remarkably high.

One of the side benefits of the RFDS was seen quite early: women, especially women with children, became less reluctant to settle in the isolated areas. Apart from the knowledge that they could call for immediate medical help, the RFDS radio network has removed what is in itself almost an illness: the oppression of loneliness and isolation which was so keenly felt by many of the early settlers. The woman who can have a brief chat with her nearest neighbour, even if she is 100 miles away, and who can listen to her child actually speaking to the School of the Air teacher in a distant base, is brought into continuing human contact with the rest of the world. On 10 November 1940 the Perth *Sunday Times* quoted Dr F. D. Sweetman of the Wyndham base of the Aerial Medical Service as saying, "At least a dozen women came to live on Kimberley stations during the two years I was there."

One of the more recent books dealing with various aspects of the RFDS is *Flying Nurse* by Horrie Miller's daughter Robin. Like her father, Robin has loved

TOP: The late Robin Miller prepares to supervise the loading of car crash victims into an aerial ambulance (*Shepherd Baker Sullivan Associates*). BOTTOM: A Beechcraft Baron aerial ambulance over Perth (*Royal Flying Doctor Service*)

flying for as long as she can remember. Her first flights as a passenger were with "Mickey Mouse Airlines" between Perth and Broome. Her mother, Mary Durack, is a famous Australian writer and a member of the Durack family which pioneered the pastoral land of the north-west. For these reasons Robin has a special affinity with the Kimberley area and the Northern Territory.

Robin trained as a nurse and obtained first her private flying licence and then her commercial licence. After an attempt to convince an airline that she could handle an airliner as well as any man—an attempt from which she withdrew graciously when she saw that the airline was running out of arguments—she was almost resigned to nursing as a profession and flying as a pastime when she found an opportunity to combine both these skills.

The Western Australian Department of Health was faced with the problem of administering Sabin anti-polio vaccine to thousands of people living in the country and outback areas of the State. To administer the vaccine as it was done in the city suburbs and larger towns, from a caravan moving from one spot to another, would have taken years and been extremely costly. "Aerial vaccination" was the answer, and Robin Miller asked for the job in the northern areas of the State.

The vaccine is administered by mouth, usually by letting the requisite dose soak into a lump of sugar and giving it to the patient to suck or swallow. As Robin Miller flew around the northern communities, from mining camps to cattle stations, the Aboriginal children gave her the name of "Sugar Bird Lady".

Robin Miller's childhood years in the Kimberley area, and all the stories which she heard of the north and its people, made her familiar with the Flying Doctor Service from her very early days. It was an easy and natural thing for her to step from her work as a "flying nurse" into employment with the RFDS—not as a nurse but as a pilot. Generally speaking the RFDS still believes in John Flynn's dictum of "one man, one job" though there have been some exceptions to this rule.

Miss Miller has described the startled reaction of some patients when they realise that the pilot is a *woman*! But apart from the thousands of miles which she has flown around the north-west and the Northern Territory she has proved her flying skill and endurance in other ways. While she was still nursing at one of the Perth hospitals she was given the opportunity to fly one of two tiny French single-engine aircraft from Paris to Australia, for delivery to an agency which hoped to promote the sale of the model in this country. The other aircraft was piloted by Dr Harold Dicks, who is well known in RFDS circles and had encouraged Robin Miller's ambition to fly when she was still training as a nurse.

The Paris–Perth flight was rather like a "Battling

Dr Harold Dicks with the Fox Moth aerial ambulance *John Flynn* at Marble Bar in the 1940s (*Royal Flying Doctor Service*)

Parer" venture all over again. The two flyers had some hair-raising experiences. Rapidly diminishing fuel forced them down on to a Greek military airfield, where they were received with great suspicion. They were buzzed by a jet fighter near the Syrian coast, and Miss Miller was so sure that it intended to shoot them down that she untied her pony tail and opened the air vents to let her long blonde hair fly loose so that the other pilot could see she was a woman. Apparently this convinced him that they were harmless, because he vanished into the distance. Next, they were forced to land in the Saudi Arabian desert, and were held prisoner by the local police for several days before being allowed to proceed. Eventually, after a flight which in its brief hops from one airfield to another was very much like those of the England–Australia pilots of the 1930s, they made it to Perth—only to find that the agency to which they were delivering the aircraft was going out of business.

Miss Miller and Dr Dicks made further flights together when they went to America to pick up Beechcraft aircraft bought by the RFDS. Their trans-Pacific flights, from San Francisco to Hawaii and then island-hopping across the south Pacific, were also full of adventure. Their aerial partnership had a pleasantly romantic outcome when they married in 1972, but ended all too soon when she died, after an illness, in December 1975.

In her book *Flying Nurse*, Robin Miller makes frequent mention of how medical treatment is much needed by Aboriginal communities in various parts of the outback, and a good deal of her flying was to take medical aid to such communities. It was a need that the RFDS recognised very early in its existence, and in 1939, for example, Dr F. D. Sweetman was making half-yearly visits to Kunmunya Mission, on the north-west coast of the Kimberley area, to hold clinics for the Aborigines there.

Twice-yearly visits may not seem very frequent,

The late Robin Miller (Mrs Harold Dicks) with Dr Harold Dicks after ferrying a Beech Baron from America for the Western Australian section of the Royal Flying Doctor Service in 1968. The newly-acquired aircraft is in the background

but they were considerably better than none at all, as was the case before the doctors began flying around the missions, stations, mining camps, and other outback communities. It was the beginning of the "patrol" system instituted by the RFDS, which was part of John Flynn's scheme. It developed from the missionary "patrols" through the outback, in which by horse and buggy or motor car they covered huge circuits: marrying, baptising, reading the burial service over the graves of those who had died since their last visit, and generally bringing spiritual comfort. Now it was the turn of the doctors, who could bring physical comfort and reassurance. The patrols became just as important as the emergency flights, because quite often the doctors were able to diagnose conditions which eventually would become urgent, and either treat them on the spot or arrange for the sufferer to be taken to hospital.

Now the patrols are almost taken for granted, and white and black Australians roll up to the clinics when the doctor and nurse arrive. Sometimes a specialist will make the rounds, and the RFDS has done a great deal to control trachoma, the eye disease which once was the curse of the outback. It is believed that the disease was introduced by Malay traders, and though it is quite easy to cure in its early stages it can blind a sufferer if it is neglected. It is extremely infectious, and on one of Robin Miller's flights she carried an eye specialist who treated an entire tribe suffering from trachoma.

Dentists also make the rounds, sometimes setting up their dental chair and other equipment in the middle of a paddock to relieve a sufferer from raging toothache. They, like the doctors on medical patrol, believe that prevention is better than cure and have saved countless people from the agony of a neglected tooth which suddenly flared into tormenting life, with the nearest doctor or dentist hundreds of miles away. It is all very different from the days when Dr Fenton was asked to extract a tooth with any tools he had handy, and was obliged to use a pair of pliers from his aircraft tool kit.

The RFDS is now so completely a part of outback life that it is doubtful whether there is any isolated homestead of any size that does not possess a Flying Doctor radio and medical outfit. The radio network is now highly sophisticated. In the recently-installed single sideband radio communications system, an emergency-call alarm arrangement uses automatic devices. With this system, if an outpost sends an emergency call when the base control station is unattended, the alarm activates circuits which switch on the transmitters for a predetermined period and then cause alarms to ring either in the doctor's residence or another selected location.

The RFDS doctors have become experts at long-distance diagnosis. A series of careful questions elicits the symptoms in the case of illness, and the doctor then decides whether it is sufficiently urgent to warrant an immediate flight or whether the patient should be treated from the medical kit. To avoid confusion, every item in the kit is numbered, so that the caller can be told, for example, "Give two tablets of number five and two more in four hours' time, and let me know how the patient's getting on when you call in the morning."

Much the same system applies in the case of accidents. Usually the doctor gives instructions for first-aid treatment to be administered before he arrives, because the RFDS principle is that it is better to go to a case which may not really need a doctor than to take the chance of neglecting one.

Many RFDS aircraft now carry a variety of equipment which would not even be recognised by Clifford Peel, who outlined the Aerial Medical Service idea when he was still a medical student. Some of them are more than flying ambulances: they are virtually flying hospitals where the patient can be treated while en route to a land-based hospital.

Apart from the emergency calls and the medical patrols, the system provides for a regular daily schedule on which people can obtain less urgent medical advice. These schedules extend into the passing of personal and business telegrams, which have become an important financial part of the RFDS, and to the School of the Air. But everything stops if an emergency call is received.

The Royal Flying Doctor Service now has twelve mainland bases and two Tasmanian bases, with about 2,350 fixed outposts maintaining regular contact and a variable but about equal number of portable transceivers also in contact. The Service flies nearly two million miles each year on about 4,000 separate flights. The medical consultations conducted personally and by radio total about 95,000 annually, and some 340,000 telegrams are handled over the radio network each year.

Perhaps the aims and activities of the RFDS were summed up best in a speech by Lord Dunrossil, then Governor-General of Australia, on 4 June 1960. He said, "This is really the Australian answer to an Australian problem, idealistic but intensely practical."

The Flying Dentist at work in the outback of New South Wales, while the aircraft pilot looks on (*Royal Flying Doctor Service*)

13

Off to War Again

Australians had to act fast to build a miniature air force into a weapon capable of fighting in a global war

Like the other democracies, Australia was poorly prepared for the war which exploded when Hitler invaded Poland in 1939. There had been plenty of storm warnings, but countless people—politicians and others—refused to consider the possibility of another world holocaust. The Australian military machine built up by 1918 had rusted into nothingness. Anyway, there was only a superficial resemblance between the wood, wire, and fabric biplanes of the 1920s and the monoplanes like the Supermarine Spitfire, the Hawker Hurricane, the Messerschmitts, and Heinkels. The later-model Spitfires were to achieve a formidable 500 m.p.h. in a dive, and the firepower of a single machine was equivalent to that of an entire flight of the 1918 air forces. The only problem was that there were not enough Spitfires. But for the dedication of a few persistent men, there would have been none at all.

The RAAF was tiny in 1939, though it had grown considerably from the time in 1934 when it had sunk to a total of 820 officers and men and about twenty first-line aircraft. Prime Minister Joseph Lyons' defence proposals of 1937 and 1938 had launched the RAAF into an expansion programme, but it was only half complete when the fateful words were uttered by Neville Chamberlain on 3 September 1939, and Australia's Prime Minister R. G. Menzies said, "It is my melancholy duty to inform you officially that, in consequence of a persistence by Germany in her invasion of Poland, Great Britain has declared war on her and that, as a result, Australia is also at war."

The RAAF was then commanded by Air Vice-Marshal R. Williams, CB, CBE, DSO, who had made his name as a flyer in the first World War. Other Australian airmen from the first war were still in the RAAF, among them Group Captain F. H. McNamara, VC, CBE, who was in London as Liaison Officer with the RAF, and Air Commodore S. J. Goble, CBE, DSO, DSC, who was the Air Member for Personnel.

The basic organisation of the RAAF was virtually the same as that of the RAF. Unlike the RAF, the Australians had little opportunity to fly the new types of military aircraft which were coming, all too slowly, off the British production lines. The aircraft principally used by the RAAF were DH Moths, Bristol Bulldogs, Westland Wapitis, Hawker Demons, Avro Cadets, and Avro Ansons. It also operated Supermarine Southampton flying-boats and Supermarine Seagull amphibians. These were all good aircraft of their day, but that day was passing rapidly and for many it was long past.

The RAAF had taken every opportunity to become involved in active flying apart from training operations. Unlike RAF pilots, who occasionally functioned under what might be termed active service conditions when called upon to bomb or strafe recalcitrant tribesmen in some part of the British Empire, the RAAF flyers made considerable peaceful use of their aircraft. Two Air Displays were held in 1934, one in Laverton to celebrate the centenary of the State of Victoria and one at Richmond for the visit of the Duke of Gloucester. About fifty aircraft took part—in fact the greater proportion of the available machines. The Hawker Demons were shown off for the first time in 1935, during an Empire Day display, and there was a Jubilee Flight to mark the Jubilee celebrations for King George V. That was the year in which the RAAF began to expand after a long, spiralling run-down, and the airfield facilities were improved and increased while new aircraft were ordered from Britain.

Peacetime flying involved more than the vanities of aerial displays. In 1936 an American explorer went missing in the Antarctic, and two RAAF aircraft, a Wapiti and a Moth, were fitted with floats and sent to join in the search. When the explorer was located they dropped supplies to him and his party. The RAAF was to play a key role in many dramatic, and often tragic, search and rescue missions. Apart from training activities, RAAF operations consisted mainly of survey work and aerial photography, spotting forest fires, and the search flights.

Whenever the various Aero Clubs put on a pageant the RAAF sent two or three machines to show how really professional flyers could perform at speeds of almost 300 m.p.h. The crowds thrilled to a now-forgotten sound: the screaming whine of the wind through the taut rigging of a Bristol Fighter as she nosed down in a mock attack. As the 1930s drew to an end there was a certain urgency about some of these demonstrations. The RAAF was now actively seeking recruits, and hoped that the displays would tempt young men with the requisite physical and mental qualities to join the service. The selection and training boards set almost impossibly high standards, and those who won their wings were the pick of the nation.

In 1938 the Australian Government invited the RAF to send an adviser to study the RAAF and submit suggestions for improvement. The Inspector-General of the RAF, Air Marshal Sir Edward Ellington, arrived in July and drew up a report which was passed to the Air Board for action. Sir Edward recommended training improvements, especially in advanced flying training; better conditions for the permanent officers; promotion by selection instead of seniority; more interchange of senior officers between the RAF and RAAF; better weather forecasting; the introduction of a scheme similar to that of the RAF Volunteer Reserve, which provided for young men to undergo an initial training period with yearly refresher courses on RAF stations; the interchange of civilian and service pilots, so that airline pilots would gain some useful background in military operations and RAAF pilots would get the feel of flying for a genuine purpose; and, last but not least, that new types of aircraft should not be put into production until they had been tested in Britain.

The latter was something of a sore point with the RAAF, which for a number of years had been trying to achieve a greater measure of independence by having aircraft designed and built in Australia instead of relying upon overseas machines.

Sir Edward's recommendation was a complete about-face from the 1918 period, when Britain had asked Australia to produce 200 fighter aircraft as soon as possible. According to the British aviation magazine, *The Aeroplane*, of 28 September 1938, an Aircraft Construction Committee was set up to tackle the job, using materials available in Australia or imported from America. A series of timber tests was carried out to determine the most suitable type for aircraft construction, and it was found that nine or ten Australian timbers would be ideal. The committee decided to build DH9As, using American Liberty engines, and an assembly line was being organised when the war ended.

After the first World War, Avro and de Havilland commenced production in Australia, and H. E. Broadsmith of Avro continued the committee's timber research and built Avro 504Ks with Australian timber for the RAAF. All that Broadsmith needed was a Government subsidy and he would have established a viable Australian aircraft industry. He designed and built a big cabin biplane, with a forty-five-foot wingspan, which was ideal for Australia's infant airlines, but the money ran out and the firm folded.

Even after that, the Government could have made use of the local talent which was looking for a chance to prove itself. Its ability was demonstrated when the Aero Club of New South Wales announced a contest for locally-built light aircraft, to be held in November 1924. Captain Edgar Percival, who eventually had to move to England to produce his Percival Gull and other light aircraft, had accumulated invaluable experience by modifying two war disposals Avro 504Ks to run a joyriding and charter service. H. E. Broadsmith designed a machine for him, Denzil Macarthur Onslow backed him financially, and Onslow and Percival built the aircraft for the contest.

Wackett was working hard on his Warbler, fitted with the Wizard engine of his own design, and Captain Robert Reid and L. J. R. Jones also announced entries, but the test was won by Percival's machine. As far as arousing Government interest was concerned, the contest was like a stone dropped in a pond.

L. J. R. Jones had been building Australian aircraft since about 1910. His airframe designs were good but he could not find suitable engines to lift the aircraft off the ground. Even the automobile was still in its infancy, and attempts to adapt the heavy motor-car engines to aircraft were fruitless. Motorcycle engines were too low in power although they were light in weight. So Jones fitted one of his aircraft with a high-speed de Laval turbine, driven by steam. It raised the aircraft off the ground, but then the rotor burst and Jones was wounded by flying fragments. Fortunately he was able to walk away from the crash that followed.

Other designers and builders were forging ahead. General Aircraft Co. Ltd, a wholly Australian-owned

TOP: A DH86 Dragon Rapide used as an aerial ambulance by the RAAF in the early years of the second World War (*RAAF Museum*)

BOTTOM: This Avro Cadet was one of the somewhat mixed bag of aircraft used by the RAAF in the early 1930s (*RAAF Museum*)

business, designed and built the three-seater Genairco in 1929 and this model flew very successfully throughout Australia and New Guinea. The firm became Tugan Aircraft Co. and built six metal Moths for the RAAF. Lawrence Wackett designed and produced the Gannett, a Fokker-type aircraft with a welded steel-tube fuselage and tail and plywood wings, for Tugan. It was an excellent aircraft for its period.

Metal was beginning to take over from timber in airframe construction, and metallurgists had produced aluminium, duralumin, and various alloys which were light enough to take over from fabric. Junkers was among the first to design aircraft clad with metal, giving his aircraft the strange "corrugated-iron" look which characterised them for many years. In 1929 L. J. R. Jones designed and built, almost single-handed, Australia's first aircraft constructed entirely of welded steel tubing, though he covered it with fabric. Carrying a pilot and four passengers, it cruised at 110 m.p.h.

Jones called it the *Wonga*, and then decided to design an all-Australian machine for the 1934 England–Australia air race. It was to have two Australian-built Harkness Hornet engines moulded into the wings, and in terms of aircraft design of that era was one of the most modern designs yet produced. Airframe construction went rapidly ahead, but Jones struck a snag with the engines. Tests showed that they were too unreliable for the stress of the long-distance flight, and they could not be improved in time for the aircraft to take part in the race.

A Mr Wikner built a low-wing monoplane and a small high-wing monoplane with a cabin, but as usual the prophet was not without honour save in his own country and he had to go to England to make any progress. His cabin monoplane, the Wicko type, was built and flown successfully in the United Kingdom.

Larkin, of course, was forging ahead with aircraft building just as he was in his struggles with civil aviation authorities and the other airline operators. He was a man who deserved a great deal more success and recognition, but somehow he did not manage to survive his various battles and by the end of the 1930s he had faded almost into obscurity. His Larkin Aircraft Supply Co. Ltd produced what was probably the first Australian all-metal aircraft, the Lascoter: a five-seater, with a payload of 1,640 pounds, which could cruise at ninety miles an hour. The Lascoter went into operational use in 1930, and was followed in 1933 by the Lasconder, a tri-motor, seven-seater monoplane with a payload of 2,300 pounds and a cruising speed of ninety-five miles an hour. WA Airways and Qantas (as recounted) also built a number of aircraft, though of British design.

Apart from this work on powered aircraft, the glider clubs were very serious about their own projects. Engineering students of the University of Sydney formed a gliding club in 1929 and built their own training glider, developed under the guidance of T. D. J. Leech. The students carried out valuable research work which included the development of a special lightweight short wave transceiver radio, fitted into the glider and used for instructing a pupil while he was in the air. Both Sydney and Melbourne universities had courses in aeronautical engineering, and in 1930 the Australian Institute of Engineers formed an Aeronautical Engineering Branch.

Some Australian politicians began to take the threat of war seriously in about 1936, and the Australian Aeronautical Research Committee was formed. H. E. Wimperis, an English aviation expert, was invited to visit Australia in 1938 to advise the Government. By the time he arrived the Commonwealth Aircraft Corporation had begun work at Fisherman's Bend, Victoria.

Australians with all this design and production work behind them did not take kindly to Sir Edward Ellington's notion that Australian-produced aircraft should not be used by the RAAF until they had been tested in Britain. However, the old Imperial prejudice died hard. The feeling by successive Australian Governments that the mother country would somehow look after them died equally hard. As *The Aeroplane* said in 1938, "The history of aeronautical engineering in Australia is one of lost opportunities and lack of foresight by the authorities. One by one those whose contributions could have been greatest were forced out of the industry. No continuity has been possible and every effort has had to start from nothing to build up some organisation against apathy and even against obstruction. Now it seems that all that is past."

But when the second World War broke out it was only in the very recent past. Australia did have an air force, but like the other armed services it was little more than a nucleus of trained men and a pitifully small amount of equipment. Luckily there was no lack of men like Norman Brearley, who may have been past their prime so far as front-line service was concerned but who were eager to re-enlist for any useful work, whether it was "flying a desk" or helping to train the young men who offered themselves in such hordes that the enlistment rolls in the eastern States had to be closed. The existing structure of the RAAF could neither equip nor train them. Group Captain Keith Isaacs, now retired, joined the army when he was disappointed in his desire to get into the air force. When his unit was shipped to Western Australia he found that vacancies still existed in that State and was able to transfer into the RAAF.

On 3 September 1939 the RAAF had 310 officers, 3,179 non-commissioned officers and airmen, and 158 reservists. The aircraft consisted of eighty-two Avro Ansons, which were twin-engined "multi-

TOP: The famous Wirraway, designed and built in Australia, was a basic RAAF machine from 1939 onwards. Although it was used mainly for advanced training, the Wirraway saw service on the battle fronts (*RAAF Museum*)

BOTTOM: The smile on the face of the pilot demonstrates the relaxed handling qualities of the Wackett Trainer, developed to bridge the gap between the Moth and Wirraway trainers (*RAAF Museum*)

TOP: A Wackett Trainer and a Wirraway on the ground during the second World War (*Commonwealth Aircraft Corporation Pty Ltd*)

BOTTOM: The Commonwealth Aircraft Corporation designed and built the Woomera but only a limited number became operational (*RAAF Museum*)

purpose" machines theoretically suitable for many duties from reconnaissance to bombing; fifty-four Hawker Demons, elegant fighter aircraft but already obsolete; twenty-one Supermarine Seagull V amphibians, a machine which by that time had been discarded by the RAF but was still a useful aircraft so long as no one had to fight in it; the first seven Wirraways built by the Commonwealth Aircraft Corporation; and a gaggle of training machines including DH60 Gypsy Moths, DH82 Tiger Moths, and Avro 643 Cadets.

The RAAF squadron numbers ran up to 23 Squadron. This brings to mind Churchill's statement that, "We called it the 8th Army to make the Germans think that we had seven more armies." In 1938 there were only eight operational squadrons: 1, 2, and 21 at Laverton in Victoria; 3, 4, 5, and 22 at Richmond in New South Wales; and 23 at the recently-established base at Pearce near Perth in Western Australia. By 1939 the squadron numbers 10, 11, and 12 had been added to the list. There were two flights from 12 Squadron at Darwin and one from 23 Squadron at Brisbane.

Some of the 10 Squadron crews were in England, preparing to fly recently-purchased Short Sunderland flying-boats back to Australia. These big, long-distance aircraft were to play an important part in convoy escort and anti-submarine work, and the British sensed this all too well. They arranged for the 10 Squadron men, and their machines, to be incorporated into RAF Coastal Command.

To replace them, the RAAF used the time-honoured expedient of commandeering whatever might be required to fight a war. First they took over three of the Qantas Short Empire flying-boats, for 11 Squadron to use on coastal patrol. Hudson Fysh, with first World War flying behind him, was able to foresee this as the world rolled towards war. In 1938, after seeing Germany's preparations for the war in the air, he wrote in his diary, "Never since the days of Napoleon has our future been so challenged. We can only win out through leadership and action, and the making of sacrifices. There is no easy way."

He had tried to work out a war plan with Imperial Airways, and asked Short Bros to quote on fitting long-range fuel tanks in the Qantas flying-boats. Also he obtained maker's drawings for the installation of machine-gun mounts in the aircraft.

His idea was that the flying-boats might have to

X-12

A2-3

These two pages show some of the aircraft of the 1930s and 1940s. All photographs provided by the RAAF Museum, Point Cook

FACING PAGE, TOP: A Vought-Sikorsky OS-2U Kingfisher seaplane

FACING PAGE, CENTRE: A Dornier DO-24K flying-boat

FACING PAGE, BOTTOM: The famous Walrus amphibian flying-boat

THIS PAGE, TOP: Douglas Dolphin amphibian flying-boat

THIS PAGE, TOP CENTRE: The Percival Vega Gull, a fast and popular all-purpose light aircraft developed during the late 1930s, shown here with RAAF markings

THIS PAGE, BOTTOM CENTRE: The Percival Proctor

THIS PAGE, BOTTOM: The Percival Prince amphibian flying-boat

fly roundabout routes in order to maintain the air link between Britain and Australia, and that they should be prepared to help themselves.

Apart from taking over their machines, the Government soon asked Qantas for a different kind of help. The RAAF wanted to buy new Consolidated PBY Catalina flying-boats from America in 1940, when the USA was still neutral. The Government felt that it would be better for a civil organisation to collect the aircraft, and turned to Qantas. Lester Brain, the chief of operations, took charge. Consolidated factory pilots would deliver the machines to Hawaii, and then Qantas pilots would fly them across the south Pacific to Rathmines for delivery to the RAAF. The first one to arrive was only the third aircraft to make a direct flight across the vastness of the south Pacific, in succession to Kingsford Smith's *Southern Cross* and *Lady Southern Cross*. P. G. Taylor (later Sir Gordon Taylor) had flown with Smith in the *Lady Southern Cross*. Now he was helping to ferry the Catalinas.

The first was delivered in January 1941, and after that the ferry flights continued steadily. The war began to hit Qantas hard. After recovering from having their aircraft commandeered, two of their remaining flying-boats were sunk in Darwin harbour on 19 February 1942. The *Centaurus* and *Corinna* went down in the welter of more than forty sinking, burning, or damaged ships.

But Qantas executives were dreaming of new uses for the ubiquitous Catalina flying-boats, even though they carried the RAAF roundel. The air link to Britain was cut when the Japanese swarmed into Singapore, but even before this the flight across half the world had presented many hazards. The British had suggested that Qantas might fly to Perth, Cocos Island, and Ceylon, and P. G. Taylor liked the idea. In 1939 he had flown exploratory flights across the Indian Ocean for the Commonwealth Government, with the specific aim of selecting and surveying emergency air bases for wartime use, and he suggested that the 3,513 miles from Perth to Ceylon could now be flown direct. The new Catalinas just had the range to swallow the distance in one gulp.

Hudson Fysh approached A. B. Corbett, Director-General of Civil Aviation, with the idea, and was told that it was out of the question. So he wrote to the Governor-General, upsetting Corbett considerably. "I feel very disturbed that you should have taken such action," he wrote to Fysh.

P. G. Taylor flew to London to talk about the new route, and found that General Sir Archibald Wavell, commanding the India and Burma theatres of war, was keenly interested. In 1943 Hudson Fysh also went to London to talk to the British, who agreed to provide RAF Catalinas. Qantas prepared to start the new service across the Indian Ocean, flying from Koggala in southern Ceylon to Crawley Bay on the Swan River near Perth. On 10 July 1943 Captain R. B. Tapp took off in a Catalina and landed in Western Australia on 11 July. The fuel load needed for the long run allowed the aircraft to carry a payload of only 852 pounds, but it was enough for mail, official documents, and the occasional VIP passenger. Qantas had broken the wartime blockade and was back in the business of running an overseas airline—even if it was not flying its own aircraft.

The Catalinas bought from America were part of the re-equipment programme which the RAAF and the Government began with urgent haste within days of the declaration of war. Forty of the stubby Wirraway fighters had been ordered from the Commonwealth Aircraft Corporation in 1938, but initial production was slow. Few tradesmen in Australia had the skills necessary for building all-metal aircraft, and CAC had to begin with a training programme. But Wackett could see the war clouds gathering, and he was the sort of executive who seemed to be everywhere at once: explaining, cajoling, ordering, arguing over the telephone with laggard suppliers, and having his say at meetings. By 1939 CAC was a small but efficient modern aircraft plant which could turn out one complete aircraft a week—yet it seemed that, like many other Australian aircraft factories, it was doomed just as it moved into full swing. R. G. Casey (now Lord Casey), the Minister for Supply and Development, told CAC that there would be no more Government orders when the forty Wirraways had been delivered.

Events soon enmeshed Casey as they did everyone else. In September he telephoned Wackett to ask how soon he could increase production to three Wirraways a week, and how much it would cost. On the following day he was on the phone again. Now, the Government wanted six aircraft a week. Wackett had the answer ready. Yes, it could be done, and it would cost £150,000 for extra factory equipment. "Agreed," Casey answered. A mere £150,000 was nothing in wartime. Ten years earlier a tenth of that sum would have rescued any of the Australian aircraft manufacturers which were sinking into oblivion, and the nation would have had a vigorous aircraft industry. Unfortunately, history has repeated itself since the second World War.

Wackett put all his immense energy into the task of supplying the RAAF with Australian-built fighters and other aircraft. By the end of the war the CAC had built 757 Wirraways, and Wackett believes that production in Australia saved the Government £1 million.

The Wirraway was only one of the aircraft built by Australian assembly lines. The Wirraway was basically a training aircraft, to teach men who would then fly British-built machines. They did their basic training in DH82 Tiger Moth biplanes, but there was need for an aircraft to bridge the gap between the rather simple Moth and the more sophisticated Wirraway. Wackett responded by designing a low-wing monoplane fitted with a DH Gypsy Six engine. It was a pert machine with a noticeable North American Aviation influence in its design, and this became more pronounced when Gypsy engines ran short and the Wackett Trainers were fitted with Warner Super Scarab seven-cylinder 165-hp radial engines instead. In all, 200 Wackett Trainers were built for the RAAF.

Even Wackett's amazing thrust and energy were not enough to supply the aviation demands of Australia in the second World War: a war which for the first time brought the enemy to her gates and saw the first attack on Australian soil. The Government Aircraft Factory was set up at Fisherman's Bend to build Beaufort bombers, and CAC established an aero engine factory in Sydney. De Havilland, by that time the oldest aircraft manufacturer in Australia, though still a subsidiary of the British parent company, increased its already extensive capacity.

Lawrence Wackett designed a bomber with twin radial engines, and two prototypes were built. He then designed a twin-engined anti-tank aircraft, but it did not leave the drawing-board. The Government was demanding fighters: fighters to combat the Japanese divebombers and twin-engined strike aircraft based as close to Australia as the Celebes islands (today Sulawesi). From their main base in Rabaul, the Japanese were attacking the Australian posts in Papua New Guinea. Darwin and Broome had felt the devastating impact of their bombs.

Within thirteen weeks the Boomerang fighter had been designed, built, and flown. It was the only wholly-Australian fighter produced during the war, and was fitted together from pieces of Wirraway airframe, engine installation parts adapted from the Beaufort bomber, and a massive twin-row Wasp 1,200-hp radial engine. It had a top speed of 296 m.p.h. and a ceiling of 29,000 feet.

The Boomerangs began to come off the production line, but just too late to be used as interceptors in the critical year of 1942. They had been designed and assembled so hurriedly that the pilots discovered many problems, but when these had been eliminated the Boomerangs proved to be useful RAAF–Army co-operation aircraft during the jungle fighting in Papua New Guinea. A total of 250 were built for the RAAF.

The Boomerang had been a good effort, but not quite good enough. The RAAF asked why fighter aircraft designed overseas could not be built by the CAC. Why not the Spitfire? That was a fighter which had proved itself countless times in the hands of RAF and RAAF pilots.

But Wackett had a stubborn loyalty to North American's designs, and he insisted on his preference

FROM TOP DOWN: Vultee Vengeance; Miles monoplane; Miles Magister; Fairey Battle with wheels partly retracted. (*All photographs from RAAF Museum*)

TOP: The well-named Lockheed Lightning was used as a fighter, a photo-reconnaissance aircraft, and on ground-strike missions and other operations by the Allied air forces including the RAAF (*RAAF Museum*)

Apparently this Lockheed Vega was taken over by the RAAF during the second World War. In this photograph she still carries her civil aviation registration (*RAAF Museum*)

This Waco YQC-6 was one of a variety of American light passenger aircraft which appeared in Australia during the late 1930s. The above photograph was taken in 1942 (*RAAF Museum*)

The Auster, specially designed for army co-operation work (*RAAF Museum*)

for their P51 Mustang. Activity at the sprawling CAC plant had to be stepped up yet again, for the Mustang was a far more complex machine than any previously produced in Australia. The assembly line was in full swing when the atomic bomb brought an end to the war, and 170 Mustangs had rolled from beneath the factory arc lights into the sunshine outside the Melbourne plant. Many of them were still in service when the Korean War began, and were flown by RAAF 77 Squadron during that campaign.

While CAC technicians were working on the Mustang, the designers were planning a superb new piston-engine fighter. The first design was for a machine with a Mustang-like tail and a cleanly-cowled Pratt and Whitney R2800 engine, but then the Rolls-Royce Griffon 125 liquid-cooled engine became available. It was the world's best reciprocating engine and the design of the new CA15 was modified to use this mighty 2,035-hp power plant. Some of those who saw the prototype CA15 believe that she was the most beautiful piston-engined fighter ever to fly. Her straight, tapered wings were the new laminar-flow type and she achieved a speed of 442 m.p.h. Unfortunately she was too late to prove herself in action, and was not test-flown until March 1946.

The air war was greedy for every type of aircraft, and in 1942 the CAC designers began work on the basic local design of a four-engined transport with a range of 1,500 miles. She was to have had a somewhat similar appearance to the Douglas DC4, and if she had ever progressed further than the drawing-board she could have been adapted for Australia's domestic airlines when the war ended. Unfortunately she was never approved, and the blueprints were filed among the "might-have-beens" in the drawing office. The CAC, under Wackett's vigorous guidance, could

have been the beginning of a solid Australian aircraft industry, but the time was not yet ripe. Some of its efforts might even have been forgotten if Ian H. Ring, chief engineer of CAC, had not recalled them at a Royal Aeronautical Society meeting in Melbourne in 1968.

Other men were trying as hard as Wackett to keep Australians flying. Arthur Butler, the bustling airline operator from New South Wales, told the Commonwealth that he would gladly carry out a free survey of the quality, type, and condition of the men and machines available in New South Wales. He would then submit a report to the Government. He said, "I flew to Canberra during September 1939 and submitted our offer to the Minister of Civil Aviation, Mr J. V. Fairbairn. He wasn't very interested, and seemed to think I would be more use as an instructor in the RAAF. The ensuing argument was interrupted by Mr R. G. Casey, who was Minister for Supply and Development. He was so interested in our plan he invited me to his office and requested me to confirm my offer in writing."

But it came to naught. Butler abandoned the idea, but found a practical way to help the war effort by re-organising his workshops to repair RAAF DH82 Tiger Moth trainers after they had been bounced and battered by trainees.

The whole problem, especially in the early days of the war, was that Australia was still psychologically linked with Britain. The feeling was that somehow, at whatever cost, the Royal Navy and Royal Air Force would preserve Australia from any real harm. The British Empire was still referred to as a living entity,

FROM TOP: The Vickers Valette transport; Northrop Delta light transport; Ryan STM trainer. (*All photographs RAAF Museum*)

The Republic P43 Lancer, at top, has a certain resemblance to the Brewster Buffalo in the bottom photograph. The Brewster Buffalo was supplied to Commonwealth air forces during the early years of the second World War. The aircraft was used on ground strafing and bombing operations against the Japanese assault on Malaya and Singapore but was no match for the enemy fighters (*RAAF Museum photographs*)

and there had been repeated promises to Australia that in case of any threat to the country the British fleet would be in Singapore within seventy days. As late as mid-1939, the Canberra correspondent of the London *Times* cabled that, "The Commonwealth Government are satisfied with the assurances they have received repeatedly from Mr Chamberlain and the Admiralty." Subsequent disasters, from Dunkirk to Singapore, flowed from the complacency of Whitehall planners in those vital pre-war years.

Their failure, or refusal, to see the writing on the wall is still astounding. In the Adelaide *Express and Journal* of 21 January 1939 there is a photograph captioned "A Record With Bombs". It shows an Italian Air Force crew which had just set a new record by flying a total of 12,000 miles with a bomb load of two tons, at an average speed of 293 m.p.h. In the same newspaper, a lengthy article is headed, "Air Lessons of Spanish War". The lead-in states: "The swift machines of today call for greater skill and longer training, and make large reserves of skilled pilots urgently necessary."

The article made the point that the huge increases in speed since the first World War made accurate bombing and gunnery far more difficult. The article also stated that Spanish Government pilots needed twelve months' training before they were allowed to fly their "Moscas", an adaptation of the Boeing P26, in combat. Bombing was so inaccurate that the insurgent forces of General Franco, using twin-engined Heinkels flown by Luftwaffe pilots, took nearly two years to score a direct hit on the petrol storage tanks at Barcelona, even though the tank farm had no anti-aircraft protection. The Germans fighting for Franco also used Messerschmitt 109 fighters, while the Italians on the same side employed Fiat CR32 biplanes. The Government forces had some Junkers 52 three-motor bombers and the Russian twin-engined bombers which they called "Katiuskas", but they had lost aerial supremacy. The German and Italian air staffs found that the Spanish Civil War of 1936–39 was an ideal proving ground for their Heinkels, Junkers, Bredas, Messerschmitts, and Savoias.

Both the Supermarine Spitfire and the Hawker Hurricane were fully operational when that report was written, and Spanish Government pilots said that their "greatest ambition was to fly a Hawker Hurricane. They were filled with enthusiasm at the idea of what they could do to the Fiats and Messerschmitts if they had a few Hawkers."

The standard model Spitfire was revealed to the public for the first time at the Paris Aeronautical Salon in 1938, and, to quote the *Express and Journal* once more, "The 8 Gun Fighter is Fastest Warplane . . . with a maximum level speed exceeding 350 m.p.h. it is the fastest aircraft yet in service with any of the world's air services." The aircraft on display "attracted the admiration of the man in the street, who recognised the beauty of its streamlined form, and of the expert, who was quick to realise how cleverly low head resistance had been married in its design with military efficiency."

"Another quality difficult to appreciate from examination of the aeroplane in its place on the Royal Air Force stand is the ease of control which makes the Spitfire and its contemporary in the service—the Hawker Hurricane—safe and docile vehicles for the newly trained pilot and swift in aerial manoeuvre at all speeds."

The article described at length how the Spitfire had endured gruelling tests, including a high-speed belly landing, and dilated upon its superb design in which the riveting was absolutely flush with the outer skin to eliminate drag at high speeds. The aircraft had been designed for comparatively quick and simple assembly, with rapid replacement of prefabricated parts such as the tail plane and wingtips.

Extensive planning and the manufacture of many tools and jigs were necessary so that this magnificent fighter could be turned out at assembly-line speeds. It was as well that this initial work was under way, for apart from the planes contracted for by the "parent" company, 1,000 had been ordered from the new aircraft factory which Lord Nuffield was building in Birmingham.

And yet this supreme fighter, whose designer R. J. Mitchell had been at pains to create an aircraft suitable for mass production as well as domination of the skies, was *not for Australia.* There is no practical reason why, after a tooling-up period, it should not have been produced in this country. The reasons must be traced back to the "colonial syndrome" which linked so much Australian political and strategic planning with Whitehall and allowed Australian leaders meekly to accept such dicta as the RAF's decision that Australians should fly only in British-built aircraft. Too many of the British planners still thought in trench warfare terms and too many decisions were made under the spur of immediate military necessity. There can be no other explanation for the fact that 645 Hurricanes and Tomahawks were sent to help the Russians in 1941, while Malaya had to be defended with 161 obsolete bombers and fighters. The *Repulse* and the *Prince of Wales*, the British battleships sunk by the Japanese off Malaya, had no air support of any significance from land-based aircraft, and the aircraft carriers *Hermes* and *Indomitable* were not even within reach. The price, of course, was paid by the fighting men ashore and afloat—and by the flyers who battled with their obsolete Buffalos and twin-engined Hudsons against the fanatical assaults of the Japanese.

14 Winged Victory

Dark blue RAAF uniforms became a familiar sight in beleaguered Britain. Australians flew in every theatre of war including Russia, Burma, and the Battle of the Atlantic

The first RAAF flyers to serve in any theatre of the second World War were those of 10 Squadron who had been sent to pick up the Short Sunderland flying-boats ordered from Britain. They arrived just in time to be taken over by the RAF, and thereafter flew with Coastal Command. As the months passed by, the "phoney war" erupted into the German invasion of Scandinavia, Holland, Belgium, and France. The forerunners of the RAAF in the northern hemisphere were joined first by a trickle and then by a flood of men in the dark-blue RAAF uniform, which many Britons thought was that of some kind of naval air service.

The rapidly escalating numbers of RAAF personnel on active service were the result of one of the better ideas of pre-war planning, even though it came only just in time. This was the Empire Air Training Scheme, generally known as EATS. The basic purpose of the scheme was to protect the motherland and provide air support for British expeditionary forces. The plan was based on the assumption that Germany, and perhaps Italy, would be the principal enemies, and that most of the main campaigns would be fought in Europe. Britain was woefully short of trained men and combat aircraft, and staff planning aimed at building up a massive force of both as quickly as possible. Initially it was thought better to train Australians, Canadians, and other imperial flyers to fight with the RAF, rather than to raise and train separate squadrons, although this thinking began to change quite early in the war. To begin with, the majority of RAAF personnel in Britain served as "individuals" in squadrons of the RAF Fighter, Bomber, or Coastal commands, but as the war progressed so RAAF squadrons became fully operational from British bases. Even after that a number of RAAF men served in RAF aircrews. It was not unusual for the crew of a Lancaster bomber to contain an Australian and a Canadian together with the British crew.

The EATS scheme had various advantages. One was that it enabled aircrew to be trained without the distraction of German attacks on their airfields. Another was that the vast open spaces of Canada, Australia, and South Africa gave plenty of room for trainees to manoeuvre without the problem of cluttered cities and countrysides beneath them. The dominions of the great British Empire provided plenty of space for massive training centres, and their classrooms, workshops, simulators, and airfields were used by RAF trainees as well as men from many other parts of the world.

Canada became the centre of the EATS, but Australia made a heavy commitment to the scheme. Her first EATS trainees began work in Somers, Victoria, on 29 April 1940, and the first Canadian training centre opened on the same day. Australia's initial commitment called for her to supply 16,000 fully-trained aircrew by March 1943, followed by 10,000 in each subsequent year. Together with the first aircrews, 25,000 men were to be trained as ground staff. The cost to Australia was estimated at more than £60 million.

In September 1939 Australia had only one flying training school and sixteen instructors. A number of the pilots taught to fly in the early months of the year were, much to their disappointment, appointed

TOP: The famous Curtiss P40 Kittyhawk (*RAAF Museum*)

CENTRE: The RAAF did not obtain the deadly Supermarine Spitfire until quite late in the war (*RAAF Museum*)

BOTTOM: A very early photograph of the Hawker Hurricane, taken in the late 1930s (*RAAF Museum*)

The Victa Aircruiser was one of the very few wholly-Australian aircraft, designed and built in this country, to be produced since the end of the second World War. See text for full story. TOP: The Aircruiser prototype at Moorabbin airfield, Victoria, on the day of her successful test flight. From left are C. Tuttleby the test pilot, H. Millicer the designer, and K. Langford of the Department of Civil Aviation. (*photo: H. V. Millicer*) BOTTOM: Another chapter in the Aircruiser story. Members of the Tariff Board inspect Aircruiser VH-MVR on the tarmac at Bankstown airfield, New South Wales, when the Board was considering the need to protect the Australian light aircraft industry against overseas imports. (*photo: H. V. Millicer*)

These Sabres were among the new aircraft supplied to the RAAF during the various stages of reorganisation and re-equipment which occurred during the 1950s and 1960s. The pace of technological progress has been so fast that the RAAF has been faced with difficult decisions over its choice of aircraft during the past ten years. (*Department of Air*)

76

A "gunship" of 9 Squadron RAAF, the helicopter unit of the RAAF component in South Vietnam, flies over Phuoc Tuy Province. The gunner mans his weapon in readiness to return fire from Viet Cong ground forces. Most of the helicopter gunners were airfield defence guards who volunteered for flying duties. (*Defence Public Relations*)

instructors instead of being sent overseas. They passed on their knowledge to a rapidly-increasing number of young Australians, and while these recruits were flying their "circuits and bumps" the groundwork was being laid for schools in radio and air gunnery, observation from the air, aerial photography, aerial navigation, air/army co-operation, and seaplane and flying-boat flying. The ground staff, without which any squadron would soon become non-operational, were to be catered for in schools of mechanical and electrical engineering, aircraft rigging, and aircraft armament. Together with these, all those less glamorous but equally necessary branches of the service—medical, catering, and administration—had to be provided for.

The whole operation was organised, and began to grow, with that astounding speed which seems possible only in wartime. As soon as men passed out of the primary flight training schools they proceeded to a central flying training school and then operational training units. These were designed to put a sharp edge on the weapon forged by earlier schooling. The young men who had been taught to fly were now tutored in the trade of killing, and trained in the fine arts of handling fighter aircraft in combat, bombing selected targets by night or by day, and flying as part of an operational unit. Side by side with this ran the dry-run aerial tuition of navigators, wireless operators, and air gunners.

The training of airmen never really came to an end. New aircraft were introduced from time to time, and each of them was more complicated than its predecessor. If the war had lasted for another year then the fighter pilots who were trained originally to fly Spitfires or Hurricanes would in all probability have been retraining to fly jet fighters. New techniques of bombing and aerial combat were being developed all the time and everything learnt on the fighting fronts was passed back to the operational training units.

By 1942 RAAF pilots who had flown Spitfires and Hurricanes in the European and Middle East theatres were returning to Australia to be regrouped for the battle against Japan. Some of these men with many hours of combat flying had to be retrained in the American P40 Kittyhawks, the fighter aircraft most freely available to the RAAF.

Kittyhawks had already been used by RAAF men in the Middle East, and opinions varied as to their quality. Some pilots who had flown British fighters felt that the Americans suffered badly by comparison, but others were enthusiastic. In a 1974 ABC television programme, several ex-RAAF men were interviewed and questioned as to their opinion of the Kittyhawk. "Sluggish, but plenty of power," was the verdict of ex-Group Captain Clive "Killer" Caldwell. Squadron Leader John Waddy, who shot down two Messerschmitt 109s and two Junkers J52s in one engagement when flying a Kittyhawk, said that she was "A very strong aircraft. Mine had 114 holes in her but I still got home." Squadron Leader Bobby Gibbs felt that the Kittyhawk was a good gun platform, "but you had to be able to shoot". The consensus of RAAF opinion seems to have been that the Kittyhawk lacked the grace and agility of Spitfires and Hurricanes, but that she was a burly machine capable of taking plenty of punishment—and, in the right hands, inflicting a lot of damage.

The privilege of handling Spitfires and Hurricanes was at first limited to Australians who served in the northern hemisphere. There they blazed their name through the skies as they flew the British fighters over Britain, Europe, and the Mediterranean theatre of war. A number of them took part in the Battle of Britain, in which Flying Officers R. Glyde, J. Cock, and J. Curchin and Flight Lieutenant P. Hughes shot down more than thirty German bombers and fighters.

The first RAAF Spitfire squadron, 452 Squadron, was formed on 12 April 1941 and flew its first operation on 2 June. Under Flight Lieutenant Bungey, from Glenelg in South Australia, the RAAF squadron had to forge a reputation comparable with Spitfire squadrons which had flown in the Battle of Britain. It succeeded magnificently. By the end of its first operational year it led the list of British fighter squadrons in confirmed destructions of enemy aircraft, and in three successive months it shot down more than any other squadron in Britain. Its exploits had been rewarded with a DSO, five DFCs and three Bars, and a DFM.

Bungey was succeeded by Wing-Commander Brendan "Paddy" Finucane, the much-decorated

These Spitfires were flown by RAAF pilots serving with the RAF in 1940 (*Australian War Memorial*)

RAF Spitfire pilot who soared to fame during the Battle of Britain. When he was shot down over France, 452 Squadron came under the command of another famous ace: Pilot Officer Keith W. Truscott, known always as "Bluey" because of his flaming red hair. In his first year with the squadron he had thirteen confirmed kills. He returned to Australia in 1942 to take part in the war against Japan, and took a vigorous part in the Milne Bay battle where the Australians gave the Japanese the first defeat in their headlong progress towards Australia.

Keith Chisholm, another 452 pilot, was as determined on the ground as he was in the air. With six confirmed kills on his record he was shot down in the English Channel, picked up by the Germans, and sent to Lamsdorf prison camp. He made two attempts to escape, one of them in company with the redoubtable Douglas Bader, but was recaptured each time. His third escape led him on a long and dangerous track through occupied Europe until he reached the American forces in France in August 1944. Chisholm was one of many Australians whom the Germans and Italians found extremely hard to handle when they put them behind barbed wire.

Another Australian Spitfire squadron, 457, was formed late in 1941. It was led by the famous RAF ace Squadron Leader (later Air Commodore) Peter Brothers, who is now living in Sussex, where he still talks fondly of his "days with the diggers". During 1942 his squadron gained a reputation to equal that of 452, but by then "the diggers" were looking anxiously homeward. The first great Japanese assault had been rebuffed but there were still many campaigns to be fought. In January 1943, 452 and 457 Squadrons arrived in Australia as part of No. 1 RAF Spitfire Wing, which was posted to the air defence of the north-west. The third squadron in the Wing was 54 Squadron RAF, and the combination of British and Australian Spitfire pilots who had learnt their trade the hard way over Britain and Europe was a nasty shock for the Japanese. On 2 March 1943 the Spitfires slammed into sixteen enemy bombers and shot down three, and on 15 March all three squadrons scrambled to intercept a raiding force of forty-four Betty bombers and Zero fighters. The snarling dogfight ended with seven of the enemy shot down and seven damaged, four missing Spitfires and one damaged. The raid on Darwin did not take place as planned by the Japanese.

Apart from the men who served as "individuals" with the RAF, squadrons of RAAF bombers attacked Germany from 2 September 1941 onwards. On that date twin-engined Hampden bombers manned wholly by RAAF aircrews took part in a raid on Berlin. During the first eight months of 1943, RAAF bomber squadrons flew five times to Berlin. On 23 August 1943, in the first of the monster attacks which reduced the German capital to rubble, a formation including RAAF Lancasters was led by Group Captain H. I. (later Sir Hugh) Edwards, VC, DSO, DFC, of Fremantle. He was the first Australian "air VC", winning the Cross for an attack on Bremen in July 1941. He led fifteen Blenheims in a low-level attack on the port, which was still burning after a night raid. It was an incredibly daring attack in which the Blenheims flew almost at rooftop height, weaving through the cables of a thick balloon barrage and actually zooming under electric cables strung between tall pylons. When they had dropped their bombs they machine-gunned every target that seemed to merit a "clobbering", then turned and wave-hopped back across the North Sea. Four of the Blenheims were shot down and three had had to break off the attack and limp back to base.

Edwards had already won a reputation for aggressiveness and daring by the low-level attack on an enemy convoy, in June 1941, which won him the DFC. In December 1942 he led five squadrons of RAF and RAAF bombers in an attack on the Philips radio factory at Eindhoven in Holland: an important strategic target because it manufactured a large percentage of the radio equipment used by German armed forces. Although thirteen bombers failed to return, the raid was successful and left most of the factory area in a flaming shambles.

"Hughie" Edwards, who later was promoted to air commodore and was appointed ADC to Queen Elizabeth II, held a commission in the RAF which he joined in 1936. As soon as the war broke out an increasing number of Australian flyers appeared in Britain, the first of them being the 10 Squadron RAAF pilots who had been sent to collect a squadron of Sunderland flying-boats ordered by the RAAF. The flying-boats were taken into the RAF and the Australians stayed with them, to fly with Coastal Command on convoy protection, rescue missions, and patrols of the thousands of miles of wild water comprising the vital Western Approaches to the British Isles.

They were very soon followed by many more, and in 1942 their activities were summarised in an article headed "Ubiquity" in the Australian War Memorial publication *These Eagles*. The writer said: "Almost everywhere there has been air fighting, there Australians have been found. In Australia, and over the islands near it and the seas which girdle it, Australians fly and fight.

"They are fighting in Britain, over Germany, Denmark, Holland and France, they are fighting over the Atlantic in the Battle for Ships, and they patrol the seas which gird the British Isles.

"They are posted in Iceland and have fought side by side with the pilots of the Red Air Force on the Russian Front.

"They have fought and are fighting over the

FROM TOP: The mighty Mustang was among the last of the great piston-engined fighters, and of much the same vintage as the Bell P39 Aircobra shown in the centre photograph. The Beechcraft 17, in the bottom photograph, was an ancestor of the sleek Beechcraft often seen in Australia today. (*RAAF Museum photographs*)

Top: This Curtiss Shrike saw service with the RAAF during the second World War. Bottom: An aircraft seldom remembered nowadays; the Noorduyn Norseman (*RAAF Museum photographs*)

Mediterranean and the Bay of Biscay, they are strafing Rommel's forces in the Western Desert, and are flying in defence of Malta. They have flown in Greece and Crete and in Syria.

"Men of the R.A.A.F. are to be found in India and have flown over Burma, Malaya, Sumatra, and Java, and the islands and seas about them.

"In Canada and Rhodesia they are training, and every now and again comes news of an Australian pilot having performed some remarkable feat in some out-of-the-way place on the earth's surface."

The article quoted Sergeant Pilot George Sayers, of Melbourne, as an example of the "ubiquity" of Australian flyers. Sayers, an EATS trainee, was twenty-nine in 1942, and by that year had 900 flying hours in his log and had flown on forty-three operations in the Middle East, Britain, Russia, and south-east Asia.

Other pilots mentioned in the article are Pilot Officer J. P. M. Haydon, of Canberra, who flew in the paratroop and commando raid on the German radar station at Bruneval, France, on 27 February 1942; and Flying Officer Selwyn Clark, Pilot Officer Mark Sheldon, and Sergeant Pilots Arthur Gould and Barton Campbell, all of whom flew with the RAF Hurricane wing sent to Russia in 1942. The Hurricane squadrons spent two months of operational flying in Russia, under the arduous conditions of one of the most bitterly-fought campaigns of the whole second World War. On Flying Officer Clark's first operational flight his squadron encountered five Messerschmitt 109s and a Henschel, and shot down the latter and four Messerschmitts without loss to themselves.

Australians were heavily engaged in the defence of Malta. The island was under round-the-clock attack by the Italian and German air forces, and RAAF men flew against them, and against Italian Navy warships around the island, in Spitfires, Beauforts, Hudsons, and other aircraft.

One RAAF Hudson crew, led by Pilot Officer J. G. Cowan Hunt of Sydney, won "one of the war's most amazing aerial duels". In June 1942 the Hudson was patrolling off Malta when it was attacked by three Junkers 88s. Two dived to attack from astern, with one of them pulling up steeply to shoot up the Hudson from below. The RAAF gunner fired at this one, and saw bullets enter its nose, and then both Junkers turned away. One made for home, and the remaining two prowled around the Hudson until they judged the moment ripe for another attack.

The three aircraft manoeuvred at high speed as the Junkers tried to outfox the Hudson. Occasionally, Pilot Officer Hunt flew her only ten feet above the water and waves splashed against the fuselage. The whole encounter was fought out at less than 400 feet.

At last the two Junkers decided to make a head-on attack. The Hudson flew unwaveringly towards them, and at 600 yards the gunner gave one Junkers a tremendous burst. Flames and dense smoke poured from its engine and it broke off the attack. It was last seen losing height, with its sole remaining companion hovering anxiously near by. The Hudson returned safely from patrol.

One of the most spectacular dogfights over Malta took place when RAAF Hurricanes attacked a force of Cant bombers escorted by twenty Macchi 202 fighters. A Sydney pilot described his remarkable part in the engagement: "As we were going for the bombers, the Macchis dived on us. I counted 16, and as I went after them, I saw the rest of the Hurricanes keeping engagements of their own. One of them was among six of the enemy, four of whom were on his tail. I put a few bursts into the one I was after, and his tail began to smoke. Then another attacked me from astern. I turned and saw this chap and flew straight at him. I expected him to turn, as the Italians usually do; but he didn't, he came straight on. Perhaps the pilot was killed and the Macchi was flying itself; anyway, I flew right into him and the Macchi broke up in the air. There were pieces all over the place.

"I am told that one of my wings broke off, but I don't know, for I didn't see the Hurricane again. I was in a spin and didn't need to jump, I just opened the lid and fell out. I was about 2,000 feet up at the time, and as I came down over the island I saw that I was going to land on a village. I hit the wall of a house, and landed on a flat roof right on top of a dog, which let out a terrific yell, jumped off the roof, and belted up the road."

RAAF Boomerangs fly in line abreast (*Australian War Memorial*)

A completely different type of service was performed by RAAF men stationed in Iceland. They had the dreary, dangerous job of patrolling the North Atlantic convoy lanes, which during that period of the war ran through far northern latitudes. Others were engaged in the "ferry service", flying new aircraft across the Atlantic from the USA to Britain. Two RAAF flyers, with a RCAF wireless operator, set a new ferry record early in 1942. They flew their Hudson across in seven hours forty minutes, thirty minutes less than the previous record held by a Liberator bomber.

The flight was even more remarkable because all three men were recent graduates of the EATS scheme. The scheme expanded enormously after the first aircrew began training at Toronto, Canada, in April 1940. The demands and pressures of war were far greater than anyone had anticipated, but the EATS scheme responded magnificently to the challenge. By the end of 1940 double the number of aircrew had been trained and sent to Britain than the original scheme had planned for. Extra training schools were opened far ahead of schedule, and by 1943 it was apparent that the aircrew and ground staff being produced by the EATS scheme, and the aircraft pouring from the factories of Allied countries, would combine into an overpowering weapon.

The British air forces in the Mediterranean area won their aerial superiority against what, to begin with, seemed like overwhelming odds. The situation in the sky was exactly the same as on land and sea. The RAF was still using such antiques as Gloster Gladiators; the Allied land forces, including the Australians, were short of armour and heavy artillery; the Royal Navy with the support of a few Allied warships was outnumbered by the great fleet which Mussolini had built up during the 1930s.

The outcome of the clash belongs to history. The Italians rolled forward to conquer Egypt, and were rapidly beaten into submission. Their fleet fought its first and last "big ship" battle with the Royal Navy at Matapan. Their airmen learnt to give a wide berth to aircraft flown by British, Australian, South African, and Polish pilots.

The situation changed overnight when Hitler sent his panzer divisions into Greece in the northern spring of 1941, and at almost the same time dispatched Field-Marshal Erwin Rommel, with the hastily raised Afrika Korps and supporting Luftwaffe units, to put some stiffening into his Italian allies. From victory, the British were forced into retreat, and the Mediterranean campaigns fluctuated between galloping advances and helter-skelter withdrawals until Montgomery gained the decisive victory of El Alamein.

The entry of the Germans into the Mediterranean war brought a new dimension into aerial combat. The RAF developed the first of what became known as Tactical Air Forces, to work in close co-operation with land operations and to strike at Rommel's supply line across the Mediterranean. When the fortress island of Malta had been secured and the Germans and Italians decided that it was not worth expending any more men and machines in attempts to bomb it into submission, the island became a "floating aircraft carrier" which eventually made it impossible for any enemy vessels except hospital ships to cross the Mediterranean. The light bombers, fighters, and fighter-bombers operating along the North African coast struck constantly at German and Italian movements of troops, armour, and vehicles, although they now had to battle against Messerschmitts instead of Italians.

The first complete RAAF operational unit in the Mediterranean area was 3 Squadron. Equipped with Gloster Gauntlets, and Lysanders, they flew their first dry runs in late 1940 and became operational in January 1941. At first they were used mainly for army co-operation, to provide air cover for the land forces and to observe artillery fire. Their aircraft were almost obsolete and the Lysander could fly as slowly as 50 m.p.h., with a top speed of only 230 m.p.h.

The squadron performed well during the invasion

The elegant Airspeed Oxford (*RAAF Museum*)

The second World War gave a tremendous boost to the production of four-engined aircraft. TOP: The Avro York, an "all-rounder" employed for many purposes, especially freight and passenger transportation, during and after the war. CENTRE: The Boeing B29 Superfortress. BOTTOM: The Avro Lincoln. Several of these were converted for Australian overseas passenger services after the war (*All photographs RAAF Museum*)

of Libya and the capture of Bardia, securing air superiority that enabled the British and Australians to advance without fear of air attack. During this period they were re-equipped with Hurricanes. The new aircraft came only just in time. In April 1941 the Germans came roaring into North Africa and proved a very different enemy from the Italians. They hurled themselves full-tilt at the Allies in Cyrenaica and forced them into a fighting retreat eastwards through Libya, while Messerschmitts and Junkers snarled through the skies abandoned by the Italians.

With RAF and other Allied units, 3 Squadron fought back hard. It was a difficult and dangerous time because a pilot often returned from a sortie to find the ground crews frantically packing up equipment to move the airfield, which was in any case no more than another patch of dusty ground another 100 miles or so to the rear. It is said that at one time the pilots were ordered to burn their machines because a German spearhead was approaching during the night, while it was too dark to risk take-off. The pilots refused, and while the squadron vehicles rumbled off in the darkness they stood by their Hurricanes, prepared to risk an unlit take-off rather than surrender or burn the fighters. But the Germans did not appear and the pilots took off at first light.

When the withdrawal was completed the squadron was re-equipped again, this time with Tomahawk fighters. The new aircraft arrived in time for the squadron to join the invasion of Syria, planned by the Allied staff to clear out this pocket of Vichy French in the eastern Mediterranean. It was expected that resistance would be slight, but the campaign developed into one of the hardest fought by Australian air and land forces. During the four weeks of fighting, 3 Squadron shot down twenty-four enemy aircraft and destroyed or damaged about fifty on the ground. Confirmed kills rose above ninety.

From Syria the squadron returned to the Western Desert, to take part in a new offensive. They found that the Allied air forces had been built up to overwhelming strength. British, Australian, Canadian, Rhodesian, South African, and New Zealand aircrew, many of them the products of the Empire Air Training Scheme, were being welded into a force that would use the Germans' own tactics against them. The Germans, learning from British and Australian tactics in the first World War, had developed close air support into an essential part of their blitzkrieg tactics. Bombers and fighters flew ahead of advancing troops to break up enemy formations on the ground, destroy vehicles, armour, supply and ammunition dumps, rail junctions and other strategic points, and blast the opposition into such a demoralised condition that they broke and ran before the ground forces.

In North Africa, the Tactical Air Force performed the same role. Fighters, fighter-bombers, and bombers kept up a round-the-clock assault on everything they could reach, including German and Italian bases and airfields far behind the lines. The long streams of trucks ploughing through the sand, to bring reinforcements and supplies to the front line, were pounced on by fighters which raked them from end to end with bombs and bullets. Many a panzer grenadier perished in a blazing truck, with Nazi dreams of world domination fading to the scream of aircraft engines. The bombing of such bases as Mersa Matruh became so common that Allied pilots referred to it as the "milk run".

The de Havilland Albatross flew for Imperial Airways and might have superseded flying-boats on the England–Australia route if the war had not intervened (*Hawker Siddeley photograph*)

Sleek, handsome, and deadly. RAAF air defence and ground attack squadrons are currently equipped with the Mirage III-O. (*Department of Air*)

Macchi aircraft, used for basic training by the RAAF, are manufactured in Australia under licence. The other basic training aircraft are the Winjeel and the HS748. (*Department of Air*)

It was far from a one-sided battle. The Germans have proved themselves to be cunning and aggressive combat pilots since the days when Rittmeister von Richthofen first swung his cavalry boots over the leather-padded rim of an Albatros cockpit. They fought back savagely and 1,132 RAAF men were killed in the Middle East.

Not all the Australians there flew with 3 Squadron. Others flew with RAF units, and by June 1941 it was estimated that Australians in the Middle East had accounted for 195 of the enemy. 450 Squadron RAAF, entirely composed of EATS trainees, joined the fray in mid-1941. They were equipped with Kittyhawks, and 3 Squadron had been re-equipped with that fighter. Despite the varying opinions expressed as to its combat qualities, it chopped into the enemy like an axe. The Germans and Italians did not like them at all. 3 Squadron Kittyhawks celebrated New Year's Day, 1942, by shooting down five enemy aircraft, and soon after that the squadron charged into fifty German and Italian machines, shot down seven, and put the rest to flight. In another battle, RAF and RAAF fighters shot down twenty of a flight of thirty of the enemy and damaged all the remainder.

It is no exaggeration to say that Allied air superiority was a critical factor in the defeat of the Afrika Korps. Only one example is the way in which the flyers smashed Rommel's attempt to recapture Tobruk in May 1942. The "Desert Fox" had his tactics worked out to a strict timetable. His troops were to drive round the British left flank, capture Bir Hacheim, roll up all the positions in the Allied south and rear, and march victoriously into Tobruk on 30 May.

His plans were upset in what some people called the Battle of the Gap. The Gap was a corridor which the Afrika Korps cut through British minefields and in some places was only 1,000 yards wide. As the German and Italian vehicles and armour rolled through the Gap they were attacked by Kittyhawks, Tomahawks, Kitty-bombers and Hurri-bombers, Hurricanes, Boston medium bombers, and Skua divebombers. Flying short hops from their airstrips, and returning again and again despite enemy counter-attacks, the Allied flyers destroyed hundreds of vehicles and threw the advance into chaos. At the same time, Rommel's rear bases and airfields were bombarded by Wellington and Albacore bombers. Together with Allied resistance on the ground the aerial assault upset Rommel's timetable, and though Tobruk fell a month later the British command had at least gained a breathing space to rally for yet another savage defence of what were then the strategically vital areas of Egypt and the Suez Canal.

The list of decorations awarded to RAAF flyers in the Mediterranean theatre is an impressive record of courage and flying skill. Wilfred Arthur from Queensland, awarded the DFC after shooting down four of the enemy in one day. "Killer" Caldwell, who was attacked by two Messerschmitt 109s and shot down one of them even though he was wounded and his aircraft damaged. Thomas Price of Adelaide, whose Seagull amphibian was attacked by three Italian fighters when he was on reconnaissance for H.M.A.S. *Sydney* off Bardia. The three fighters chopped up his slow machine with the gunnery but he still managed to bring her down to a safe landing. Hugh Throssel of Perth, serving with a South African squadron, who destroyed five of the enemy during twenty-nine sorties over the German lines. And so the list goes on.

During the struggle for aerial supremacy in the

A Lockheed Lodestar of the RAAF on an airfield in northern Queensland in 1944 (*RAAF Museum*)

Top: The Douglas Boston (*RAAF Museum*)

Centre: The ground crew prepares a Halifax bomber on an English airfield for a night strike on Berlin (*Australian War Memorial*)

Bottom: Prelude to the invasion. The Halifax at lower right was one of a strike force attacking German military concentrations in the Pas de Calais, northern France (*Australian War Memorial*)

Mediterranean, a different type of air war was being fought over Europe. Australians served in the first RAF raid on Berlin, on 25 August 1940, and just over a year later the first bombers completely manned by Australians in the European theatre, twin-engined Hampdens, joined in the raid on Berlin on 2 September 1941. From then on RAAF crews flew Lancaster, Wellington, and Halifax bombers on scores of raids on targets throughout Germany. Despite Goering's boast that if the British bombed Berlin "You can call me Meyer" (a name held by many Jews) the German capital was assaulted with mounting fury and frequency by RAF and RAAF units.

Berlin was one of many targets. During those wartime years the British radio news bulletins listed an almost daily string of strategic points blasted from the air. "The railway marshalling yards at Hamm" were mentioned so frequently that the phrase became a joke. The bombers concentrated on railways, factories, harbours, oil refineries, and a huge variety of similar targets, with the aim of crippling German war production.

Since the second World War it has become fashionable to deplore the aerial bombardment of Germany and to say it had little effect on the outcome of the war. But those who remember the war will recall the gigantic menace poised by Nazi Germany over world freedom. The German High Command stopped at nothing and had already killed thousands of European and British civilians in their bombing of cities from Rotterdam to Liverpool. If British aircraft had not sought out the rocket programme, then it is highly likely that London would have suffered a rain of rockets and a holocaust of destruction as bad as that inflicted by Allied bombers on Dresden.

The factual results of the bombing have long since been obscured by a cloud of deplorers and apologists, but it is absurd to say that it was ineffectual when it forced the Germans to concentrate so much of their war effort in attempts to fight it off. And the effect on morale was enormous. The air crews were the first to prove that the Germans could be beaten: first in the Battle of Britain and then during the bombing raids which were pressed home night after night, week after week, month after month, despite everything the Germans could do. The people of conquered Europe took new heart when they saw German cities burning; the people of Britain learnt with delight of the first 1,000-bomber raids which carried the war deep into the homeland of their enemy. Attitudes change, and it is now difficult to realise that one actually felt pleased to hear that a city was being bombed into fiery death. But war itself is insane and the emotions of wartime are hard to justify in the peaceful years that follow.

The famous North American B25 Mitchell bomber flown by the RAAF (*RAAF Museum*)

And, of course, the bombers did not have it all their own way. The news bulletins which reported the raids usually concluded with some such words as "four of our aircraft are missing". These bland words conveyed nothing of the immense physical effort, dangers, courage, and fear which were all part of a raid on Germany.

The first bombing raids on German strategic installations began in May 1940. After the Battle of Britain the raids grew steadily in size and number though they were still somewhat restricted because the bomber squadrons were equipped mainly with Wellington medium-range bombers. Nevertheless these aircraft pounded every target they could reach. In May 1942 a total of 1,047 bombers hit Cologne.

In the meantime the British aircraft factories were working on the production of Lancaster and Halifax bombers, and when these began to reach the squadrons in 1943 the raids were stepped up enormously. In those days the new bombers seemed to be gigantic. The Lancasters weighed ten tons and consisted of 70,000 separate parts: a miracle of modern technology hurried into production under the impulse of war. They carried up to 22,000 pounds of bombs, while the Halifax bombers carried up to 13,000 pounds apiece.

Bomber Command became a weapon more formidable than anyone had dreamed of in 1939. It consisted of about 1,600 heavy bombers, mostly Lancasters and Halifaxes together with a few obsolescent Wellingtons and Stirlings. The whole was divided into groups, each responsible for about 350 aircraft divided into squadrons of eighteen bombers apiece. Each group commanded a number of stations. A station was the airfield and administrative centre for one or more squadrons: sometimes bomber squadrons alone, sometimes bomber and pathfinder squadrons.

The British bombing force comprised about nine per cent Australian aircrew. These men flew either as "individuals" in RAF crews, which also contained Canadians, Rhodesians, New Zealanders, and other products of the Empire Air Training Scheme; as complete Australian crews in bombers attached to RAF squadrons; or as personnel of complete RAAF squadrons flying out of RAF stations. For example a station might have two RAF squadrons and one RAAF squadron.

In the winter of 1943–44 the quantity of bombs dropped on Germany increased steadily, and in May 1944 the record of 30,000 tons of bombs was dropped on strategic targets connected with the forthcoming invasion. This may seem a puny amount compared with the subsequent devastation wrought by atomic bombs, and then the gigantic tonnages dropped by the Americans in Vietnam, but in those days it was awe-inspiring.

The whole business of strategic bombing, which by 1943 had become a regular fact of life, began at top level. Bomber Command made the decisions on data supplied by staff planners who wanted strikes on this, that, or the other target: U-boat pens, a ball-bearing factory, trains moving troops into France, aircraft plants, steel works, railway junctions, and so *ad infinitum*. German-occupied Europe became one huge target because it was all armed and aimed at total war.

These decisions filtered down through the groups to the stations to the squadrons until finally they were implemented by the aircrews. Most of the men who flew the bombers were very young, often as young as eighteen, rarely as old as thirty. The aircraft they had been trained to fly were the most complex ever designed until that time, but they gave little comfort to their crews. They were built purely as killing machines to carry a maximum load of bombs to a target, drop them with as much accuracy as possible, and, hopefully, return to base.

Many did not return. Their paper-thin fuselages could be pierced by a rifle bullet, to say nothing of the jagged shards of metal spat out by anti-aircraft shells. Their massive Merlin engines were held in place by a complex series of aluminium alloy struts which could be torn apart by a fragment of shell. The hydraulic systems which operated the undercarriage, tail, ailerons, and gun turrets were vulnerable to bullets and flak. When the heavily-laden machines lumbered into position for take-off they carried hundreds of gallons of high-octane gasoline and tons of explosives—not only their bomb load but the ammunition for their own guns. The entire aircraft was a flying bomb, and a flaw of wind, a pilot error, or some mechanical fault which escaped the meticulous checking of the ground crew could turn aircraft and aircrew into a roaring fireball at the end of the tarmac. Homeward bound it could be worse. Often the aircraft had suffered damage over the target and had to land with one or more engines feathered, some of the crew dead or wounded, and the survivors at an extreme pitch of physical and emotional exhaustion after many hours in the air. It was easy to make a mistake, and a mistake was often fatal.

When a squadron was warned that it would be

This RAAF Boomerang was christened Sinbad II by her pilot (*RAAF Museum*)

flying that night, each aircrew went to check its aircraft and take her up for a test flight. Now all the planning was concentrated into groups of seven men: pilot, co-pilot, navigator, bomb aimer, wireless operator, rear gunner, mid-upper gunner. Each crew was a tight little unit, almost a family, of its own. They all knew and depended upon each other. Within the monolith of service discipline each crew had its own little democracy in which ability and personality were more important than rank. The pilot was in overall command and, though he might be a sergeant while others of the crew were officers, his orders had to be obeyed. The pilot was usually "Skipper". Distinctions of rank were forgotten and they called each other by their Christian names. After a few weeks of active service together they developed an almost superstitious feeling for the link which bound them and their aircraft together. If one of their number was removed by death or wounds, or they had to fly in a different aircraft, they felt uneasy with the replacement until they had had time to adapt. The link grew tighter as they survived one mission after another. A tour of duty consisted of thirty completed missions; the crew knew that the chances of survival grew progressively smaller with each mission and the old hands were regarded with some awe by the newcomers.

The Lancasters awaited them on the tarmac, each with her attendant ground crew in the charge of a sergeant. Aircrews and ground crews also had a special feeling towards each other. The flyers relied utterly on the mechanics, armourers, and fitters to keep the aircraft in first-class condition. The ground crews had a pride in their machines and aircrews which helped them to give added devotion to their duty.

The aircraft, which were huge compared with the bombers of the first World War, were painted a dull black on the lower surfaces and camouflaged green and brown on the upper surfaces. The white circles of the RAF and RAAF roundels, which in those days were the same, were painted black. The aircraft soon lost the immaculate condition in which they were delivered from the factories, and the paint became scratched and dented by ground crew working on the engines and fuselage.

The ground crew sergeant reported the aircraft's condition to the pilot, who made a careful ground check to assure himself that everything was in order. Then he led the aircrew up the ladder through the rear hatch and they took their places. It was a tight squeeze. Pilot and co-pilot sat side by side up in the nose, with the navigator, bomb aimer, and wireless operator immediately behind them. The mid-upper gunner was stationed about halfway down the fuselage and the rear gunner sat in his "dustbin" in the tail. Of all of them he was the most vulnerable and exposed.

With the aircraft on the runway the aircrew took about twenty minutes for their pre-flight checks of engines, instruments, and equipment, and then at last she gathered momentum. Among those who watched the great ominous machine trundle along the tarmac there were plenty who could remember the days when the only bombs dropped from aircraft were tossed overboard by the observers; when

Wirraways in close formation (*Australian War Memorial*)

In the original of this photograph, the words 'Mosquito . . . holds record for prop driven A/C Sydney to Brisbane' can just be deciphered on the board propped against the undercarriage (*RAAF Museum*)

warplanes were flimsy affairs which could not even fly as fast as the Lancaster's take-off speed of some 100 knots.

The pre-raid flying test included a simulated bomb run, drill against fighter attack, and tests of guns, engines, and radio gear.

The tests were followed, later that day, by the briefing. It was attended by a great crowd of flyers, usually at least 100, often more than 200, depending upon the number of squadrons flying out of the station that night. As soon as the crews entered the briefing room, a glance at the huge map told them their target for that night. A ribbon pinned to the map showed the courses they would take from the airfield to the town, city, or installation that was to be "clobbered".

The briefing was far more than a simple "Go get 'em, lads!" The station commander explained the reason for striking the target, the intelligence officer gave the latest information on the enemy defences, the navigational officer and meteorological officer told the flyers about the courses they should fly to the target and back and the weather to be expected en route. Usually the squadron or squadrons represented at the briefing were only one part of the strike force, and the movements of all other units of the group had to be co-ordinated.

"Time over target" was an important factor. The first aircraft to arrive could be luckier than the follow-up squadrons, because the enemy defences might not have been fully alerted. In the later stages of the war the enemy was on full alert all round the clock, and the use of pathfinder aircraft, which dropped huge flares to mark the target area, left no doubt as to the bombers' destination.

The pathfinder force had developed from an idea formulated by the Australian pilot D. C. T. Bennett, of Toowoomba. Bennett made a name for himself before the war as author of two books on aerial navigation and as pilot of the Mercury Pick-a-back aircraft. This was evolved as a solution to the problem of running a regular trans-Atlantic airmail service. One aircraft carried a smaller aircraft for part of the way, so that the "passenger" could conserve its fuel. At a given point the smaller aircraft took off to complete the Atlantic crossing while the carrier returned to base.

Bennett found an answer to the dangers of a "bomber's moon". Early in the bombing campaign, aircraft had to fly under ideal weather conditions which enabled them to see the target. This meant, of course, that the enemy could see them equally well, and also that a number of missions had to be aborted because of poor visibility.

Bennett suggested that a force of expert navigators should be trained to find targets under any conditions, and to mark them in such a manner that they would be unmistakable. He flew on the first pathfinder missions in order to develop and perfect the techniques and the crews soon became adept at finding and marking targets. They flew in twin-engined Mosquitos, the all-purpose fighter-bomber of laminated plywood which de Havilland developed during the war, or in modified Lancasters and Halifaxes. With the use of conventional navigation aids, radar, radio direction-finding, and some of the earliest radio beam guidance systems, they became so expert that they could pick out a target despite fog, cloud, or industrial smog. The "back-room boys" of the British Ministry of War Production developed a wide variety of flares to be dropped by the pathfinders. Some were of multi-million candlepower and others were designed to float above cloud for long enough to show the target area to the bombers. The types and patterns of the flare markings were changed frequently, to foil German attempts to deceive the bombers by setting up false patterns over deserted country.

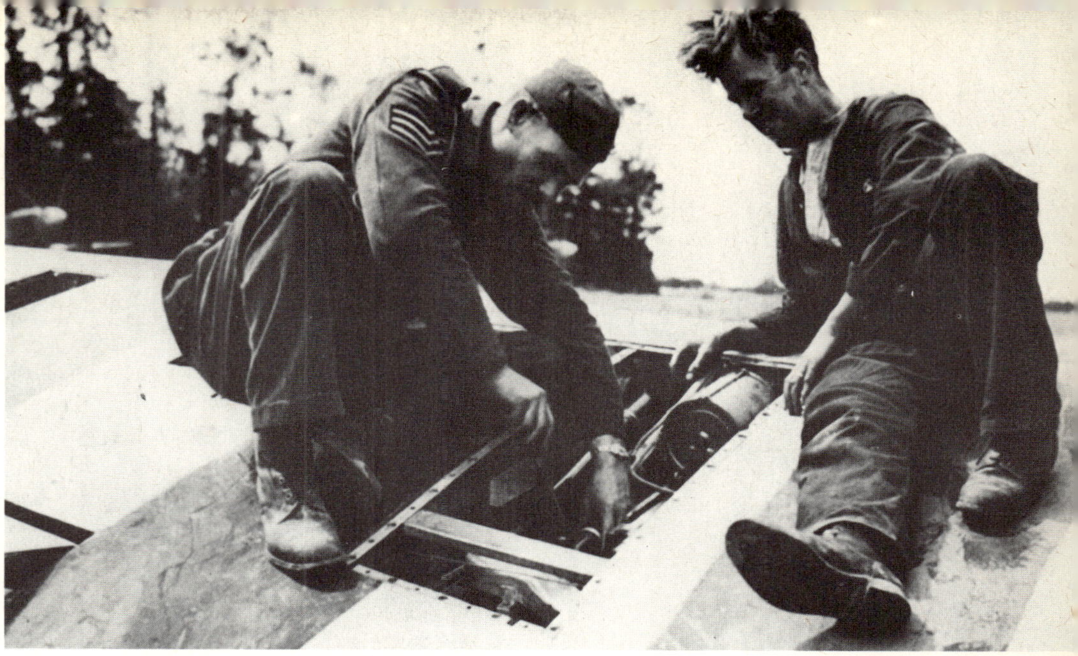

Top: The finishing touch: armourers completing overhaul of an aircraft's weapons before a strike (*Australian War Memorial*)

Bottom: Spitfire pilots are "given the gen" by the Intelligence Officer a few minutes before take off (*Australian War Memorial*)

TOP: RAAF winners of the Victoria Cross. *Left* Wing Commander H. I. Edwards. *Right* Flight Lieutenant William Ellis (Posthumously Awarded).

BOTTOM: Group Captain C. R. "Killer" Caldwell, whose aircraft markings recorded the destruction of 17 German, 3½ Italian, and 7 Japanese aircraft, in 1945 (*All photos Australian War Memorial*)

When the briefing concluded there was little to do but wait. Some men wrote letters: others played cards, chatted, read. A few managed to sleep. All of them thought about what lay ahead: some with apprehension because it was their first mission, others with stark knowledge because they had been there before. They did not discuss their fears. The accepted practice was to make light of the whole affair—to treat it, in some way, as though it was a huge sporting event. There was really no alternative. If the weather worsened, or the operation was cancelled for some other reason, they did not know whether to feel relief or disappointment. Perhaps the worst periods were when a long spell of bad weather kept the aircraft grounded for days on end, so that aircrew relaxed into a false sense of security. It was almost easier to keep going at full stretch all the time, with mind and body so occupied that there was little time to think.

A meal was served to the aircrew before take-off, and often it consisted of such items as bacon and eggs, a luxury in wartime Britain. Some men ate heartily: others found it hard. Then at last they were donning their flying suits, flying boots, gloves, helmets, oxygen masks, and the lifejackets which early in the war were christened "Mae Wests" after the buxom American film star. The bombers flew at high altitudes through freezing winter nights, and most flyers bundled themselves in heavy scarves and jerseys under the flying suits.

The flyers were intensely conscious of each other as partners in an elite experience: the sharp edge of the sword wielded by Bomber Command.

Clumsy in their flying gear, and carrying their parachute bags, they climbed into the vans which took them to the aircraft. The Lancasters, massive shapes in the chill winter dusk, stood in their dispersal areas round the airfield. One by one, as the aircrew boarded and settled themselves into position, the bombers moved out on to the perimeter track and then on to the runway, to form a long queue of roaring machines. As each of them saw its take-off signal blinked from the control van it bellowed off into the night, until the station was left silent and empty to await their return.

At airspeeds of less than 200 knots the bomber flight from English airfields to the German heartland was a long haul, especially when the target was Berlin or one of the more distant industrial cities. And, once the narrow North Sea was crossed, the whole distance was flown over hostile territory, through belt after belt of searchlights and anti-aircraft defences. The battle began as soon as the bombers crossed the coast of Holland, but there was little they could do to fight back.

The Germans had brought their anti-aircraft defences to a high pitch of efficiency. Radar, the British invention and closely-guarded secret which

In the cockpit of a Hudson (*Australian War Memorial*)

helped to win the Battle of Britain, had now been developed by German scientists and was used to direct searchlights and anti-aircraft fire. The flak artillery units had been provided with heavier cannon which could fire to the extreme ceiling height of the Lancasters. Once the searchlights and radar-directed cannon had locked on to a bomber they could be evaded only by flinging the aircraft wildly around the sky. To some extent there was "safety in numbers", because so many bombers were in the air that the flak could not deal with all of them, but some of them had to suffer.

The flak was bad enough but the night fighters were even worse. They were like tigers leaping from the ambush of the dark; their identification lights, switched on to protect them from their own flak, like vicious fiery eyes. Gunners watched for them tensely, with trigger fingers so itchy that they had been known to fire at imaginary shapes or even at another bomber looming up out of cloud.

The night fighters were specially-adapted Messerschmitts manned by a pilot, a radar operator, and a rear gunner. Their crews possessed their full share of German skill, courage, and aggression. Strict training and discipline and plenty of active service had welded them and their machine into a single slashing weapon. In speed, power, and manoeuvrability they had a great advantage over the lumbering Lancasters, especially because they had learnt to strike at blind spots out of reach of the bomber gunners. Often the best defence was violent evasive action. When a gunner shouted a warning through the intercom, the pilot wrenched the bomber into a violent corkscrew turn and then rolled, dived, climbed again, and corkscrewed once more in a manner hardly dreamed of by the aircraft designers.

One point sometimes forgotten is the brute strength required to handle heavy military aircraft of second World War vintage. Ivan Southall, the famous Australian writer of children's stories, was a flying-boat pilot during that war. In his book *Fly West* (published in 1975) he comments upon the cumbersome weight and sluggishness of the Short Sunderland flying-boats used by the RAF and RAAF. They were

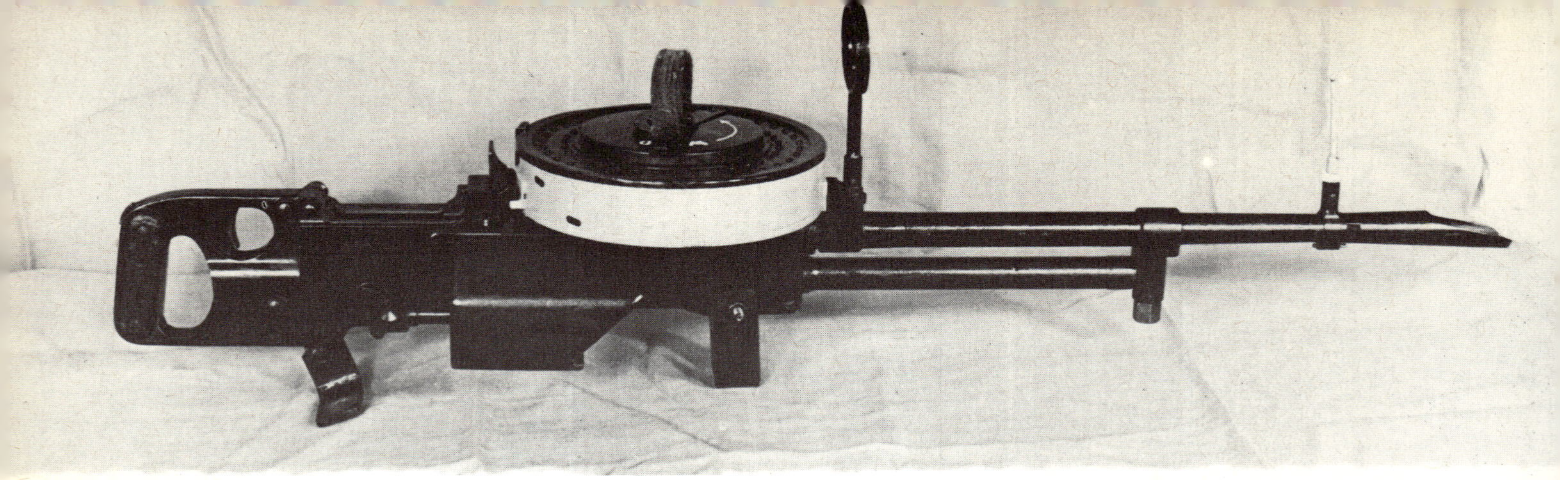

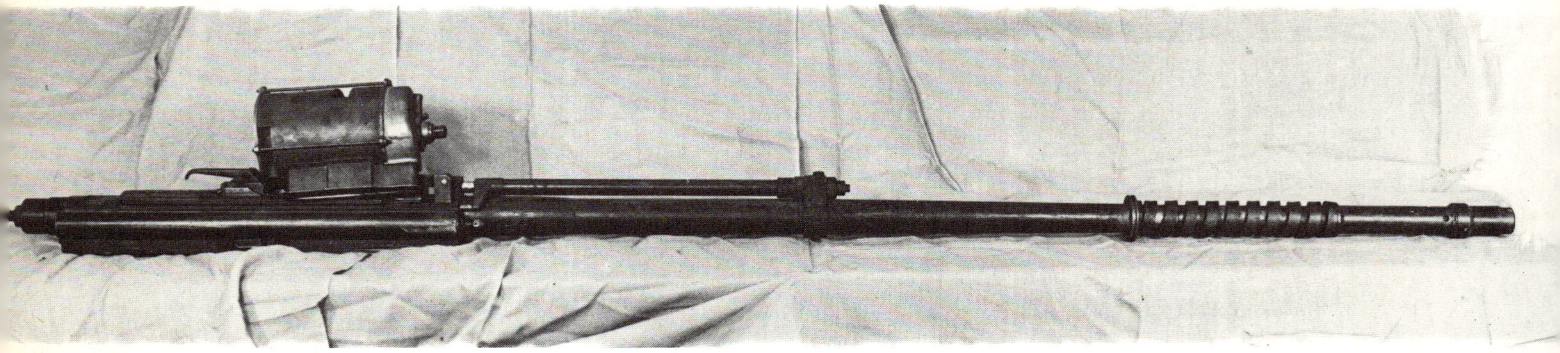

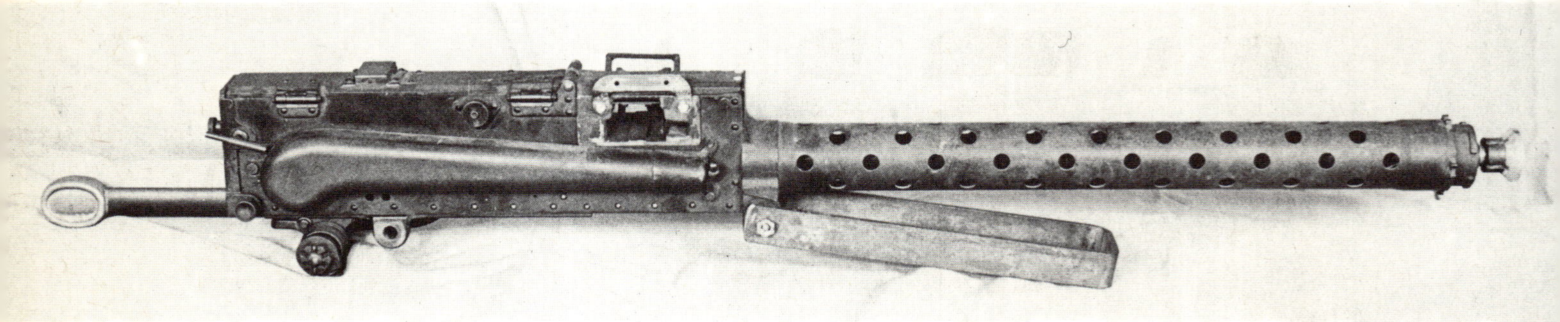

RAAF armaments. FROM TOP: Vickers gas-operated Mark I .303 calibre machinegun fitted between the wars and in the second World War; 20 mm Hispano cannon of the wartime and post-war periods; Vickers Mark V .303 calibre machinegun of the second World War (*RAAF Museum photographs*)

well suited for their original purpose—long-distance ocean patrols and courier runs—but when pilots had to take off or land under difficult weather conditions or engage in wild evasive action, the sheer bulk of the machine imposed a heavy physical burden. And of course it was simply impossible to fling the boat-hulled craft around the sky with the agility of a Spitfire.

The Lancasters, Wellingtons, Halifaxes, and Stirlings used against Germany were somewhat more manoeuvrable, but the pilots still had to fly gross loads of well over 50,000 pounds. Despite hydraulic assists, the operation of the controls was a man-sized job and some pilots returned from sorties in a condition close to physical exhaustion.

When the target was in sight the time for evasive action was at an end. The aircraft had to fly straight and steady as she lined up for the bomb run. This was the only reason why all the risks had been taken, and for most flyers it was a time of incredible tension as the aircraft flew to the directions of the bomb aimer. For the pilot, it was like being a motorist driving against a stream of heavy traffic at night, except that the dazzle came from searchlights and flak explosions. It took brave men to hold the aircraft steady in the air.

The sight above the target was unforgettable. In big raids, the fires on the ground lit up the interior of the aircraft with sombre ruddy shadows, shot with flashes of different colours from the pathfinder flares, the searchlights, and the exploding shells. Some of the cannon shells drifted lazily up in a series of incandescent blobs, deceptively slow until they suddenly accelerated as they came closer and either tore past

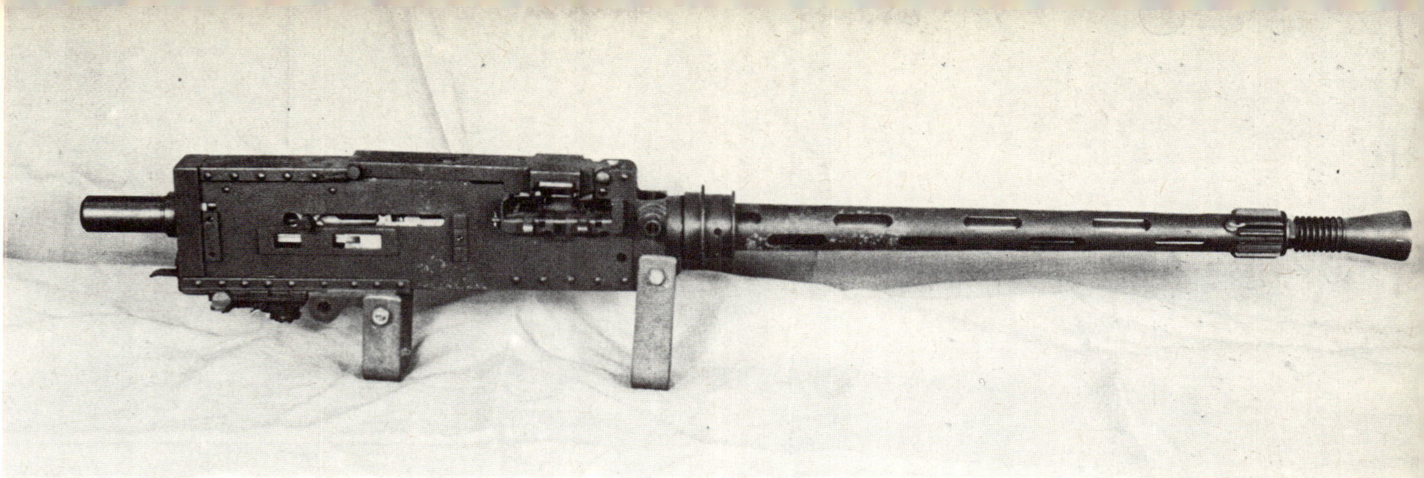

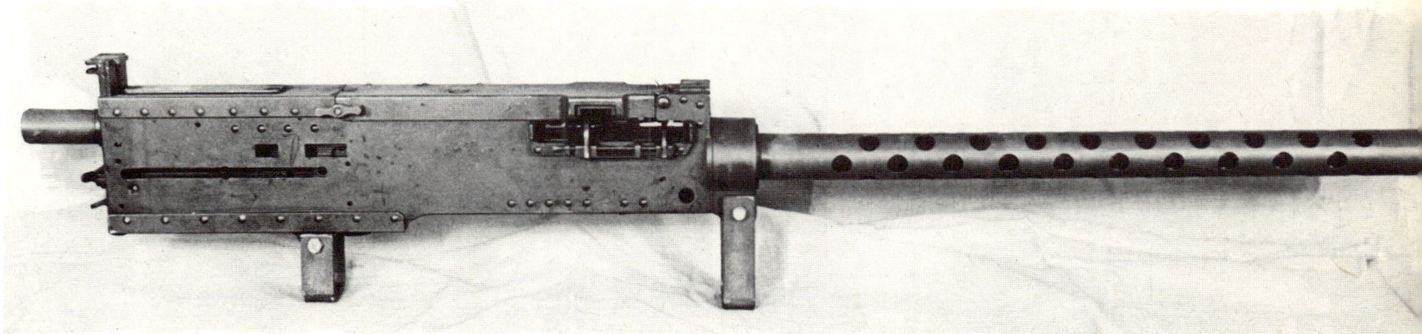

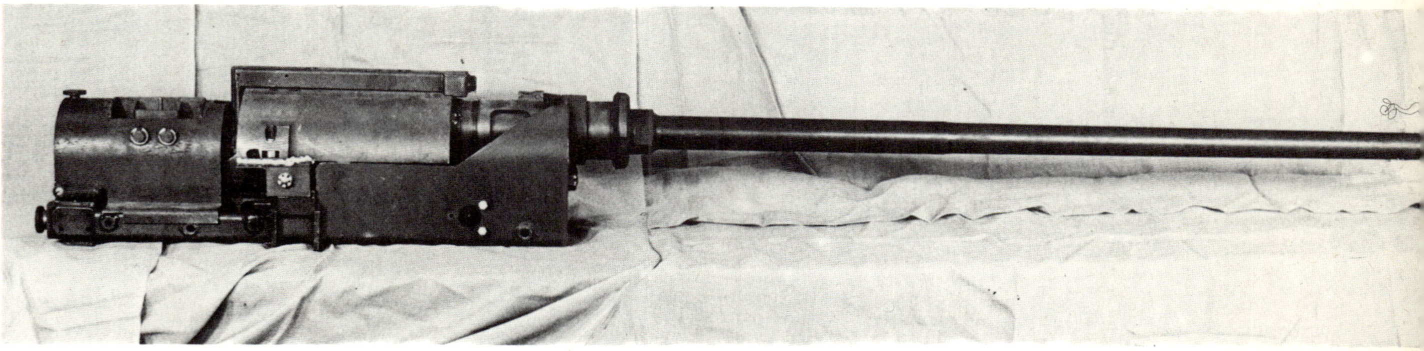

RAAF armaments. FROM TOP: Browning .303 calibre aircooled machinegun of the second World War; Browning .50 calibre machinegun used in the second World War and in Korea; 30 mm Aden gun fitted in RAAF Sabre jets (*RAAF Museum photographs*)

or hit the bomber with a vibrating crash. Other shells were invisible until they exploded in a cloudy orange flare and a booming thud that could be heard above the engines.

For most of the distance covered from the airfield the other bombers had been seen only as occasional flicking shapes, or by the glow of their exhausts: above the target they could be seen clearly. Sometimes too clearly, as they came close and sheered away. There were instances of bombers colliding over the target, and at least one case of falling bombs striking an aircraft flying below. There would be the sight of a bomber suddenly shuddering violently, as though she had collided with an invisible barrier in the sky, then dipping, sliding downwards; flames licking from engines over wing on to the fuselage, billowing astern as she began to spin uncontrollably, and maybe just a glimpse of mushrooming parachutes shining in the light. Or the sight of a bomber exploding with a direct hit: bombs, fuel, ammunition tearing her apart in a swiftly-falling uproar of flame.

Then there was the sudden lift as the huge bombs dropped, the shudder as the bomb doors closed, the rapid ninety-degree turn, and the course set for home. The aircrew were sweating profusely under their thick layers of clothing. From the pilot, the whole affair demanded computer-like reactions and sheer physical strength. The rest of the crew had to trust him utterly. When the bomber was flung in wild gyrations through the sky they could only hope he knew what he was doing as they clung grimly to their places.

And it was not over yet. They had to run the gaunt-

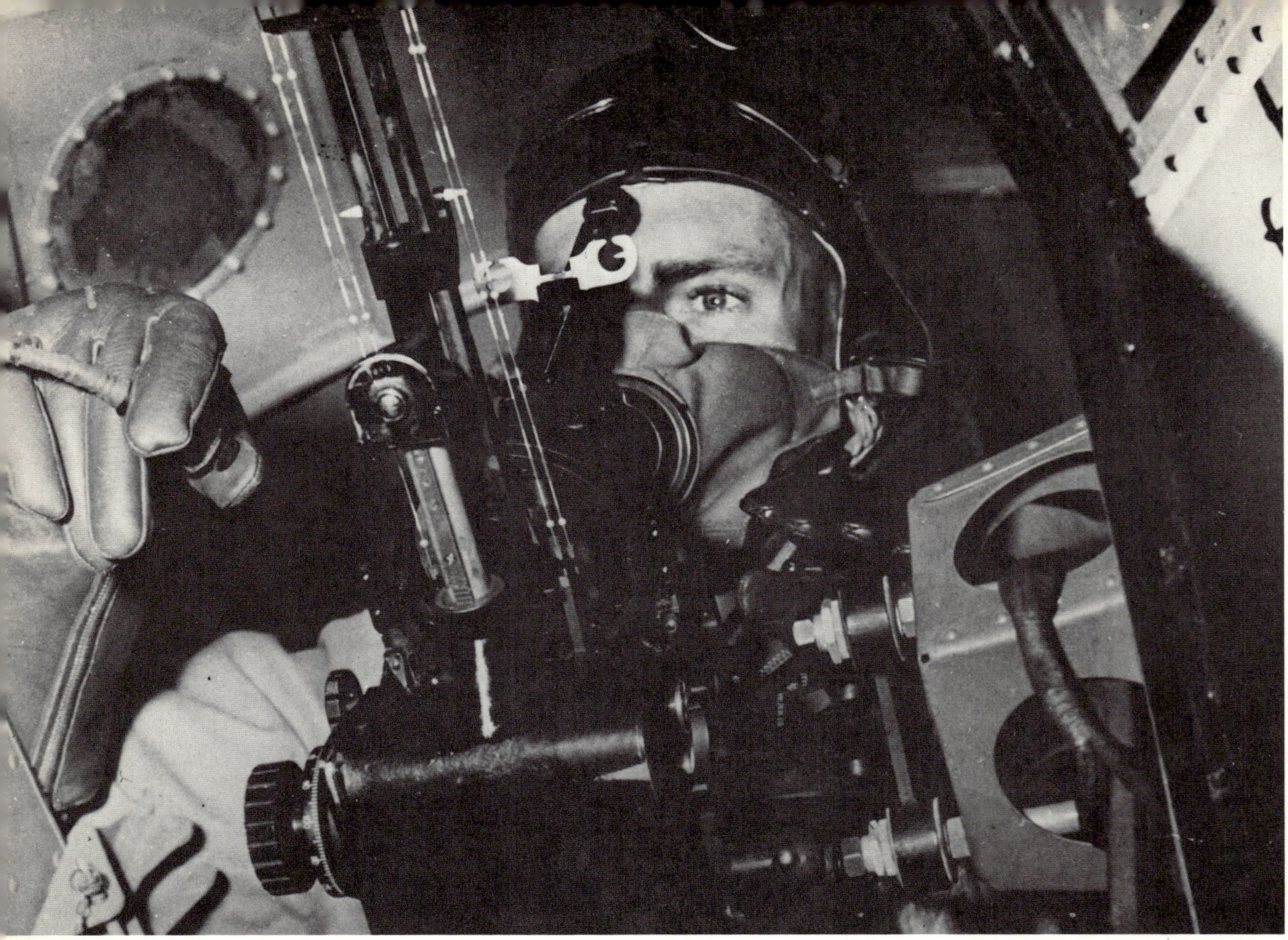

TOP: Bombs away! The bomb aimer "presses the tit" (*Australian War Memorial*)

LEFT: RAAF missiles of two wars. 250 lb. depth charge used in the second World War; 112 lb. general purpose bomb of the first war; 250 lb. and 100 lb. anti-submarine bombs of the second war; and 100 lb. smoke bomb (*RAAF Museum*)

let of the flak belts yet again. If they had been unlucky, then their aircraft was wallowing along with the slip-stream screaming through great gashes in the fuselage, wounded men to be cared for, and the dead to be moved aside. Sometimes the pilot had to fight desperately to maintain height, with the aircraft sinking lower and lower until at last he was forced to give the order to jump. Some bombers crashed before they reached the Dutch coast: others came down "in the drink" and the survivors paddled their rubber dinghy in the cold North Sea. Sometimes they were picked up.

A typical story of a Lancaster flight is told in *RAAF Saga*, published by the Australian War Memorial in 1944. It is that of Flight Sergeant G. C. C. Smith, of Sydney, an air gunner who was the only Australian in a RAF Lancaster. He and his crewmates had made five Berlin sorties apart from other raids, and on 15 February 1944 they were en route to Berlin again. Smith was acting as rear gunner as they passed safely through the flak belts, and Berlin was in view when he sighted the identification lights of a night fighter on the port beam. He and the fighter pilot fired simultaneously, and both hit their marks. The fighter exploded, but a two-second burst from her four machine-guns and two cannons had torn through the Lancaster from mid-section to tail. Immediately after that, a second fighter attacked from astern but the bomber pilot took violent evasive action and lost her in the darkness.

Smith's right leg was shattered by a cannon shell and machine-gun bullets. The hydraulic line to his turret was cut and his parachute cover was on fire. The attack by the two night fighters had left the Lancaster a flying wreck. The mid-upper gunner's leg was broken, the rear wheel was shot away and the starboard wheel was damaged, the flaps had dropped and could not be raised, the bomb doors were jammed, the instrument panel was shattered, one engine had been pierced by a cannon shell, and the aircraft was on fire amidships from a broken oil pipe.

The fire was extinguished and the wounded mid-upper gunner was laid out on the floor of the aircraft. As the pilot turned her for home he told Smith, on the intercom, that the rest of the crew would help him out of the rear turret. Smith knew that the mid-upper turret was out of action and refused to be moved. He continued to operate his turret manually, and despite appalling pain from his shattered leg he continued to keep alert for fighters as the Lancaster headed for home.

By that time the aircraft was almost unmanageable, and she strayed offcourse into a flak belt. The pilot took such wild evasive action that the crew thought he had lost control, but eventually he escaped the flak and reached the North Sea.

The crew chopped away the bomb door release and let the bombs fall into the sea, then tried to extricate Smith. The door of the turret was frozen hard, and after they chopped it open with the crash axe they found that Smith's wounded leg was caught in the ammunition belt and controls. He was still conscious as they worked for an hour to free him, then gave him morphia and laid him on the aircraft floor.

FROM LEFT: 40 lb. general purpose bomb; 100 lb. USAF practice bomb; 25 and 11½ lb. practice bombs; 20 lb. Cooper bomb of the first World War; 100 lb. USAF general purpose bomb of the second war; 4.5 inch reconnaissance flare; USAF Mark 41 marine marker. In front is a 4½ lb. incendiary bomb of the second war (*RAAF Museum*)

The pilot headed for the nearest airfield in England, with the intention of making a belly landing, but then found the bomb doors could not be closed. He decided to attempt a landing with the damaged undercarriage, and brought the aircraft down safely with the bomb aimer and wireless operator lying on either side of Smith to protect him if they crashed. The whole crew was brought out without further harm, though it was found that Smith had a badly frostbitten face in addition to his shattered leg, which had to be amputated. For his devotion to duty he was awarded the Conspicuous Gallantry Medal.

Despite the rapid growth of the RAAF during the first two years of war, and the audacity displayed by Australian flyers in the European and Mediterranean theatres, the whole organisation remained to some extent subsidiary to the RAF and to overall British strategy. Then, with devastating suddenness, the Australians had a war of their own to fight. Instead of sending their young men into battle for the mother country they needed them to defend the homeland against the onslaught of Japan. Australians continued to fight with the RAF, either as individuals or in squadrons which remained in the northern hemisphere, but an increasing number partook of their share of the war a lot closer to home. It was the beginning of a severance from RAF domination, however willingly that domination might have been accepted in earlier years, and from then on the RAAF grew

Lockheed Hudsons were the first American-built aircraft to be used operationally by the RAF. Australians with the RAF used these fine machines on all types of operations from anti-submarine work to general bombardment and eventually they were flown by the RAAF (*Australian War Memorial and RAAF Museum*)

steadily into a wholly Australian air force which made its own decisions.

The first significant RAAF clashes with the Japanese occurred in Malaya, Burma, and the Dutch East Indies (now Indonesia). When the Japanese attacked Malaya, Group Captain J. P. J. "Black Jack" McCauley, RAAF, who was renowned as a stern disciplinarian, was the senior RAAF officer in command of 1 and 8 Squadrons, equipped with Hudsons, and 21 and 453 Squadrons equipped with Buffalos. Both these types of aircraft were at a disadvantage against the Zeros which had foolishly been mocked by pre-war comedians, and against Japanese fighter pilots whom propaganda had portrayed as being too short-sighted to be effective. Nevertheless the Australians fought their outclassed aircraft with self-sacrificing fury. They used the Hudsons to blast Japanese invasion barges at sea and on the beaches, and Japanese troops and vehicles moving inland. They flew the Buffalos against Japanese bombers, and when fifty-one crated Hurricanes were landed on 13 January 1942 they worked twenty-four hours a day to assemble them into a force that would give the Japanese a bloody nose.

But by that time the battle for Malaya and Singapore was only a month away from the conclusion which has been a matter of controversy ever since. The RAF and RAAF bomber and fighter pilots did the utmost with the weapons at their disposal, a total of only 161 outmoded aircraft, and these were whittled away in the air while the airfields were being overrun on the ground.

Most of the aircraft that could still fly were withdrawn to the Dutch East Indies. Black Jack McCauley set up his headquarters at Palembang on the island of Sumatra. Now in command of six RAF units and the two RAAF Hudson squadrons, he directed attacks on Japanese forces on the mainland and then on enemy vessels invading the island. At the same time RAAF 2 and 13 (Hudson) Squadrons were fighting from Ambon and Timor. Mr Bruce Bowley of Adelaide, then a corporal with McCauley's headquarters, recalls that Black Jack was a stern but just commander, and says that his handling of the situation undoubtedly saved the lives of many Australians.

It was a grim period in which every day whittled down the number of pilots and aircraft a little more. The Japanese attacked with their usual fanatical determination: the British and Australian fighter and bomber pilots fought back with equal skill and courage. Unfortunately the outcome of the campaign was virtually a foregone conclusion. The enemy was attacking in overwhelming force. The Allies, who had taken the threat too lightly for several years beforehand, were at a moral disadvantage and also had to consider the effects of a last-ditch stand upon the civilian population of Singapore and other cities.

Some of the facts about the tremendous battle put up by RAAF units during the first phase of the Japan-

TOP: Lockheed Hudsons (*Australian War Memorial*). CENTRE: Low level bombing raid on Japanese shipping at Wewak on 2 September 1943 (*RAAF Museum*). BOTTOM: Near misses ahead of an enemy convoy escort during a skip-bombing attack by RAAF aircraft (*RAAF Museum*)

"I made a pass." A group of fighter pilots watch Flying Officer J. H. Smithson demonstrate how he chased a formation of Japanese bombers, and shot down two, near Darwin on 11 October 1943 (*Australian War Memorial*)

ese attacks are still coming to light. In February 1974 the Herald group of newspapers published an article called "Doolan, Hero of Ambon", by the well-known journalist Denis Warner.

Denis Warner wrote of the way in which the 2/21st Battalion, 2nd AIF, literally fought to a finish against the Japanese invaders of the island of Ambon. Out of 1,300 men who defended Ambon, only 363 came home. The others were killed in the battle, executed after capture, or died in captivity. Driver W. T. Doolan, who killed an estimated eighty Japanese soldiers before he was shot by a sniper, became a hero to the Ambonese and a symbol of endurance to the surviving Australians.

In his article Denis Warner mentioned 13 Squadron RAAF, which did its utmost until forced to evacuate the island. This evoked a letter to the Adelaide *Advertiser* from Kym Bonython, ex-RAAF pilot and now a celebrated entrepreneur. His letter, with later amendments supplied for this volume, said "I refer to Denis Warner's excellent article on the dark wartime days of Ambon in the Dutch East Indies. Mention was made of the heroic part played by Number 13 (Hudson) Squadron of the RAAF. No mention was made of the equally meritorious part played by its sister, 2 (Hudson) Squadron.

"Based in Koepang, Timor, units of 2 Squadron were operational in the same area from their airfield at nearby Namlea. Both squadrons worked in close co-operation.

"There are as many stories of heroism by members of 2 Squadron that your correspondent seems to have overlooked. Many of 2 Squadron crews lost their lives in this area—such as Flt-Lt. Parker Hodge and South Australians Flt-Lt. Rob Burns Cuming, FO Fred Robilliard, and Sgt. (Air Gunner) Jack Mawdsley.

"One of the greatest tales of devotion to duty was provided by 2 Squadron pilot, Flt-Lt. Bill White. Pleading for 'one more crack' at the enemy, he suffered such damage to his plane and crew that he was unable to take off from Laha Airfield at Ambon. He was taken prisoner and shortly afterwards was beheaded by the Japanese.

"Another member of 2 Squadron, with close Adelaide connections, FO Robert Law Smith, made

Top: The mighty Hercules. The RAAF currently employs the Hercules C130A and E, Caribou, Dakota, Mystère 20, HS748, and BAC111 as transport aircraft. (*Department of Air*) Bottom: A Lockheed SP-2H Neptune of 10 Squadron, which is stationed at RAAF Base Townsville. These aircraft, together with the Lockheed P-3B Orions of 11 Squadron, at RAAF Base Edinburgh, form the long range maritime reconnaissance element of the RAAF

The controversial F111C. When this aircraft was ordered by the RAAF it was regarded as a leap into the future, but the manufacturers encountered many problems before the first machines could be delivered. See text for the story. (*Department of Air*)

Squadron Leader K. W. "Bluey" Truscott, D.F.C., and Squadron Leader L. D. Jackson at Milne Bay, New Guinea, in October 1942, after playing a vital part in fighting off the Japanese invasion of Milne Bay. Squadron Leader Truscott died later in the war in an aircraft crash in Western Australia (*Australian War Memorial*)

an historic evacuation from Namlea carrying an incredible 39 passengers in a plane which normally held a crew of four. This was the last plane out of the area.

"Tattered remnants of both squadrons re-formed in Darwin under the joint leadership of Wg-Cmdr. (later Air Vice-Marshal) Frank Headlam, and Wg-Cmdr. John Ryland (later general manager of TAA).

"From Darwin, these ill-equipped crews continued to harass the enemy on Ambon, again with further heavy losses. Pilots such as Adelaide's FO Phil Brooks and Perth's FO John Venn were among those to crash into the waters of Ambon Harbor in the weeks following the fall of that outpost.

"No doubt there were many stories of bravery on land, sea and air, but the joint efforts of Numbers 2 and 13 Squadrons deserve recognition. In fact, both squadrons subsequently received the Presidential Citation from America for their work in those perilous days."

The fighting retreat slowly but inevitably became a rout. Of the RAAF aircrew and ground staff still alive a number managed to escape by dangerous and arduous routes and eventually made their way back to Australia, while others fell into captivity.

North of Malaya, Australians were helping to counter the Japanese attacks on Burma. Americans in P40 Tomahawks and RAF men in Buffalos had broken up the first Japanese air raids, and in January they were reinforced by RAF 17, 135, and 136 (Hurricane) Squadrons, which included many Australian pilots. RAF 113 (Blenheim) Squadron included sixteen Australians, and most of the aircrew in RAF 45 Squadron, which also flew the Blenheim bombers, were Australians.

They fought as hard as their mates in Malaya but were defeated for much the same cause: the rapid depletion of essential supplies and the loss of airfields during the chaotic military situation on the ground. They made their presence felt before the survivors withdrew to India. In one attack on a Japanese airfield, 45 Squadron destroyed thirty aircraft.

The RAAF and its ancillary units were hard hit very early in the Japanese assault on Australia and the islands along the northern coasts, and there were outstanding examples of heroic devotion to duty among both civil and military flyers. An exemplar of such men was Goff (Godfrey) Hemsworth, who had been Ray Parer's companion during the Centenary Air Race from Britain to Australia in 1934. His story is told in the book *Battling Parer* and in *Front Line Air Line*, the story of Qantas during the second World War. Hemsworth joined Qantas as first officer in one of their flying-boats running between Sydney and Singapore, and when some of these were taken into the RAAF he went with them, with a commission as flight lieutenant. He was stationed at Port Moresby, the base for 11 and 20 (Flying-Boat) Squadrons, from September 1939 onwards. In 1942 he was awarded the Air Force Cross, for which his citation reads in part: "Before his appointment to the RAAF, Flight Lieutenant Hemsworth had completed approximately 5,500 hours civil flying, and since his appointment has been engaged for approximately 2½ years on flying-boat duties in the New Guinea and Solomon Islands areas. In the course of his service duties prior to the outbreak of the war with Japan, he completed long hours of flying, both by day and night, amounting to approximately 1,300 hours; carried out extensive reconnaissance patrols, and took a leading part in the establishment of aircraft bases in the New Guinea and Solomon Islands areas. When base facilities for aircraft were being arranged in the area, this officer displayed considerable foresight and initiative, which proved to be of great assistance in the completion of the work . . . Following upon the outbreak of the war with Japan, Flight Lieutenant Hemsworth has been continuously employed as captain of an aircraft in operations against the enemy. On one occasion, after an attack over Japanese-occupied territory, his aircraft was severely damaged by enemy action. He landed the aircraft safely at an emergency landing base, effected repairs sufficient to take off, and returned to base on one engine after a flight of approximately six hours' duration."

Flying-boats on coastal patrol. The top photograph is of a Martin PBM Mariner and the other of a Consolidated PBY Catalina (*RAAF Museum*)

Hemsworth was one of the real old breed of Australian flyers; not only an expert pilot but one who could keep his machine airworthy with his own hands. But on 6 May 1942 he took off on his last flight. As captain of a Catalina flying-boat, with a crew of eight, he left Port Moresby to search for a Japanese invasion convoy believed to be sailing through the Solomon Sea between Woodlark Island and the Louisiade Archipelago. The convoy had been glimpsed by other flyers but bad weather had prevented proper reconnaissance.

Hemsworth flew his RAAF Catalina number A24.20 over the mountains, jungle, and swamps of the Papua New Guinea coast and offshore islands until he was above the Louisiade Archipelago. There, spread out on the glittering sea three miles south of Misima Island, he saw the enemy convoy. It was heading for the Jomard Passage at the northern approaches to the Coral Sea.

At the height he was flying he could not see the exact composition of the convoy. He sent a sighting signal to Port Moresby, then began to let down for a closer look. During the next few minutes he gave his wireless operator details of the number of troop ships, landing craft, and warships including an aircraft carrier which made up the convoy.

It was one of the most important messages of the war, because Hemsworth had seen part of the force which the Japanese were mustering for an assault on Port Moresby. If that Allied base had been overwhelmed then the Pacific war would have taken a very different turn, but Hemsworth's message prompted the American counter-attacks which culminated on 8 May 1942 in the Battle of the Coral Sea: the first full-scale defeat of the Japanese navy and air force.

At 2.20 p.m. on 6 May, Hemsworth sent his last message: "Attacked by eight enemy planes. Sky full of them am on fire have hit two." Nothing more was heard of Hemsworth and his crew, though some post-war enquiries led to a belief that they were picked up out of the sea and clubbed to death on the deck of the Japanese destroyer which "rescued" them.

Qantas aircraft were very quickly involved in the war which flooded across the south-west Pacific area. The big-bellied Qantas flying-boats were ideal for the task of flying ammunition and supplies to the front-line garrisons on the widely separated island bases. The Singapore route was abandoned shortly before the surrender, and after that the flying-boats flew to the Dutch East Indies. For a number of people they were literally lifesavers. On their return trips to Australia they carried wounded servicemen and women and children who would otherwise have been left to the mercy of the Japanese.

The flights became shorter, and the dangers greater, as the enemy pressed closer to Australia. By the end of January 1942 the Japanese fighters were ranging over the Timor Sea, and when Captain A. A. Koch

The de Havilland Sea Hornet, shown here with Royal Navy markings, was flown by some Australians serving with the British forces between 1944 and 1946 (*RAAF Museum*)

attempted to fly the *Corio* on yet another mercy trip to Soerabaya the flying-boat was attacked by seven Zeros. With two engines blazing, the *Corio* was forced down to water-level and Captain Koch zigzagged wildly as the Zeros swooped repetitively to administer the *coup de grâce*. At last he was forced to drop the aircraft on to the water, five miles from land, and the impact threw him through the windscreen. The *Corio* began to sink but five other men struggled free and one of them helped Koch, whose leg was broken, to swim ashore at Timor.

While some Qantas flying-boats flew on these dangerous missions, others maintained communications between Australia and Britain. Prudent pre-war forethought had pioneered a new route to Colombo, via Cocos Island instead of Singapore, so when that island fortress fell the flying-boats continued to fly mail and VIP passengers across the width of the globe. Another of their number was lost when the *Circe* disappeared late in February 1942, homeward bound with evacuees from Java, and it was presumed that she was shot down. Others provided invaluable services immediately after the disastrous Japanese air raid on Darwin, when they provided rapid communication with the southern ports of Australia.

The RAAF, like most other people and organisations in Darwin on 19 February 1942, was caught badly on the hop. There were no RAAF fighters operational in Darwin at that time, and the only aircraft available to fight off the swarm of attacking fighters and bombers were eleven P40s of the US Army Air Force 33rd Pursuit Squadron. They fought gallantly against overwhelming odds but with little or no effect on the outcome of the raid. The first Japanese raid struck hard at the main RAAF base and when all the P40s had been destroyed either in the air or on the ground they strafed and divebombed the airfield and its installations almost at their leisure, though small groups of men including Group Captain Scherger, the North-West Area commander, tried to hit back at them from the ground. One of them, Wing-Commander Archibald Tindal, was the first RAAF man to be killed in action in Australia. He died as he fired a Vickers machine-gun at the attackers.

The secondary RAAF airfield, at Parap, was also

These Wirraways were among the earliest off the production line when Australia began to produce her own fighters. The photograph was taken in 1939 (*Commonwealth Aircraft Corporation Pty Ltd*)

the civilian airfield which had been used by all the record-breaking Australian flyers from Ross and Keith Smith onwards, as they flew in or out of Australia. Dutch KLM airliners also used the field. In February 1942 it was the base for a flight of RAAF Wirraways and it also housed DCA communications facilities. Most of the installation was badly damaged or destroyed during the first raid.

Far worse was yet to come. When the second wave of attackers arrived, a few minutes before noon, they headed straight for the main RAAF base and deluged it with bombs. Everything that had survived the first raid was blown skywards on the volcanoes of flame and debris. The RAAF hospital, in which every bed was occupied by sick and wounded who had been flown in from the islands, was badly damaged. Fortunately, the patients had been carried out to lie in trenches as soon as the second air raid warning was received. For a few hours after the raid the base was completely disorganised and a number of personnel headed south, either because an order was misinterpreted or because of the general chaos of the whole Darwin area. By the following day the long task of clearing up the wreckage and rebuilding the base had begun.

On 3 March 1942 there was a less publicised but equally disastrous raid: that on Broome. It was particularly tragic because sixteen Dutch and Australian flying-boats were moored in the harbour, and eight of them were packed with Dutch women and children who had just been rescued from Java. The flying-boats were attacked by about six Zeros which sank them one after another. As they vanished in blazing lakes of petrol the Zeros turned on the airfield and destroyed five bombers on the ground.

For the RAAF in Australia, the first three months of 1942 were the low point of the war. After that there was nowhere to go but up—and they climbed fast. They were helped by the rapid response of the Americans to the challenge of the Pacific war. Slowly at first and then in almost extravagant quantities the Americans provided the aircraft with which their own air forces and the RAAF could strike back at the Japanese. These aircraft included the Liberator bomber, used by the American Fifth Air Force Heavy Bombardment Unit. Australians learnt to fly the Liberators by flying on missions with the Americans. A number of the RAAF men to be retrained in Liberators were veterans of the bombing war over Europe or the operations against the Afrika Korps, and they found the Japanese anti-aircraft defences to be but a pale replica of the German flak belts. Eventually the first RAAF Liberator squadron was formed and began to strike the Japanese over a great semicircle which involved round-trip flights of nearly 3,000 miles.

Among the men who had returned from the northern hemisphere was Bluey Truscott, and he earned fresh laurels as he joined in the battle for Milne Bay. This was one of the decisive battles of the Japanese endeavour to conquer the islands north of Australia as a preliminary to invading the mainland. They planned to conquer the vital base of Port Moresby by an overland thrust from the north coast of Papua together with an invasion from the sea. For the latter, they needed to secure Milne Bay as a pivot for the invasion force and to protect their rear.

The Australians and Americans just beat them to it. By working at almost frenzied speed in the dense jungle of Milne Bay they managed to hack out the

The famous Liberator, used by the RAAF for long-range bombardment in the latter years of the war. After the war, some of these mighty aircraft were converted for overseas passenger services (*RAAF Museum*)

first airstrip by 20 July 1942, and by the end of the month the Kittyhawks of 75 and 76 Squadrons and some Hudsons of 6 Squadron were based on the strip. They took the offensive immediately and fought off raiding and reconnaissance aircraft during the weeks while the Japanese were preparing the invasion.

It came on 24 August. Hudsons attacked barges laden with troops, and Kittyhawks became fighter-bombers to strike at the invasion convoy. The first Japanese landed at Milne Bay on the night of 25 August, and as soon as the Kittyhawk pilots could see to fly on the following morning they attacked the enemy trenches, barges, and supply dumps. It was the worst possible flying weather, with thick cloud at treetop height and almost continuous rain, but at least the pilots had not far to go. The Kittyhawk undercarts splashed up water like paddlesteamers as they took off from the flooded airstrip, and then it was a matter of gaining a little height, flying a couple of miles over the jungle, and pouncing on the Japanese at the water's edge.

But the Japanese were determined to die for their Emperor and they ignored their losses as they pressed closer to the airstrip. For ten long days it was a strange and terrible battle fought among the swamps and jungle, with the Kittyhawks being used almost as artillery. It was perhaps the earliest example of really close ground/air co-operation during that war. Australian veterans of the Battle of Britain flew so often against enemy positions that they lost count of their missions, and now and again had to fight off Japanese attempts to bomb the airstrip. RAAF ground crews had an equally important task as they worked in torrential rain to service and re-arm the Kittyhawks when they returned to the strip.

RIGHT TOP: RAAF ground staff prepare a Liberator for a strike over Timor (*Australian War Memorial*). RIGHT BOTTOM: RAAF Liberators over Balikpapan, Borneo, on 1 July 1945, after bombing from a height of 7,000 feet (*RAAF Museum*)

The Lockheed Ventura (*RAAF Museum*)

One of the "biscuit bombers" which supplied Australian troops in the jungles of Papua New Guinea. A number of RAAF aircraft and aircrew flew on such operations but the aircraft in the photograph is a Douglas C47 of the USAF (*Australian War Memorial*)

The battle for Milne Bay ended on 7 September 1942 when the Japanese withdrew their troops, but it was only the beginning of the long campaigns in Papua New Guinea. Australian servicemen and civilians had suffered badly during the first onslaught on the islands, and Kevin Parer, the younger brother of Ray Parer, was among those who lost their lives. Kevin was one of the civilian flyers who had maintained air communications between the isolated settlements before the war, and he continued with this work after the Japanese attacked. One day he was about to take off from Salamaua in his twin-engined DH84 Dragon when Zeros swooped on the airfield, and he jumped out of the aircraft and dashed for a trench. A few moments later he changed his mind and ran back to the aircraft, apparently with the idea of getting her into the air in an attempt to save her from destruction. He was machine-gunned before he reached her.

The Japanese landed at Lae, Salamaua, and other places along the northern coasts between March and July 1942. The whole huge area north of Australia, from Broome in the west to Thursday Island, became a battleground. Land operations were limited to the speed at which men could hack their way through the jungles of the islands, sea operations were a constant thrust and parry between Japanese invasion convoys and American fleet units, but the sky was literally the limit for the Australian, American, and Dutch airmen who fought in the area. It was the new kind of war, in which aircraft were deployed as weapons against an enemy hundreds of miles away.

The RAAF struck back hard and fast. 11 and 20 Squadrons, of Catalina flying-boats stationed at Port Moresby, were constantly in the air on reconnaissance missions and they suffered badly, including the loss of three boats sunk at their moorings by a raid on the port. They were supported by Hudsons and Kittyhawks which carried out a number of raids on the enemy footholds along the northern coasts, and as the weeks passed these aircraft were reinforced by American fighters and bombers together with more RAAF units. Despite the damage they inflicted on enemy shipping, aircraft, and shore bases they could not prevent the steady build-up of Japanese forces which in July began to advance through the Owen Stanley Ranges and down the Kokoda Trail. They were the northern part of the pincer movement of which the Milne Bay invasion was the southern prong.

RAAF and American aircraft supported the desperate and eventually victorious defence and counter-attack of 2nd AIF troops and Australian militiamen, although many of their actions had to be little better than guesswork. The ground forces were hacking at each other under the dense screen of the massive rain-forest trees and could rarely be spotted from the air. An important part of the airmen's work was that of acting as "biscuit bombers", a nickname derived from the hard army biscuits, packed in sealed tins, which were among the supplies dropped into forest clearings. Unglamorous though this task may seem, it was another example of the vital importance of air supremacy. One of the reasons why the Japanese had to retreat was because they had outdistanced their supply facilities. As the Australians advanced they received steady replenishment of food, medical stores, and ammunition dropped by the biscuit bombers.

Soon, the RAAF and USAF solved another problem of jungle war: that of delivering fighting-fit troops to the front line instead of men exhausted by battling through the jungle. A massive airlift was organised to back up the offensive on Japanese positions along the northern coast. RAAF Hudsons, USAF Dakotas, Australian commercial airliners, and B26 bombers carried soldiers, stores, ammunition, and artillery on to airstrips cleared by advance parties. When the attack began it was backed up by Bostons, Beaufighters, and A20s, while RAAF Wirraways spotted for the artillery.

The Allies encountered little Japanese resistance in the air, because their strategy had included enormous raids by USAF B17 and B26 bombers and RAAF Beauforts on enemy airfields and supporting facilities. The Japanese air force was virtually eliminated on the ground.

A brilliant example of air supremacy used to speed the land campaign was demonstrated in the battle at Wau, the goldmining town with the "uphill airfield" where Ray Parer and his friends had made so many difficult landings in the 1930s. Kanga Force, an Australian Army group, had been fighting in the Wau area for several months and was entirely supplied from the air. In January 1943 this force was driven back towards Wau by a Japanese attack, and its commander asked for reinforcements. These were assembled in Port Moresby and flown the 150 miles to Wau. The C47 Dakotas which carried them had to land on the uphill airfield and take off again in a hurry to allow others to follow. The troops jumped from the aircraft and went straight into action against the Japanese, who by that time were only fifty yards from the edge of the field. Next day the Dakotas landed field guns, while Beaufighters strafed the enemy. In a two-day battle the threat to Wau was averted.

RAAF Boston attack-bombers fly at "zero feet" as they swoop down in a whirlwind strike against the Japanese (*Australian War Memorial*)

Allied strategy in that period of the war was concentrated upon the plan to beat the Germans and Italians before turning a final effort on the destruction of the Japanese war machine, but American and Australian forces in the south-west Pacific area maintained a steady forward pressure. The bulk of the Allied forces in this theatre of war were American, but RAAF squadrons played a substantial part. The Allies had learnt very fast that the best tactics for the jungle and island war called for close co-operation between air, land, and sea forces and the flyers were used as the spearhead of combined operations designed to knock out one Japanese base after another. As usual, the RAAF ground units were as important as the flyers whom they kept in the air. Aircraftmen of airfield construction units were close on the heels of the first assault waves of American and Australian troops who swarmed ashore from landing craft. Their task was to reconstruct Japanese airfields which had been blasted by B17s, Kittyhawks, Beaufighters, and Bostons, so that they might be used for another step forward. RAAF craftsmen and tradesmen set up their workshops on the edge of the jungle, and medical personnel including RAAF nurses took care of the sick and wounded. An innovation prompted by the jungle war was an air-ambulance unit which flew casualties of all services from the battlefield to base hospitals.

A hard-working and little-noticed war was fought by the RAAF equivalent of Coastal Command. Japanese submarines penetrated as far as Sydney, and thirteen squadrons were formed to fly anti-submarine (A5) and reconnaissance patrols. A number of men lost their lives in crashes, and there was no recorded submarine-sinking by RAAF aircraft in Australian

FACING PAGE TOP AND BOTTOM: Bristol Beaufighters. The aircraft in the bottom photograph was an Australian-built machine flying over New Guinea (*RAAF Museum photographs*)

Top left and bottom left: RAF Beaufighters manned by RAAF aircrew strafe a train in Burma; Australian-built Beauforts release a bomb load over New Britain (*Australian War Memorial*)

Top right: A Beaufighter scatters the crew of a grounded Japanese bomber (*Australian War Memorial*)

285, 286

Centre and bottom right: An RAAF Bristol Beaufort; and groundcrew "bombing up" a Beaufort (*RAAF Museum*)

The famous DH Mosquito was widely used in the European and Mediterranean theatres of war by RAAF aircrew flying with their own or RAF squadrons (*RAAF Museum*)

waters, but the flyers provided an early-warning service which helped to keep the enemy at bay.

Another jungle war was being fought far from Australia. In Burma the 14th Army, which comprised British and Indian divisions, was slowly grinding its way forward against desperate Japanese resistance. The 14th Army was called "The Forgotten Army" because its exploits were frequently overlooked by the media, in those days crammed full with glamorous exploits closer to hand. But under the command of General William Slim the 14th was showing the Japanese some new tricks in jungle fighting. Like the Australians in Papua New Guinea the troops were helped by air drops, close air support, and the tactical bombardment of targets in and behind the enemy lines. About 1,000 Australians flew with the RAF squadrons on these missions and in some squadrons they made up more than half of the aircrew. The majority were men trained under the Empire Air Training Scheme.

They flew Liberators and Wellingtons on heavy bombing missions, Blenheims, Bisleys, Beaufighters, Mosquitos, Vultee Vengeance divebombers, and Hurri-bombers on low-level strafing attacks, and Spitfires as fighters. In all these aircraft they gained a resounding name for themselves and helped to gain domination of the air in the India–Burma theatre. The targets were often different from those attacked in Papua New Guinea because Burma had seaports, railways, roads, bridges, and industrial plants such as oil refineries. The Japanese were using these and the RAAF men came to excel in "beating up" railway trains and riverboats as well as ammunition dumps and troop entrenchments. One Beaufighter squadron destroyed or damaged 104 locomotives, 128 motor transports, forty-seven steamers and barges, and 837 sampans during one six-month period. The Spitfire squadrons gave the Japanese some horrifying experiences.

By that time the war affected almost every country in the world outside of Central and South America, and Australians were flying in most of them, whether as trainees, transport pilots, or on active service. Although a number of units had returned to Australia to take part in the battle against Japan, a great many Australians still flew as individuals in RAF squadrons in the northern hemisphere or in RAAF squadrons in the European theatres.

These included 451, 454, and 459 Squadrons, which took a very active part in the savage campaign being waged for possession of the eastern Mediterranean. The Italians surrendered in September 1943 and the Afrika Korps had beaten an inglorious retreat from North Africa, but the Germans held on grimly to their outpost fortresses in Crete and other Grecian islands and to their conquests in Greece, Yugoslavia, and the Balkans. The three Australian squadrons were engaged in reconnaissance, in the battles for the islands of Cos, Samos, and Leros, in bombing German bases, and in attacking the ships that supplied the island garrisons. It was a rough corner of the war in which the Australians pressed home low-level attacks against enemy convoys despite fighter support flown from islands and the mainland. The Baltimores of 454 Squadron located the convoys and kept track of them until the Marauders and Beaufighters of the other two squadrons came roaring in with rockets, bombs, and cannon fire.

By 1944 the Beaufighters of 451 Squadron comprised the only Australian fighter squadron remaining in the Mediterranean area. They were moved to Corsica, and from this island base they struck at the Germans over Italy and escorted Mitchells and Marauders on strafing and bombing attacks. Earlier in the hard-fought Italian campaign, Kittyhawks of 3 Squadron RAAF which had fought in North Africa were used to slash at enemy transport winding along the narrow mountain roads, and they took part in many other operations including the fearful battle for Monte Cassino. Australians also served as aircrew in the RAF squadrons of Wellington, Halifax, and Liberator bombers pounding the Germans as the war inched its bloody way up the mountainous leg of Italy.

It would have been hard to find a British airfield on which at least one Australian accent could not be heard. Australians flew in RAF Coastal Command, Bomber Command, and Fighter Command, and then in the special squadrons of Albemarle glider-towing aircraft formed to carry troops on the invasion of northern France. A total of 6,500 Australians served as individuals in ninety RAF squadrons. Besides these, the RAAF squadrons based in Britain displayed innumerable examples of what official citations called "contempt for danger". The "shipbusters" of a Beaufighter squadron became experts at smothering the fire of flak ships guarding German coastal convoys, by such fierce and repetitive low-level attacks that the gunners were killed, dazed, or frightened away from their weapons, while RNZAF Beaufighters swooped in to torpedo the merchant ships. On one of these operations the two squadrons hit nine ships and destroyed or seriously damaged all nine.

Before and after D-Day, 6 June 1944, air operations in the European theatre rose to a crescendo. There was a grim irony in the fact that the Luftwaffe, with which the Germans had expected to blast an entry for their invasion of Britain, was now thrown entirely on the defensive. The only air offensive left to the Germans was the one on which they were working frantically behind the lines: the V1 pilotless aircraft known as "flying bombs" and the V2 guided missiles. Otherwise the greater part of the Luftwaffe machines

TOP: RAAF groundcrew overhauling a Halifax bomber. CENTRE AND BOTTOM: Two photographs taken during the same attack by rocket-firing Beaufighters on a German convoy off the Dutch coast (*Australian War Memorial photographs*)

An RAF Sunderland flying-boat flown by RAAF aircrew straddled this U-boat with bombs and depth charges (*Australian War Memorial*)

and personnel was devoted to fighting off the round-the-clock bombing assaults on Germany. RAF and RAAF heavy bomber squadrons attacked almost every night and American Flying Fortresses were overhead almost every day. Apart from these massed attacks, swarms of fighter-bombers struck at every target within their reach.

Allied domination of the air over the English Channel and northern France was secured well in advance of the invasion. Early in 1944, what was virtually a shuttle service of fighter-bombers flew to and fro across the Channel to strike at the communications supplying German coastal defences. The French railways in the areas were pounded into a shambles of twisted iron and broken bridges, to be pounded yet again when they had been patched up by the Germans. Convoys of troops, armour, and supply vehicles dared not move in the daytime, and had to be widely scattered because any concentration drew instant attack. Just as important as these attacks were the photographic missions which enabled the Allied staff to keep a daily check on enemy troop movements and dispositions. When the invasion fleet left southern England it was preceded by wave after wave of aircraft carrying paratroops, towing gliders, or carrying the huge bomb loads to be dumped on German fortifications.

Fighter Command was reorganised into what was known as the 2nd Tactical Air Force, or 2 TAF. This was formed as the result of experience in North Africa, where 1 TAF had been formed in 1943. A combination of fighters, fighter-bombers, and light bombers, it was the most flexible and devastating weapon yet to be developed during that war of movement in which staff planners had to learn fresh lessons every day. 1 TAF was used with deadly effect against German and Italian units moving across the empty desert. 2 TAF, much larger and with newly-developed aircraft to fly alongside battle-proven Spitfires and Beaufighters, wrought havoc among an enemy struggling to manoeuvre amid the farmlands of northern France. Even the most minor movement was likely to draw destruction from the air, as Field-Marshal Erwin Rommel found when he was recalled to Germany for consultations. As his staff car sped along one of the winding roads it was spotted by a 2 TAF fighter which pounced with cannons blazing. The car ended in the ditch though Rommel escaped with comparatively minor wounds.

451, 453, and 464 Squadrons RAAF played a prominent part in 2 TAF operations. The squadrons flew fighter-bombers in attacks on targets ranging from radar sites to V-weapon installations. 464 Squadron took part in one of the most expert demonstrations of precision bombing during the war: the attack on Amiens prison in north-west France. The Germans, goaded by the rapidly-increasing harassment of their occupation troops by French partisans, who were supplied from the air, had rounded up 700 Frenchmen who were accused rightly or wrongly of being members of the Free French resistance forces. Whether guilty or innocent their fate was certain: either dispatch to the German slave camps or death by firing squad.

The predicament of these men was known in London because of the secret radio network which continued despite the most savage German reprisals. The Mosquitos of 464 Squadron had already been used to blast Gestapo headquarters in various European cities, but this called for something new. It was necessary to hit the gaol in such a way that the resistance men could escape without suffering in the raid.

464 Squadron was one of three squadrons chosen for the attack. The others were RNZAF and RAF units. The pilots were given a careful briefing which included a demonstration, on a rough scale-model of the gaol, of the exact point where their bombs should be placed. The whole attack was planned with split-second timing, with some pilots given the task of flying air cover and others that of smothering flak emplacements near the gaol. The aircraft which were to hit the prison were armed with 500-pound bombs.

On a grey day in February 1944 the three squadrons

An Australian-built DH military glider used for carrying troops and equipment (*RAAF Museum*)

swept low over the Channel and flashed across northern France. They hit Amiens so quickly that the Germans had no chance to respond. The first Mosquitos to strike were those of the New Zealand squadron, dropping their bombs squarely against the outer wall of the prison. The dust and fumes were still billowing above the great holes blasted in the walls when 464 Squadron Mosquitos slashed overhead. Their bombs shrieked down to explode against the end of the main cell block, and as the walls came tumbling down the French prisoners took advantage of this incredible last-minute reprieve. Those who could scramble out of their cells dashed for the holes before the astounded Germans could rally to stop them. About 260 Resistance fighters got clear of the prison. Unluckily, a number were recaptured and shot out of hand, but the majority managed to find secure hiding places.

Such a moment as this was a highlight on what became a day-by-day succession of attacks on every tactical target that could be found. After the invasion bridgehead was established the three RAAF squadrons moved to France, where Flying Officers Harry Turner and Mervyn Partridge had another opportunity to display the almost unbelievable accuracy of Mosquito bombing. They were told to hit a German Army headquarters at Choret, and with another Mosquito they hit the building squarely with 500-pounders on their first pass at the target.

Other RAAF men in 2 TAF units flew Typhoon "tankbuster" aircraft. The Typhoons were armed with eight rockets each having the explosive power of a six-inch shell, and they provided invaluable tactical support to advancing ground troops. Whenever these were threatened by the deadly German Tiger tanks, either in open attack or dug in as artillery, they whistled up the Typhoons. One pass by a Typhoon meant goodbye to a Tiger. Two of the rockets would blast one of the dreaded Tigers into a flaming mass of crumpled steel.

Typhoons became "flying artillery" with overwhelming speed and mobility. The Germans learnt very quickly that it was a waste of time to assemble a couple of dozen Tigers for one of those hardhitting counter-attacks that had been so devastating to British armour fighting in North Africa. A group of tanks was certain to attract the Typhoons, as the Germans found to their cost when they mustered upwards of thirty Tigers to attack the Americans at Mortmain. Typhoons with RAAF pilots destroyed six as the

Tigers rumbled forward, and cleaned up most of the remainder during the tank battle with the Americans.

The distinctive silhouette of the Mosquito and the sudden flashing appearance of the Typhoon tank-buster became all too familiar to the Germans in northern Europe. The Australians in 2 TAF would have a go at anything, mobile or immobile. As the Allies advanced along the invasion front, one of the most frequent targets for the airmen flying ahead of the troops were the strange "ski-run" installations set up by the Germans. They were launching sites for the V1 flying bombs which the Germans sent against London, and in the counter-attack against this "secret weapon" a number of Australians hit the "doodle-bugs", as they were christened by Londoners, while they were still over the Channel or the southern counties. These RAAF men were flying with the group known as Air Defence of Great Britain, or ADGB: a formation instituted at the same time as 2 TAF. The latter struck at the enemy on his own ground: ADGB made sure that he couldn't strike back.

ADGB had had a fairly peaceful time for a few months prior to the invasion, but when the doodle-bugs began to swarm the fighter squadrons had to "scramble" in a manner reminiscent of the Battle of Britain. The stubby-winged, almost innocent-looking, little pilotless aircraft, with their exhausts flaming astern as they drove steadily towards London, were no joke. They were winged rockets, each loaded with 1,874 pounds of high explosive. A total of 10,500 of these weapons, officially known by the Germans as Fieseler FZG 76A, was sent against Britain. About 4,000 were destroyed by the defenders, but those which got through did enormous damage and caused many civilian deaths.

The fighter pilots of ADGB had to develop new tactics to counter this threat. When they destroyed a V1 in the air they found themselves buffeted by the blast of the explosion and in danger from flying fragments, so if they could catch a doodlebug over the sea they used a technique first developed by Flying Officer K. R. Collier, RAAF. He was flying a Mark XIV Spitfire, a refinement of the now-famous aircraft with modifications that included a five-bladed propeller. Collier, with two other Spitfires, had attacked a V1 and knocked out its engine, but it had begun to glide towards a coastal town. He edged his Spitfire closer and closer to the flying bomb until one of his wingtips slid underneath its wing, so that he was able

The first aircraft to land on Kamiri Strip, Noemfoor Island (*Australian War Memorial*)

to divert its glide and send it down to explode harmlessly away from the town.

The flying-bomb attack was serious enough, but it would have been a lot worse if ADGB and 2 TAF pilots had not hit the launching sites and the bombs themselves. When the Germans switched to V2 rockets, the grim precursors of space-age warfare, there was nothing that conventional aircraft could do except try to find the launching sites and destroy them.

At last the war in Europe came to its chaotic end and the Allies began to switch all their efforts towards the elimination of Japan. Australia now had its own 1 TAF, formerly 10 Group, commanded by Air Commodore A. H. Cobby. 10 Group was formed at Noemfoor as part of the island-hopping strategy which drove the Japanese from one stronghold to another, and when it was re-formed as 1 TAF the force comprised two fighter wings, an attack wing, and two airfield construction wings. The latter units were of extreme importance in the island war.

For a few months morale in 1 TAF was rather low. As the iron ring tightened around Japan, and the Americans poured men and material into the area, the RAAF pilots began to feel that they were fobbed off with minor missions and unworthy targets. Under the direction of the US 13th Airforce they were given the tasks of cleaning up Japanese units which, in General Blamey's phrase, had been left to "wither on the vine". The RAAF men flew hundreds of missions against enemy coastal shipping, defences, troop positions, and other targets on the scattered ring of islands north-west of Australia, and a steady stream of men lost their lives on these attacks, but the feeling was that they were not really necessary. The really big actions were taking place further to the north, where the Americans were recapturing the Philippines and overwhelming Okinawa and the other islands which would be springboards for the final assault on Japan. Even if enemy units in scattered island outposts were left completely unscathed this could have little effect on the final outcome, because it was obvious that they were being cut off from supply and reinforcement and were powerless to mount a counter-attack of any significance.

Senior RAAF officers threatened to resign and the situation seemed near crisis point, but began to settle down when the promise of large and important action loomed on the horizon. This was the assault on Tarakan and Borneo to the north-west and New Britain and Bougainville to the north-east.

Australian soldiers, sailors, and airmen took a substantial part in these operations and 1 TAF dominated the air. The RAAF men flew Liberator bombers and many types of fighters and fighter-bombers as they took part in the combined operations that cut up the Japanese piecemeal in one jungle battle after another. The tactics of close infantry support developed in North Africa and Europe were pressed home in the tropics and Australians were still fighting in the air when the first atom bomb was dropped in August 1945.

Like the other armed forces of the Commonwealth the RAAF had grown gigantically during the second World War. A total of 216,900 men and women had enlisted in all branches and of this number 171,511 were still in service. 6,447 died on active service.

15

ANA, TAA, EWA, and Others

Government attempts to monopolise Australian aviation aroused furious opposition. But their new airline, and several others, soared to success in the post-war period

When the guns ceased fire on 11 November 1918, Australian aviation fell into a sharp decline from which it was rescued only by the dedication of such individualists as Norman Brearley and Hudson Fysh. Governments were simply not interested in the peacetime potential of flying machines.

But the Australian government already had shown that it intended to take over the nation's aviation when, on 6 August 1945, the atomic bomb blasted the remnants of Axis hostility into history.

Labor Party theorists of that era were convinced that anything done by capitalists could be done better by socialists. Nationalisation was in the air. The Labour Party of Great Britain, which avalanched into power in August 1945, planned to nationalise steel, railways, banks, and medical services. The Australian Labor Party had decided to nationalise airlines.

Late in 1944, Acting Prime Minister F. M. Forde told the House of Representatives in Canberra: "The government has decided that a wholly Government-owned statutory authority be formed to take over the operating and maintaining of interstate airlines . . . the effect of this decision is that the Commonwealth Government will take over all interstate operations."

At a time when the Allies were straining every nerve to pound the Axis into submission, this announcement was hardly headline news and it went virtually unnoticed by all except those who were vitally concerned. But when the war effort began to swing over towards peacetime reconstruction, the argument came out into the open.

Labor doctrinaires said it was logical, in a country the size of Australia, to establish one integrated airline to serve everyone. They claimed that the cost of operating the huge new aircraft developed during the war was beyond the reach of Australian companies, and that State and Federal governments would in any case have to become involved in the huge cost of new airfield facilities. Those who paid the piper should call the tune.

Free enterprise spokesmen said the proposal was absurd. They noted that government-owned enterprises had a sorry history of inferior service to the public, allied to massive operating losses. Look at the railways, they said. Look at the shipping companies run by governments. The only way to ensure good service was by open competition. Those who gave good service would be successful. Those who did not would fall by the wayside.

And, they asked, why rock the boat when it was sailing so well? The struggling days of the 1920s and early 1930s were long in the past. Australian airmen and airline operators could point to an enormous bulk of successful and profitable achievements. All of them were men who had backed their dreams with hearts, minds, and bank accounts. Would bureaucrats—more often than not without incentive—be able to say the same? After all, the history of government involvement in Australian civil aviation was pathetic.

Despite the turmoil of war, Australian civil airlines were in a comparatively vigorous condition and their executives were planning to re-equip and extend their services. Qantas, the "Front Line Airline", had given superb service during the war and was

Qantas Super Constellation L1049 flying over Sydney. Qantas introduced Super Constellation aircraft on to their world routes in 1954 and by the end of January 1959 the company employed sixteen of these famous aircraft. They flew on main trunk routes from Sydney to London through San Francisco and New York; to San Francisco and Vancouver; to Manila, Hong Kong, Tokyo, Johannesburg, and on the Bird of Paradise service from Sydney to Port Moresby. The "Super Connie" carried eighty-two passengers at a cruising speed of 340 m.p.h. (*Qantas Airways photograph*)

A Qantas 747B over Sydney Harbour Bridge. Qantas now employs the mighty Jumbo jets on their international services. (*Qantas Airways photograph*)

working on a scheme to use Liberator and Lancastrian aircraft, modified for passenger-carrying, on the route to England via Singapore. This service commenced in April 1946 and made an old dream come true: that of using wheeled aircraft, instead of flying-boats, on the long flight to London.

Within Australia, the scene was dominated by Australian National Airways although it had vigorous competition. The DC3s ordered by ANA in the late 1930s had logged hundreds of thousands of miles during the war. This incomparable aircraft had rarely failed its owners, and apart from maintaining domestic air services the ANA DC2s and DC3s had been commandeered from time to time to fly coastal reconnaissance and to help in the Papua New Guinea theatre of war. They evacuated refugees and participated in the airlift of troops and support equipment.

The ANA aircraft were the only long-range machines in Australia when war broke out, and from time to time they were commandeered to fly reconnaissance for troopship convoys leaving Australian ports. They patrolled as far from land as possible, while disappointed passengers were soothed by ground staff who were not allowed to tell them why the aircraft had been suddenly removed from service. For most of the war ANA Douglas aircraft averaged seventeen hours flying time a day. The Americans, who averaged only eleven-and-a-half hours a day, sent a technical team to Australia to discover how it was done. Among other things they found that ANA ground crews had reduced the time taken for a complete engine change, including the compulsory twenty minutes test flight, to under four hours. The changes were carried out between three a.m. and seven a.m.

During the war, ANA operated one of the very few DC5s ever built. This model was a high-winged monoplane designed for short-haul services in the USA. The American army brought one to Australia but found little use for it because of its short range. ANA hired it for the Tasmanian service, on which it carried twenty-four passengers in "austerity" seats.

By the end of the war, ANA had only nine aircraft in operation but the company had asked for quotes from the Douglas Aircraft Corporation and in early 1946 seven ANA aircrews were sent to America to take delivery of seven DC4 Skymasters. The company ran a recruiting programme which skimmed the cream of RAAF pilots and ground crew who wanted to stay in flying, and some of these found themselves flying or servicing RAAF and USAAF C47s. Many of these famous aircraft were available as "disposals" and the company bought several for conversion to passenger machines.

ANA had big plans for post-war expansion, including extension into overseas routes, and its DC4s were chartered by British Commonwealth Pacific Airlines to fly the trans-Pacific route from 1946 to 1948. In that year the latter company inaugurated its own service. ANA flew 255 crossings of the Pacific but these were the only regular passenger flights which it made outside Australia.

The company was a solidly integrated and independent organisation, with its own workshops capable of keeping the aircraft up to Department of Civil Aviation requirements and a national airfreight and passenger booking organisation. There was certainly nothing the government could teach it about running an airline.

Other airlines, large and small, were of a similar opinion about their own organisations. They included Ansett Airways, whose Lockheed Electras were the only aircraft, apart from those of ANA, to fly on major interstate routes; Butler Air Transport and New England Airways, both in New South Wales; MacRobertson Miller and Airlines (WA) Ltd, founded in 1935, in Western Australia; Guinea Airways, which flew an Adelaide–Darwin service together with regular flights from Adelaide to Broken Hill and South Australian country centres; and Queensland Airlines Pty Ltd. This company, founded in 1929, was a subsidiary of Butler Air Transport. From its initial service between Brisbane and Toowoomba it had extended to serve a number of coastal and inland towns in Queensland.

These were the larger airlines, and there were several smaller ones such as Inland Air Services Pty Ltd of Brisbane, later to be absorbed by Queensland Airlines, and Connellan Airways of Alice Springs. This organisation was perhaps the last to bear any resemblance to the airlines started "on a wing and a prayer" in the early 1920s. It was founded by E. J. Connellan, a schoolteacher from Victoria, who took to the air. In 1938 he was commissioned to carry out a survey of the Northern Territory to assess its grazing capacity, and the contacts he made during that period inspired him to start his own airline to serve the isolated stations and townships.

He began work in July 1939, using two Percival Gulls to fly out of Alice Springs airstrip on a fortnightly service to Mount Doreen, Granites, Tanami, Inverway, Victoria River Downs, and Wyndham. It was real seat-of-the-pants flying, with primitive airstrips and minimal ground assistance to aerial navigation, but apart from these regular flights the pioneer airline was open to commissions to "drop in anywhere". A number of these came from the Flying Doctor service, still known as Australian Aerial Medical Services, who paid Connellan a retainer of £1,000 a year. He also received an annual subsidy of £1,500 from the Commonwealth government. He flew one of the aircraft, and kept the other in reserve, and soon became famous among the outback community as he flew mails, passengers, doctors and patients, spare parts and newspapers to mining camps, cattle stations, and missions.

One of the great bomber fleet that pounded enemy targets in German-occupied Europe: an Avro Lancaster (*RAAF Museum*)

It was a rugged life but Connellan was one of the true bush pilot breed, with the skill to service his own aircraft and the ability to land on an airstrip that had been hacked out of the bush only an hour or so before he appeared. The war made him busier than ever, especially after Japanese bombers raided Pearl Harbour. He surveyed sites for new military airfields and carried out the unglamorous but important service of flying cattle tick serum to the station owners, to keep their cattle free from disease and enable them to feed the huge Allied forces massing in the area from Alice Springs to Darwin. He flew alone until 1943, when he took on another pilot and a mechanic.

Even though transworld air services had become an accepted fact, Australia was still close to its pioneer aviation days. Trailblazers like Horrie Miller, Ivan Holyman, Hudson Fysh, and E. J. Connellan were still very much in the picture. The work they had done, the risks they had taken, and the air services they had established were under threat of extinction by the Australian National Airlines Bill, voted through Parliament in July 1945 and given royal assent by the Governor-General, the Duke of Gloucester, on 16 August.

The Act spelled out government intentions without equivocation. It enabled the government to run an airline from any place in a State to any place in another State, any place in any Commonwealth territory and any place in Australia outside that territory, and any place in any Commonwealth territory to any other place in the same territory. And, when the government had established an airline "adequate to meet the needs of the public for interstate or territorial transport by air", other airlines had to stop running in competition.

These provisions would bar private enterprise airlines from interstate services and eliminate them from the Northern Territory. They would be limited to services within the borders of the State in which their headquarters lay.

Their howls of outrage filled the silence left by the guns of war. The Press thundered against another instance of creeping socialism. MacRobertson Miller, Guinea Airways, and ANA took legal advice, and wasted no time in issuing writs against the Commonwealth. On 29 October 1945 the contestants began to hammer out the case before the High Court of Australia, and in December the court handed down its decision. The judges found that the Commonwealth was empowered to establish its own airline, but that it could not take over any commercial airline nor refuse licences except on grounds of safety.

The Commonwealth decided that half a loaf was better than no bread, and continued with its plans for a national airline even though it would have to compete with private enterprise. The Australian National Airlines Commission was established, and Prime Minister Chifley chose Arthur Coles as chairman.

For a Labor administration this was a remarkable move; Arthur Coles was a symbol of private enterprise. He and his brother had started G. J. Coles, one of the most successful groups of chain stores in Australia. Nevertheless, Coles was an ideal choice and he made his attitude clear from the outset. "It has to be run as a commercial enterprise," he said. "I would not be interested in taking the chairmanship of a government monopoly. It was my opinion, and still is, that a monopoly of any kind, whether it is a government one or a private one, does fall down on its service. I am a great believer in competition for business keeping air services on their toes."

Coles' business background and forthright attitude ensured that the government airline was run from the start as a proper profit-and-loss operation, instead of being able to conceal its financial status behind the hazy "cash accounting" methods of other government operations.

Arthur Drakeford named the other members of the Commission early in 1946. They were A. C. Joyce, the Senior Assistant Secretary to the Treasury; Captain Edgar Johnston, Deputy Director of the Department of Civil Aviation; D. (now Sir Daniel) McVey, Director General of Postal Services; and W. C. Taylor, a Sydney lawyer.

They met for the first time on 13 February 1946, to start an airline from the ground up. They had no aircraft, no hangars, no staff, no name. Someone suggested Trans-Australian Airlines, and this was accepted though the "n" was dropped on 11 March. The aircraft were easy to acquire. Scores of RAAF DC3s and C47s sat on the tarmac, their occupation gone now that RAAF personnel were being demobbed as fast as the authorities could process them. Skilled pilots and ground crew were to be had for the asking, and men who had worn RAAF blue in Britain, North Africa, and many other theatres of war donned the darker blue of TAA.

Executive positions had been advertised and there was a host of applications: 53 for the job of General Manager, 166 for Secretary, and 357 who'd take anything going. Many of the applicants were among the cream of Australian aviation: men like Lester Brain, who had had years of experience with Qantas. The Commissioners didn't have to look much further when they saw his name on an application, and he took up his appointment as General Manager on 3 June 1946.

S. H. Crawford was named Secretary, Group Captain K. M. Hampshire appointed as Operations Manager, R. E. Pope became Chief Accountant, and J. L. Watkins was Technical Superintendent. Hampshire did not take up his appointment, and the position was filled by Group Captain John Hyland. He was already on the staff of TAA, as chief of the National Airlines Training School at Point Cook RAAF base. Once more Point Cook was being used as a training ground for a pioneering airline venture, following the precedent laid down by Major Norman Brearley when he selected his pilots there in 1921.

The appointment of John Hyland was a good one and he later became General Manager of TAA. John Chapman, who succeeded him at Point Cook, became Assistant General Manager.

The first executives were career men who were to stay with TAA for the rest of their working lives, but they were far from possessing what is unkindly known as the "public service mentality". They were men dedicated to flying, like John Watkins who had been one of a group of Adelaide University students which built gliders back in 1929, and flew their *Zegling* at Tapley's Hill gliding camp in 1930. Between them, the first TAA executives moulded and fashioned a new airline during the challenging years when aircraft were changing from piston engines to turbo-prop to pure jets.

Liberator bombers of RAF Eastern Air Command Strategic Air Force ready for an attack on Mandalay, Burma. Almost every aircraft of RAF 99 Squadron of this command had at least one Australian crew member (*RAAF Museum*)

Arthur Coles said, "The new airline caught the imagination and appealed to thousands of young men and women whose pioneering spirit was still very much alive after the war. They all wanted a place in this exciting adventure."

In July Lester Brain recommended a daily air service between Sydney and Melbourne, to begin early in September. "We can go twice daily later in the month," he added.

By August the crews were all kitted out in the merchant marine style blue uniforms. The captains' top salaries were fixed at a top of £1,150 ($2,300) which was about four times the basic wage in those days. Miss Helen Somerville joined as the first Air Hostess Superintendent.

On 9 September 1946 the first scheduled TAA flight left Laverton, near Melbourne, for Sydney. Captain J. A. Hepburn was at the controls of the silver DC3, named *Hawdon*, and hostess Veni Vernon looked after the VIP passengers who included Arthur Coles and Arthur Drakeford. Another group of VIPs, including the Premier of New South Wales, greeted them on arrival at Mascot.

Brain's plan for regular services in September did not come to fruition and the first regular public services began on 7 October. But even with a daily

The Commonwealth Aircraft Corporation designed this all-weather fighter in 1951–52 but the design never got any further than the model shown in the photograph (*Commonwealth Aircraft Corporation Pty Ltd*)

and then a twice-daily service between Melbourne and Sydney, and a gradual extension to other ports, the new airline could not break through the profit barrier. The first nine months of operations showed a loss of £500,000, an astronomical sum in those halcyon days when the basic wage was only about £250 a year. Critics of the government airline complained loud and long. A national poll showed that thirty-eight per cent of Australians wanted to kill the infant airline, but they were outvoted by forty-eight per cent who were still willing to give it a go.

TAA was, by mid-1948, carrying more than a third of the nation's internal air traffic. Its fleet comprised two DC4s, nineteen DC3s for passengers and four for cargo, six more DC3s undergoing conversion and five new Convairs almost ready for delivery. For the previous twelve months the airline had been operating services within Tasmania and Queensland, with the permission of the governments of those States.

The miraculous DC3 was still the prime carrier on Australian domestic routes. She is recalled with fond nostalgia by countless old-timers, who remember the coughing start of the Wright Whirlwinds, the glisten of morning sun on engine cowl and wingtip as passengers looked down to see their home town disappear beneath gently inclined aerofoils, the thin, slow stream of inky oil which dribbled back from the front of the cowls as the engines settled into their stride. The DC3 was a lady, a hardworking lady. A few veterans, now more than thirty years old, are still in service.

The DH84 Dragon, a twin-engine biplane, was another well-loved machine which flew in the TAA fleet as well as with other carriers. Her rounded single-pilot cockpit made her look, from the outside, like the world's biggest grasshopper. Her two Gypsy Queen motors, with their tiny two-bladed props, seemed inadequate to lift the monster until it was remembered that her structure was only wood and fabric. She rose from the field with great docility and climbed with imperturbable slowness and majesty. DH84s entered TAA service in 1949, though these machines had been flying on the barren Nor'-West routes and the East-West run in the 1930s.

An Australian-built aircraft, the de Havilland Drover, enabled TAA to extend its Queensland services into the remote Gulf country. The seven-passenger Drover was a real workhorse, ideal for rugged outback conditions, and was seen often on the TAA network which by the end of 1951 extended over 21,917 miles.

By that time Australian airlines were looking abroad for the big glamour aircraft to carry them into the new age of aviation. TAA took delivery of its first Convair 240 from the makers in San Diego, California, in August 1948. She was VH–TAQ, the aircraft which introduced pressurisation to Australian air travellers and represented a milestone in the growth of TAA. The first Convair was followed by others and then by a stream of new designs, each of them leaving the others far behind during an era when it seemed that every year brought something bigger and better in the way of passenger aircraft. The Vickers Viscount 700 series appeared in 1954; the Viscount 816 series, labelled "Super Viscounts", in 1959. That was also the year of the delightful Lockheed Electra prop-jet and the Douglas DC6B, which strengthened the trend away from British and toward American equipment. Lockheed had a well-established Australian office and sold successfully to the RAAF and Qantas. The Lockheed Neptune P2V-H maritime patrol reconnaissance aircraft are only now being phased out of service, giving way to other Lockheeds—the P.3B and P.3C Orions—and the graceful Super Constellations flown by Qantas for so many years make aircraft enthusiasts settle into a haze of fond reminiscence.

The constant flow of new designs made life difficult for airline operators. They'd hardly finished paying for their equipment before they had to start all over again, or run the risk of being thought obsolescent. Luckily the public was showing none of its pre-war reluctance to risk itself in the air, and revenue from freight and passengers showed a steady climb. In 1949 Drakeford announced that TAA was budgeting for a profit in 1950, but when that happy day had come the government which established the airline had been voted out of office.

The galaxy of new aircraft also made it possible for a carrier to steal a march on his competitor. Ivan Holyman did this in 1953. His plans for post-war expansion of ANA had to some extent been blunted by TAA, and his hope of an overseas airline had not materialised, although ANA machines flew migrants from Italy and the company became involved in the formation of Cathay Pacific Airways and Air Ceylon. ANA was nevertheless a growing giant. Figures released on the airline's twenty-first birthday, in September 1953, were astronomical. They showed that since ANA began operations, as Holyman Airways, it had flown 180 million miles in 1.24 million hours, and carried more than 5.5 million passengers and over 31 million pounds of mail. This was a considerable headstart on TAA, but, in that anniversary year, Ivan Holyman staggered the airline industry by cancelling a provisional order for Viscounts to re-equip his fleet.

People painted gloomy pictures of ANA fading into obscurity while TAA soared onwards and upwards in its glittering new Viscounts, until Holyman revealed that he was buying two Douglas DC6As and two DC6Bs. He and his brother Victor had been the first to introduce Douglas aircraft on to Australian routes, with their DC2s bought in 1934, and since then ANA had used DC3s and DC4s. Now, the airline turned to Douglas once again, although new DC6s were in short supply and Ivan Holyman had to buy the "A" model from National Airlines of America. New "B" type aircraft were delivered by Douglas in 1955.

Holyman told *Aircraft* magazine, "I do not, in any way, discount the Viscount, which is a very good machine. I am of the opinion, however, that the public will like these DC6s better, and that their economics, utility and versatility will make them a better buy, viewed in relation to our continuing

A Douglas C47 Dakota with RAAF markings. This aircraft was the "grand old lady" of civil and military aviation. A number were converted for use on Australian domestic airlines (*RAAF Museum*)

policy of developing both passenger and freight traffic side by side.

"One of their strongest features in our view is that their long range, combined with a high payload and high cruising speed, will give them a lasting and important defence value. We cannot, in the light of our experience and Australian sentiment, forget that aspect."

He was referring to the times during the second World War when ANA aircraft had been used for many kinds of military purposes. Luckily the DC6s did not have to be commandeered to help combat another threat to Australia, but Holyman was right when he said that they would be popular with the public. They could carry fifty-eight passengers at cruising speeds above 300 m.p.h., using conventional Pratt and Whitney radials instead of the Rolls Royce turboprops fitted on Viscounts. The "A" types carried 3,000 pounds of freight, but the "B" types could double this payload because of their wider freight doors which enabled mechanical freight handling and the loading of larger quantities during short stops at airports.

The year in which Holyman introduced the DC6s to Australia was also that in which the Airlines Agreement Act of 1952 began to take effect. This, with subsequent amendments, and together with the Airlines Equipment Act of 1958, was designed to rationalise competition between TAA and ANA and to ensure that both airlines had comparable, though not necessarily identical, aircraft fleets.

This legislation, broadly known as Australia's two-airline policy, was introduced by the Liberal–Country Party government which succeeded Labor for twenty-three years in office. The "rationalisation" has been the subject of frequent criticism, but it has, on the whole, worked well for the public. However, one must wonder why aircraft of the two competing lines must take off within minutes of each other, when an adjustment of the timetables would provide a more flexible service.

The purchase of the DC6s, together with spares and ancillary equipment, cost £3 million, but the government helped with a loan and by releasing American dollars, then the world's prime currency, so that the deal could go through. At the same time, and as part of the Airlines Agreement Act, the structure of TAA was changed so that the airline had to pay the same taxes as a free enterprise concern. Until 1952 TAA had received £400,000 a year for carrying the mails, in return for which it had had to guarantee to give the mail priority over any other operations, but the new Act gave a share of the mail service to ANA. The new agreement was a fair modification of the Australian National Airlines Bill, and since then the two airlines have enjoyed—or at least put up with—"level pegging".

TAA had been going ahead dramatically. Naturally it enjoyed all government patronage. This was perhaps equalised by the fact that businessmen tended to favour the commercial competitor. Still, as time passed this distinction faded. In May 1947 the airline opened a service from Sydney to Canberra, Mildura, and Adelaide, and in July of the same year it began a regular service from Brisbane up the Queensland coast as far as Cairns. By 1949 it had taken over the domestic services still operated by Qantas, which then concentrated on overseas flying, and was serving other Queensland routes. Guinea Airways surrendered its Adelaide–Darwin service to TAA in 1948.

Australians were becoming slightly blasé about air travel, though the occasional bad crash sometimes made people think twice about trusting themselves to the air. Twenty-five died on 10 March 1946 when an ANA DC3 crashed at Hobart, and ANA suffered another disaster on 2 September 1948 when thirteen perished in a DC3 which strayed over 100 miles off course en route from Brisbane to Sydney. Only two months later the passengers of yet another ANA DC3 had a miraculous escape when the aircraft crashed on Mount Macedon in Victoria. The pilot and co-pilot were killed but the passengers and hostess were able to scramble out just before the aircraft exploded. On 2 July 1949 a DC3 owned by MacRobertson Miller Airways crashed near Guildford airfield, outside Perth in Western Australia, and the crew and passengers died. Many of these crashes could be attributed to pilot error, though the unfortunate pilots of the ANA DC3 which drifted so far off course had some mitigating factors. In the subsequent enquiry, the Department of Civil Aviation was rapped for ineffective control of aircraft on the Brisbane–Sydney route, especially since the DCA radio equipment had been unable to keep contact with the DC3 before it crashed.

Air disasters have never checked progress in the air, and the airlines overcame the bad publicity by increasing their efforts to improve services and to make passengers feel cared-for from the moment they booked on a flight. A new air service opened every few months and the airlines leaned over backwards to attract patrons and to please them in the air. ANA and TAA were the pacesetters, but there was plenty of vigorous competition. MMA was going strong in Western Australia, Butler Air Transport Pty Ltd took over New England Airways in 1948 and maintained a flourishing service in northern and western New South Wales, and Ansett Airways Pty Ltd flew the only interstate services apart from the two giants.

Reginald Ansett, who entered the transport industry as the driver of a hire car, was moving steadily from strength to strength. His airline had worked steadily during the war, and his Ansett Transport Industries Pty Ltd had interests in many parts of Australia. Among the air services which Ansett pioneered were Barrier Reef Airways, of Brisbane, which was among

TOP: This Bristol M31 Freighter was among the new aircraft seen in Australia soon after the second World War (*RAAF Museum*)

BOTTOM: The Vickers Viking, adaptable to either civil or military aviation purposes, was another of the postwar models (*RAAF Museum*)

the first to fly regularly to the island holiday resorts that soared in popularity after the war, and a flying-boat service from Sydney to Lord Howe Island. This continued until the 1970s and was among the last of the world's air services to carry passengers by flying-boats. Ansett's superb four-engine Short Sandringhams were a dramatic sight as they took off, across the yachting lanes, from Sydney Harbour.

Australia was still a land of pioneers, and the quarter-century after the end of the second World War was an era in which pioneers flourished. Among them was a pioneer airline: East-West Airlines of Tamworth, New South Wales.

It began in a manner which had become almost traditional with infant Australian airlines: the use of war-surplus aircraft to make a gallant attempt to break into the big time. Men had done this at the Great War's end, and in 1945 three New South Wales men teamed to make yet another aviation beginning. They were Basil Brown of Tamworth; Ted Wilkinson, an optometrist of Inverell; and Arthur Yeates, a motor mechanic who was also from Inverell. All of them were interested in flying and they thought it was time that Tamworth, as one of the largest country centres in New South Wales, was given a better aerial link with other parts of the country.

Tamworth saw its first aircraft in 1919. It was an Avro 504K two-seater biplane, flown by Captain

A photograph taken at the Tamworth Air Pageant on 19 November 1949. The tiny two-seater, original registration VH-UTQ, was known as a "BA"

Holden and Sergeant Murphy on one of the Peace Loan tours of country centres. 1,500 people turned up to watch a display of aerobatics, but then the Avro flew away again and Tamworth returned to the railway era. On 23 June 1947 a blue and silver Avro Anson landed near the seven-furlong barrier on Tamworth racecourse, then turned and taxied back across the field to the buildings left by the RAAF which had used the space as an airfield during the war. East-West Airlines was in business.

It began as a fairly primitive service. John L. Rentell, the company's first pilot, told the *Northern Daily Leader* when he retired in 1964 that, "It wasn't unusual for me to have to make out the tickets, weigh the passengers and load the plane and then fly it." The Avro Ansons carried only seven passengers and had no cabin heating, so it was a cold flight in winter. The airport facilities consisted only of the temporary buildings left by the RAAF, and passengers huddled round a stove before they took off on a frosty morning flight to Mascot.

John Rentell had been flying since 1935. His war-time duties as an RAAF pilot included a spell as Field-Marshal Sir Thomas Blamey's personal pilot on the Papua New Guinea area. When he was demobbed he worked in an electrical warehouse for a while, but he was delighted to be given the chance to fly again. He was particularly happy with East-West's choice of aircraft: four Avro Ansons.

The Anson was one of the all-time greats of British aircraft construction. It had been designed by Sir A. Verdon-Roe who, like so many of the early flyers and designers, had been interested in aircraft from childhood and spent most of his boyhood playtime in making and flying model aircraft. His Anson first took to the air in 1933. In 1960 the Director-General of Civil Aviation, Mr D. G. Anderson, said, "I regret to see the passing of what was in its day a fine aircraft" when he announced that Ansons would be grounded for safety reasons and removed from the aircraft registry on 1 July 1962.

Thousands of Ansons flew in peace and war. The twin-engined monoplanes were built in traditional style with wooden frames and fabric fuselages, but nevertheless had an amazing strength and resilience. At Bendigo in 1959, two Ansons were flying together when one of them dropped a little too low and crashed into the top of the other. The two aircraft locked together and the pilot and co-pilot of the lower aircraft bailed out in the face of what seemed to be inevitable disaster. The co-pilot of the upper aircraft also hit the silk, and the pilot was about to follow him when he realised that, by using his own controls, he could fly both aircraft together. He coaxed them into a landing and walked away from the result.

East-West Airlines prospered rapidly. In 1950 they added Lockheed Hudsons to their fleet, and in 1953 the company bought its first DC3 from Qantas. In 1959 they bought their first Fokker Friendship from Holland. The aircraft was flown from Holland to Tamworth by EWA pilots, in an operation organised by John Riley, the present general manager of the company. Friendships became the backbone of the EWA fleet.

The history of East-West Airlines gives a vivid portrayal of the way in which a private enterprise airline could be built into a solid organisation during an era in which the cost of operations escalated enormously and, in many parts of the world, governments were probing more of their fingers into the aviation pie.

The clues to the success of EWA must lie in two directions: strong and imaginative management and strong local loyalty. This has been a two-way loyalty in which EWA has served the area well. The airline had been established for only three years when the north-west of New South Wales was swamped by the disastrous floods of 1950. EWA's Ansons and Lockheeds dropped fodder to starving stock and relief supplies to isolated homesteads. One Lockheed flew out accumulated wool, to catch the Sydney sales.

A similar relief operation was carried out in 1955, when EWA had to make a hurried evacuation of its original airfield, threatened by rising floodwaters, and move to the new field which was still being prepared. For ten days in February of that year the EWA aircraft flew around the clock to carry mail, passengers, medical supplies, and food to communities isolated by the floods.

The growth of EWA has been a multilateral operation and the airline has worked closely with the Department of Civil Aviation in the improvement of airfields over the routes which it now flies. The airline serves thirty-one ports from Hobart in the south to Maroochydore in Queensland, and ranging as far west as Alice Springs. It has its own apprenticeship scheme and hostess training establishments, and a complex of workshops and administration and service buildings in various ports.

The value of the airline as a "commuter" service between New South Wales country centres has been a mainstay of its development, but apart from this service it has become solidly involved in holiday travel, especially to the Gold Coast, Tasmania, and

The Avro Anson was one of the sturdiest and most generally useful of the light twin-engined aircraft built during the 1930s and 1940s. After sterling wartime service, a number were converted for use on Australian internal airlines (*RAAF Museum*)

Alice Springs. Over one ten-year period, from 1963 to 1973, most of the airline's operations took a huge upward leap. The number of passengers carried increased from 11,723 to 18,109, net profit rose from $75,340 to $188,264, the mileage flown rose from 1.7 million to 3.3 million, air cargo and air mail increased from 2.3 million pounds to 4.3 million pounds. These figures reflected the general steep escalation in Australian airline successes.

Apart from its passenger and cargo services, EWA expanded into other fields. The airline had always accepted charters for any of its aircraft, and so when fishermen opened up the prawning grounds of the Gulf of Carpentaria, EWA aircraft sped their catches to southern markets. The Commonwealth Scientific and Industrial Research Organisation sent scientists aloft in an EWA DC3 to carry out cloud physics research. The NSW Ambulance Transport Board wanted rapid transport of patients from outlying areas, and EWA filled the bill with Queenair machines. In 1973 the forty-three-year-old Adastra group of companies came under the wing of EWA. Adastra has one of the longest histories of continuous operation in Australian aviation history, and since 1936 has concentrated on aerial survey and photography.

Like all Australian airlines, EWA has a fine record for safety, though it has had its troublesome moments. The first came in 1948, when an Avro piloted by John Rentell crashed on take off from the old Tamworth airfield. All the passengers walked away from the crash but Rentell had to be cut out of the cockpit, with one leg injured so badly that it had to be amputated. When he got used to a "tin leg" he followed the example of Douglas Bader and took to the air again.

In November 1957 Captain McKillop landed a disabled DC3 on the Lakes Golf Course, near Sydney. He brought her down to a perfect belly landing on the course, and she bumped and slid over greens and rough until her nose splashed into the lake. Out of a total of twenty-six aboard, nineteen suffered minor injuries but all survived.

So far there has been only one hijack attempt on Australian domestic airlines, in which an apparently crazed man died after a shoot-out with police on Alice Springs airfield, but on 6 November 1969 an EWA aircraft came very close to a hijack. Three of the passengers aboard a scheduled Fokker Friendship flight out of Sydney were prisoners bound for Grafton Gaol. Handcuffed to prison officers they sat near the front of the aircraft, but they had apparently discussed an attempt to take over the aircraft because, at nine-twenty a.m., when the Friendship was bucking through some turbulence, one of the prisoners jumped to his feet and shouted, "Now!"

With his handcuffed hand he hit the prison officer in the face, and one of the other prisoners put his handcuffs round the neck of the officer sitting next to him and pulled them against his throat. The third prisoner shouted, "No!" and took no part in the attempt.

The struggle went on for four minutes, with one passenger helping the officers, until the two prisoners were overpowered. The officers sat on them until the Friendship reached Grafton.

Early in the 1970s, EWA began to plan a move into the future by re-equipping with jet airliners. At first they favoured the British Aircraft Corporation's One-Eleven 475, and in 1970 made a provisional decision to buy this short-haul jet to enter service in 1972. An increase in price from $3.6 million to more than $4.3 million made them think again, and they looked at the Boeing Advanced 737 instead. Specially designed for short trips and small airfields, the Advanced 737 was used first by All Nippon Airways of Japan and proved ideal for short hauls in that mountainous country. It seemed equally appropriate for EWA flights, especially since it could give comfortable landings on gravel airfields.

The decision of the EWA administration to turn away from the British aircraft was, again, a part of the Australian trend towards American equipment, and the products of Douglas and Boeing now dominate Australian domestic passenger services. In fact the majority of Australian aircraft, both military and commercial, originate overseas, and this is a story of its own.

16

Struggle for Survival

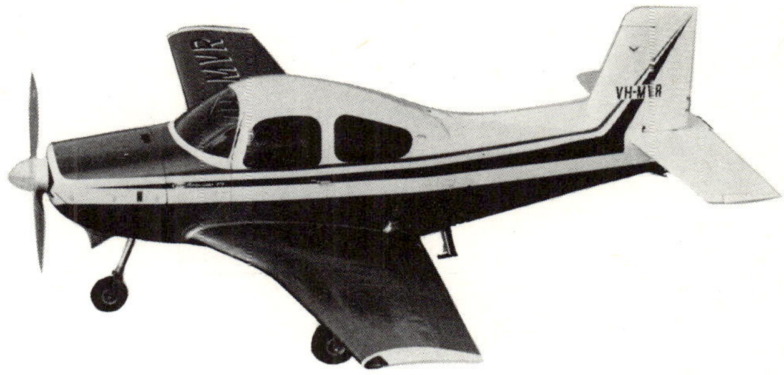

The Australian aircraft industry has always had to battle against overseas competition and official disinterest. Australian aircraft designers showed their mettle with such projects as the Nomad and the Victa

In September 1953 a crowd including Prime Minister Robert Menzies peered into the skies above the Commonwealth Aircraft Corporation airfield at Avalon, Victoria. The crowd was watching one of the new breed of aircraft: an Avon-Sabre jet fighter. Built by the Commonwealth Aircraft Corporation, and flown by Flight Lieutenant W. H. Scott on its "first official flight", the Avon-Sabre performed in a manner which would have made even the famous Spitfire seem ponderous. The sleek fuselage with its sweptback wings and gaping air intake slashed over the heads of the crowd so fast that their ears were not hammered by its engine noise until the machine was dwindling into the distance. Few in the crowd had seen anything like it. They were awed by this revelation of speed and power which spelt obsolescence for piston-engined aircraft.

The Avon-Sabre was the first of seventy-two ordered by the RAAF. In May of the same year the first twin-jet English Electric Canberra built by GAF for the RAAF had completed a successful test-flight above Avalon. Thirty-seven more were on order, and a huge hangar with a floor space of 60,000 square feet had been built for final assembly of the jet bombers and fighters.

The Canberra seemed to prove that Britain still held a lead in the design and manufacture of military aircraft. Production models flown by RAF crews had created four new speed records from England to Australia in 1951, 1952, and 1953. One of them had flown the distance in an incredible nineteen hours and twelve minutes. The bombers had flown demonstrations over Australian cities, where street crowds gaped up at the batlike monsters which seemed to float overhead, dark trails of jet exhaust floating behind, until they vanished out of view with amazing speed.

The Avon-Sabres were built under licence from the North American Aviation Corporation, but powered with Rolls Royce Avon turbojet engines. The Canberras were built under licence from English Electric. The RAAF order for these two types seemed to promise clear days ahead for the Government Aircraft Factory.

Several government aircraft factories were established during the second World War, though the principal establishment was and is the GAF at Fishermen's Bend. Between May 1941 and August 1944 they built 700 Beauforts for the RAF and RAAF, and between 31 May 1944 and 31 January 1946 the production lines turned out 364 Beaufighters for the RAAF. Before the end of the war, GAF began work on seventy-three Lincoln bombers ordered by the RAAF and more than fifty had been completed before it became obvious that this type of aircraft was becoming obsolete.

Not far from the GAF buildings stands the Commonwealth Aircraft Corporation. Between the time of its founding in 1936 and the time in the early 1950s when jet aircraft began to dominate the scene, CAC built nearly 1,400 aircraft and more than 1,600 engines. Most of the aircraft had been Wirraways and Boomerangs, but the list had included nearly 200 Mustang fighters for the RAAF. It had been the principal RAAF fighter in the immediate post-war years, until the first Vampire jets were bought in

1950, and when the Korean War broke out RAAF 77 Squadron flew Mustangs into action.

The first Australian designed and built jet aircraft were the Pika and Jindivik, which flew in 1949 and 1952 respectively. Both were designed by Ian Fleming and built by the Government Aircraft Factory, and were inspired by a joint British-Australian decision to produce a high-speed pilotless target aircraft. It could be seen that conventional anti-aircraft artillery would be supplanted by ground-to-air missiles, and that their operators would need something more sophisticated than targets towed by aircraft to teach them how to operate the missiles. An appalling new vista of warfare had been opened by the flying bombs and V-2 rockets introduced by the Germans at the end of the second World War, and the old pattern of finding ways in which to destroy the destroyers was already being repeated.

The Pika took off in October 1949, with John Miles at the controls. After successful test flights, the designers began on their next stage: that of replacing the pilot with a complex series of instruments and gyros which would enable the small jet aircraft—re-named, without the pilot, the Jindivik, or "Hunted One"—to give split-second responses to radio signals. Radio controlled aircraft were not new, and piston-engined models had been built by many people from model aircraft enthusiasts to scientists. But none of them had needed to achieve the speed and manoeuvrability of a guided missile target propelled by a jet engine.

All the problems were solved and the Jindivik was a complete success, so much so that orders came from abroad. The Jindivik continues to have a dynamic and valuable career, being still used for radar calibration, surface-to-air missile (SAM) training and normal anti-aircraft gunnery practice. Unique research in the US, using Jindiviks, is still under way.

Free enterprise was not lagging behind while the government organisations were hard at work on aircraft construction in Australia. Most of the smaller aircraft builders faded away during the 1930s, but the organisation established by Hereward de Havilland in 1928, to sell and service his company's aircraft, had moved from Melbourne to Bankstown, NSW, and gone heavily into manufacturing. When the war broke out, de Havilland had just begun to produce DH82 Tiger Moths for Australia. The two-seat Tiger Moth bi-plane was an ideal training aircraft and orders poured into the plant, not only from the RAAF but also from other Commonwealth countries. Within two years de Havilland had built more than 1,000 and had also begun to produce the Australian version of the famous Mosquito. In Britain this aircraft was built of strong softwood plywood, unobtainable in Australia during the war years, so hasty experiments resulted in production of a suitable Australian plywood. By 1948 the company had built 228 for use as trainers or fighter-bombers, but by that time all the aircraft which had beaten the Luftwaffe were being swept into obscurity and the de Havilland plant was tooling up for production of Vampire jets for the RAAF. At the same time it was producing Australia's first locally-designed and built commercial aircraft.

This was the DH Drover, an all-metal, three-engine monoplane designed by Martin Warner to replace the Dragon Rapide. Brian "Black Jack" Walker test-flew the prototype above Sydney's Bankstown airfield on 23 January 1948. Walker, whose tall, lean frame may still be seen on airfields along the Eastern seaboard, achieved a measure of fame during the second World War, when he tested the hurriedly designed and built wartime aircraft, and made the recommendations which made them safe for flyers.

The first Drover, VH-DHA, was modified in accordance with the results of Warner's test-flights and later in 1948 the aircraft was delivered to the Department of Civil Aviation for approval. DCA requested further modifications and when the Drover went into production it was with Gypsy Major Mark II engines instead of the Mark Is on the prototype, and some changes to the airframe and mainplane.

The Drover was in direct line of descent from Smithy's *Southern Cross* and the DH Hercules biplanes flown by Brearley's airline. The three engines, two set on the low monoplane wings and the third in the nose, presented a familiar silhouette though the airframe was cleaner and less cluttered than the earlier machines. Like them, it first trundled on to the airfield on a fixed undercarriage. It could carry two pilots and seven passengers, or eight if the toilet was removed.

Qantas ordered four of the Drovers for use in Queensland and later in New Guinea. In Queensland they replaced the DH84 Dragons which had flown the Qantas routes since before the war. The first production model was delivered to Qantas in September 1950, but the airline was a little disappointed when the Drover's climb performance suffered in the heat. The aircraft also suffered from propeller trouble in the centre mounting, and the airline claimed that this could not be remedied without severe effects on performance.

Qantas had ill-fortune with its trim little Drovers. One of them crashed on take-off from Mackay, Queensland, allegedly as a result of the faulty centre prop mounting, and on 16 July 1951 Captain J. W. Speire was wafting his Drover down for a landing at Lae, New Guinea, when something went wrong and the aircraft nosedived into the sea. Captain Speire and his six passengers died, and the Drover took with her a gold shipment which has never been recovered.

Fiji Airways bought two of the Drovers from Qantas in 1954, and Qantas switched to single-

TOP: The North American-CAC Sabre was one of the earliest of the "new breed" to be produced in Australia (*RAAF Museum*)

BOTTOM: The Pika, produced by the Government Aircraft Factory, was the first jet aircraft to be wholly designed and built in Australia (*RAAF Museum*)

Two illustrations of the Jindivik pilotless aircraft designed and built by the Australian Government Aircraft Factory. The one at bottom is an artist's impression of the Jindivik fitted with an air cushion landing system, used in experiments to develop the Jindivik into a pilotless air-ground strike aircraft and a low-altitude tactical surveillance vehicle (*RAAF Museum and Australian Information Service*)

engine de Havilland (Canada) Beavers from routes previously served by Drovers. But they got them back again when they bought out Fiji Airways in 1958.

In an era when aircraft design and manufacture was taking great leaps forward the Drover was perhaps a little behind the times with its cruising speed of 140 m.p.h. and maximum range of 560 miles. But it was a good sturdy aircraft which filled a gap, and it was used by TAA for western Queensland services and Flying Doctor work, by DCA for general duties, and, after it was re-engined by de Havillands with Lycomings, by independent operators for the hack work of day-to-day flying.

During the war de Havilland had produced a great variety of aircraft parts and accessories including drop tanks, metal controllable-pitch propellers, and wooden propellers. Like CAC and GAF the firm had accumulated a vast reservoir of aircraft production know-how. Either separately or in concert, the three producers could have started from scratch, and designed and built virtually any type of aircraft required for Australian civil and military purposes. CAC even had its own huge wind tunnel, perhaps a little outmoded by the dawn of the jet age, but capable of reconstruction. Wartime shortages and isolation had forced Australian aircraft designers and builders to make full use of their own ingenuity. The Australian Advisory Council on Aeronautics, founded in 1941, had done an enormous amount of research on aircraft structures and materials, with particular reference to the use of Australian materials. The Department of Aeronautical Engineering at Sydney University had carried out research into the design of wings for supersonic aircraft. The foundations had been laid for a viable Australian aircraft industry which would have made the nation at least partially independent of overseas suppliers. It could certainly have produced a range of aircraft specially suitable for Australian needs, and might even have produced something better and different from the rest of the world.

But none of this happened and the industry is still struggling. It is difficult to say exactly why. Part of the answer, of course, is the cost factor. When Norman Brearley bought his DH Hercules airliners in 1929, they cost £15,000 apiece: an enormous sum for those days but certainly not the equivalent, in any terms, of the prices charged for modern airliners. Their prices are reckoned in the millions and tens of millions, and the sophistication of the training, communications and guidance systems, workshop equipment, and everything else associated with them is a further burden on their costs. The same applies to military aircraft. The bombers of the second World War cost around £100,000, whereas the eight new Lockheed P.3C Orions recently ordered for the RAAF from the US cost about $13.5 million apiece. One cannot even blame inflation for such a rise in costs. It has been caused by such integral changes in aircraft design and construction that if a Lancaster pilot had been suddenly placed behind the controls of an Orion he would not even know where to start.

Prototype of the CA15 fighter designed and built by the Commonwealth Aircraft Corporation early in the 1950s (*Commonwealth Aircraft Corporation Pty Ltd*)

Yet, if Australia in 1950 had determinedly attacked the problems of a jet age aircraft industry, the cost factor might have been absorbed. After all, the money still comes out of the taxpayers' pockets. The pity is that it now goes to foreign aircraft manufacturers, instead of providing employment for Australians and inspiring the undoubted ability of Australians to design and manufacture aircraft.

But perhaps the cost factor was not the principal problem. A greater barrier may have been the "colonial hang-up": the remnants of an inferiority complex which made Australians in administrative positions feel that they had to look overseas for all new developments. After all, 1950 was not long after the days in which the RAAF was restricted to aircraft designs approved by the RAF.

But Australians still had high hopes of continuing to design and build their own aircraft. At the Commonwealth Aircraft Corporation, the excitement of having produced worthwhile designs during the war still simmered over the drawing-boards. The CA15 fighter, with its fantastic speed of 442 m.p.h., had never progressed beyond the prototype stage for want of government funding, though it would have been ideal for the Korean conflict in which the RAAF used Mustangs built by CAC together with some Gloster Meteor fighters. But the CAC men continued undeterred.

By 1948 CAC designs were highly advanced by world standards. Primary concentration was on the CA23 all-weather fighter. Initially it was designed to carry two crewmen, seated side by side as in the Mosquito. Power was provided by a single centrifugal jet engine, and she had two 30 mm cannon and A1 Mark 9 tracking radar for night interception.

Subsequent design proposals added a second engine, changed the seating positions to tandem, swept the wings back to forty-five degrees, and raised the tailplane until it was in a "flying" position atop the rudder fin. By 1949 the RAAF had issued a specification for an all-weather fighter (FAV) and the aircraft's design at CAC was taking final shape. It still had the flying tailplane though the final design dropped this to a lower configuration and honed the design still more. Power was supplied by Rolls Royce Avon jet engines, each with 7,000 pounds of thrust, plus reheat. High speed wind tunnel tests were carried out on models at the Royal Aircraft Establishment in Britain. A new radar relay system of interception, with a video presentation, was developed by Dr E. G. Bowen for the CA23.

By that time, work had begun on a British fighter which was to take to the air well after the CA23 project was abandoned. This was the English Electric (later BAC) Lightning. This design developed from the P.1A, which first exceeded the speed of sound on 11 August 1954. It is still Britain's prime defensive fighter, and the irony of it is that the Lightning has a strong family resemblance to the CA23.

But the CA23 design was not accepted by the Australian government. The mystique of overseas design held sway over Australians, and the RAAF was re-equipped with Gloster Meteor F.8s, which were outclassed by MiG15s in Korea, and Sabres built by CAC under licence from North American. The Sabre was reworked to make it a better aircraft than the original F.86 fighter. With Rolls Royce Avon engines, built at CAC, it was for a time the world's hottest fighter.

CAC designs continued to show a remarkable keeping of pace with international aviation developments, and yet one design after another had to be shelved. The CAC designers worked in vain on their drawings for airliners, jet trainers, and light jet transports. In the short term it was easier for the government to buy aircraft overseas or pay the licence fees for foreign designs. CAC had to content itself with designing and building such light trainers as the Winjeel, which served the RAAF well for many years.

The Department of Supply's own manufacturer, the Government Aircraft Factory at Fishermen's Bend near Melbourne, gained high skills when it produced the series of Canberra jet bombers for the RAAF but when the last of those was delivered the hum of activity began somehow to fade. Not that the GAF was idle. It became a world leader in pilotless vehicles with its production of the Jindivik and this wholly Australian product is scoring an ongoing record of foreign sales. The US Department of Defense is now modifying Jindiviks, in a multi-million dollar programme, to fit them with an air cushion landing system instead of its skid undercarriage. This would make them into self-contained aircraft. Even if this work is done abroad there will be royalties for GAF, though there have been recommendations that the work should be carried out in Australia.

Remotely Piloted Vehicles (RPVs) are becoming one of the foremost areas of aeronautical research, and RPVs were used in the latter stages of the Vietnam War and by Israel in the October 1973 war. The Jindivik was one of the first and best of these vehicles, and the GAF has since developed other RPVs such as the Ikara shipboard weapons systems and the Malkara anti-tank guided missile.

In 1972 the Government Aircraft Factory bounced back into the news in a big way. This was the result of a venture begun some years earlier and known originally as Project N. It was first presented as a design study at a Perth symposium of the Aerial Agriculture Association, at which it demonstrated the potential of an aerial agriculture aircraft powered by a single turboprop engine. This caused a flutter of interest, but professionals have seen countless designs wither on the drawing-board and Project N might have vanished into obscurity but for the enthusiasm of Alan Wrigley, Gordon Bennett, and the GAF team.

The project developed into something entirely different, and took shape as plans for a Short Take-off and Landing (STOL) machine suitable for both military and commercial aviation. It was to be a passenger and freight carrying machine that could be used under the roughest conditions of the Australian outback or on field service by the army, and as such would have a market in undeveloped areas. A prerequisite of such an aircraft would be a rugged and comparatively simple design, powerful and well-behaved.

The first aircraft was built and as Prototype 01 flew on its first test flight, from Avalon airfield in Victoria, on 23 July 1971. She was followed by Prototype 02, and during the next twelve months they flew a gruelling series of tests as Alan Wrigley and his team studied performance, solved problems, modified the design, and obtained DCA certification. One of the tests was a take off from a ploughed strip of ground in which a six-cylinder car bogged down.

During the 350 hours of test flying, the conviction grew that Australia had something really new and different to offer, and that the right place to show it off to the world was at Farnborough Air Show in Britain. Prototype 02, registered as VH-SUR and by then known officially as a GAF N22 Nomad, was

Top: This painting entitled "Helicopter Supply, Vietnam," by K. McFadyen, gives an impression of the split-second organisation required by RAAF helicopter units operating with Australian troops in Vietnam. The helicopters carried ammunition and other supplies to troops in the forward areas and picked up their casualties, often under enemy fire. (*Australian War Memorial*) Bottom: Light aircraft are now a familiar part of the Australian outback scene. Many station owners now operate their own aircraft and use them in stock mustering work and for trips to local centres. (*photo: R. V. Eussen*)

Top: Privately-owned light aircraft and helicopters did sterling work on rescue and supply operations during the inland floods of northern Queensland in 1974. (*photo: R. V. Eussen*) Bottom: A veteran flies again. Veteran aircraft of the 1920s and 1930s are eagerly sought after by aviation enthusiasts who spend a great deal of time and money restoring them to airworthiness. This fine specimen is owned by a South Australian. (*photo: Joyce Battye*)

Workshops in the Commonwealth Aircraft Factory, Melbourne. At top, craftsmen are engaged in the construction of wings for Vampire aircraft. The lower photograph shows construction of the Italian-designed Macchi jet trainer for the RAAF (*Australian Information Service photographs*)

Gloster Meteor jets were among the aircraft with which the RAAF re-equipped at the beginning of the jet age (*RAAF Museum*)

flown to Britain and in early September 1972 she was the first Australian designed and built aircraft to be put through her paces before the critical onlookers at this internationally famous event.

Hugh Field, an aviation writer for *Flight International*, flew the Nomad, "one peaceful evening during the show when the bustle of the Red Arrows' performance had ceased." He had been impressed by the way in which Stewart Pearce, the Nomad pilot, had handled STOL take offs and approaches, and when he took the controls he was convinced as to the aircraft's abilities. "I quickly felt at home with the Nomad's good manners," he wrote after his flight, and, "The Nomad was outstandingly easy to place accurately and I achieved an adequate touchdown at 50 kt."

When Farnborough was over the Nomad team took VH-SUR on a selling tour from which they returned in what Alan Wrigley called "a flood of enquiries from operators all over the world from Alaska to Peru." At a unit cost of $285,000 the new machine was reasonably priced, and production began on twenty aircraft. The first one went to the Australian Army and the second, which was the first commercial sale of a Nomad, to the famous racing driver Jack Brabham.

The Nomad has been described as the aircraft for people who want a working tool of trade rather than a status symbol. The designers were quite pragmatic in their approach to the aircraft and were less concerned with good looks than good performance. The result is a chunky, box-like aircraft with a fuselage somewhat reminiscent of a flying-boat hull, and with squared-off mainplane and tailplane, and yet the effect is very satisfying to the eye. The machine looks solid, craftsmanlike, and capable.

Everything about the Nomad 22 has been designed for hard-working, fast-flying capability under arduous conditions. The power units are the "smallest and toughest turboprop engines in the world", Allison 250-B17s. Wrigley—who has now left GAF—says of these that they show every evidence of having truly profited from well over 3·5 million engine hours accumulated by some 7,000 Series I engines flying in light helicopters throughout the world, and that, "It has been said that if a turbine can stand the life in a light helicopter—with full power applied immediately after start up, a very short flight cycle time, and another full power cycle to land followed by an immediate shut-down—it should go on for ever in a fixed-wing environment."

These engines give the Nomad 22 cruise speeds of up to 200 m.p.h., and the aircraft has a range of 1,000 miles. The rate of climb is an impressive 1,250 feet/min on cruise power, and there is enough power available to meet FAA climb requirements on a single engine even on a day when the temperature is 104 degrees and the aircraft is at her gross weight of 8,000 pounds. Another advantage of the engines is that they are remarkably quiet.

The STOL ability of the Nomad 22 is all that her designers aimed for. She can take off, fully loaded, in less than 450 feet, and land in less than 400 feet. The specially-developed wing leading-edge design

Stages in building the Nomad, the Australian designed short take off and landing aircraft which has aroused international interest. See text for complete description (*Department of Supply photographs*)

TOP: The Nomad in flight (*Department of Supply*)

BOTTOM: The Nomad displays her versatility as she takes off from a ploughed paddock which could not be negotiated by a motor car (*Department of Supply*)

FACING PAGE TOP: The Nomad soars skyward from the paddock in a cloud of dust (*Department of Supply*)

helps to give STOL with safety, and the aircraft's docile and progressive stall characteristics enable slow speed flying even in gusty, turbulent conditions.

The aircraft cabin is an "uncompromising rectangular box" with a flat full-width floor able to carry 150 pounds per square foot. Four continuous full-length seat rails allow seats to be arranged at one-inch pitching or can double as cargo tie-down tracks. The cargo space can carry 4,000 pounds of payload, and within a matter of minutes can be refitted for carrying twelve to thirteen passengers.

The list of Nomad 22 specifications seems to cover every possible use of the aircraft. Low speed loiter stability make it ideal for search and rescue, surveillance and patrol operations; a zero speed/zero altitude rocket boosted crew escape system can be fitted to military machines; the capacious cabin can be used for air ambulance work, carrying commuters over short hauls, or carting heavy machinery; even the cockpit seats have been designed to give maximum comfort in any climates. The Nomad 22 is a truly versatile aircraft, and it is wholly Australian.

In September 1975 the Federal Minister for Health announced that six Nomads had been ordered from the Government Aircraft Factory for the Northern Territory Aerial Medical Service, a government organisation. The Nomads are to replace DH Doves, of which some have been in service for eighteen years.

The Nomad project proved yet again that Australia could and should have a viable aircraft industry producing machines for domestic use and to build up export revenue, but the powers that be are difficult to convince. The aeronautical achievements of Australians, virtually without exception, have sprung from the drive and ingenuity of individuals working without regard for time-clocks or the lure of reward. They are a very different breed from the public servants who decide how the taxpayer's money is to be spent.

A project worthy of something far better than its eventual fate was that of the Airtourer, a truly indigenous Australian aircraft though it was the brainchild of Henry Millicer, a Pole who migrated to this country after the second World War.

Henry Millicer was born in Poland on 11 June 1915. He gained one aeronautical degree from Warsaw Technical University and another from the London Imperial College, and can append a string of degrees and decorations after his name. He served with the Polish Air Force until those gallant flyers flew the remnants of their squadrons out of their devastated country, and then served with the Armée de l'Air. When France fell he escaped to Britain, served for the rest of the war with the Royal Air Force, and came to Australia soon after victory. Now, as befits the Principal Lecturer in Aeronautics at the Royal Melbourne Institute of Technology, he still maintains his private pilot's licence and flies power aerobatics and gliders with the Royal Victorian Aero Club and the Gliding Club of Victoria.

In the early 1950s, when he was approaching thirty-five, he began work on a design project for

The Australians who won the 1953 Design Competition sponsored by the Royal Aeroclub of Great Britain. From left: Gordon Bennett (Stressing); Henry Millicer (Designer); Jim Tutty (Detail Drawings). (*H. K. Millicer*)

entry in a contest being sponsored by the Royal Aero Club of Great Britain. Millicer is a man with many friends. His round face and gentle manner make him seem more like a backroom boffin rather than an ex-combat flyer, and his popularity ensured the enthusiastic support of flying friends whom he had made in Australia. Two of them, Bennett and Tutty, advised him on stresses and drafting when he was drawing the plans for his Airtourer two-place light aircraft, and his whole circle of friends applauded when the Airtourer won the contest against 103 other entrants.

This group formed the Ultra Light Aircraft Association, and provided the manpower to build the first Airtourer from wood and fabric. The design of the prototype was completed in 1955, but it took another four years, mostly of night work, to develop the Airtourer to flight stage.

The prototype was ready for flight testing in March 1959. Randy Green, later to become well known as a test-pilot of rotary and fixed-wing aircraft, took the Airtourer up under the power of a small A-65 Continental engine. As he put her through her paces it became obvious that the aircraft had all the good points hoped for in the original design, although the prototype differed considerably from the prizewinning plans. They had had to be rehashed to make use of whatever parts were available, mostly donated by enthusiasts. The Airtourer had been pieced together in the old traditional style, much as Horrie Miller had built his own aircraft early in the first World War.

This first Millicer Airtourer was given the registration numbers VH-FMM and became affectionately known as "Foxtrot Mickey Mouse". Foxtrot is the phonetic code word for the letter F, but the Mickey Mouse tag was a ring-in which seemed appropriate for the pert, friendly-looking Airtourer.

However it seemed that the Airtourer might achieve no more than passing fame, as another aircraft built by aviation fanatics hooked on anything to do with flying. Millicer and his friends did not have the time or money to attempt full-time production, though Millicer made various attempts to attract sponsors.

These eventually appeared in the form of Mervin Victor Richardson and his son Garry. Merv Richardson had been one of the first to build and market rotary lawnmowers, under the famous Victa brand. His factory was near Bankstown airfield, outside Sydney, and he and his son were keen on entering the light aircraft field. They had begun work on Luigi Pellarini's design, the R-2, but it seemed that this was not ideal for their needs and they began to look elsewhere for a design. They found it in the Airtourer.

A redesigned all-metal Airtourer was produced very quickly. It began trials in 1961, with the registration VH-MVA adorning tail and wings, and with a 100 horsepower Continental 0-200-A engine. This was replaced by a Rolls Royce Continental, and in March 1963 the airframe was modified for a 115 horsepower American Lycoming 0-235. Eventually the Airtourer was produced with the option of 100 and 115 horsepower engines.

The little aircraft flew like a bird and represented a return to aerobatics in ultra light aircraft, something which had gradually slipped from the capabilities of such American machines as the Cessna, Piper, and Beechcraft. And, thus far, Australian manufacturers had given little competition to the big US companies.

There was the Cropmaster, produced originally by Yeoman Aviation at Bankstown and used for aerial spraying and topdressing, and the Transavia Airtruk: a twin-boomed agricultural aircraft designed by Luigi Pellarini. The Airtruk developed into formidable competition for foreign machines and is a credit to the Italian-born designer and his compatriots at Transavia at Seven Hills, outside Sydney. It was further proof that such things could be done well and on a large scale in Australia.

Obviously the Americans felt that such a threat must be nipped in the bud, because they stepped up their marketing activities by suddenly dropping their prices. The outcome was an increase in air consciousness among the already aviation-minded community and more and more companies, graziers, and development operators became sold on the idea of light aircraft as a solution to the Australian problems of distance, but the Australian aircraft manufacturers began to feel the economic pinch.

Towards the end of 1965 Victa Aviation joined with Transavia and Cropmaster in an application to the Tariff Board for some form of protection against US competition. They wanted a control of imports, a bounty for local manufacturers, or a combination of both.

The Americans riposted to this threat against what they had come to regard as a "traditional market" with a public relations campaign unprecedented in Australian aviation history. Representatives of foreign manufacturers bearded aviation writers to enlist their support, but the tactics of these often glib PR men succeeded only in antagonising many Australian journalists and highlighting the worth of the local product.

The hearings of the Tariff Board were a hot issue, debated bitterly by the protagonists. Men such as Victa Aviation's general manager, Colin Thomas McKenna, crusaded heartily alongside Millicer for the future of the Airtourer, and representatives of the other Australian manufacturers were forthright in defence of their aircraft.

But the Tariff Board declined to give any kind of support to the locals. Transavia decided to continue production of their Airtruk, and Victa and Cropmaster did their best to battle on but both firms had to close their doors early in 1967.

Just before the Australian government refused protection for this branch of a national aircraft industry, Victa had received a large order from a British aero club and had an option to grant a licence to a Philippines manufacturer to build the Airtourer in those islands. But the costs of running an aircraft plant are enormous and the parent company directors decided that, with the future so uncertain, Victa Aviation was no longer viable.

The Victa Airtourer project had many devotees. They included Richard Sims, a commercial pilot who had become the sales manager under McKenna. When the project died he became general manager of Hawker de Havilland's Beechcraft division. Vic Walton, a short, reddish-haired man who seems as sprightly and cocky as the Airtourer which he demonstrated at country air shows, was another. He was chief test pilot for Victa, and many sales resulted from his "barnstorming" at country shows.

In March 1967, less than three months after the Victa Aviation closedown, the parent company received a bid from Aero Engine Services Limited, of Hamilton, New Zealand, for the right to produce the Airtourer in New Zealand. Victa agreed. The jigs, together with some partially assembled airframes, were shipped to New Zealand. The first AESL Airtourer was delivered to Waikato Aero Club in May. New Zealand had been a good market for the Australian-produced Airtourers, and sales stayed good when AESL turned out the aircraft. They culminated in an order from the Royal New Zealand Air Force for Airtourers to be used as basic trainers.

Next came the most ironic twist of all. David Fairbairn, Minister for Defence in the government which had turned its back on the Airtourer as an Australian project, announced on 25 July 1972 that the RAAF would replace its Winjeel trainers with

The winning design developed into the Airtourer FMM built by members of Airtourer Group I of VCAA. The top photograph shows the machine under construction at Perfectus Aircraft Co., Newport, with Tom Webb and Trevor Sloan, members of the Airtourer Group, in the cockpit. The others show the prototype at Moorabbin in April 1959 (*E. K. Millicer*)

Top: The Victa Pellarini ready for its first test flight at Schofield, New South Wales (*Victa Limited*)

Centre: The Victa Aircruiser assembly line at the Victa factory, Milperra, New South Wales (*Victa Limited*)

Bottom: A Victa Airtourer flying over Hoxton Park, New South Wales, with Vic Walton at the controls (*Victa Limited*)

the NZ design of the Victa, called the Airtrainer. And the $3.3 million order went to New Zealand instead of to the struggling private sector of the Australian aircraft industry.

After the announcement, Millicer told reporters that he was pleased. He told the Sydney *Sun* representative that the fruits of Australian blood, sweat, and tears were returning to Australia.

But he kept the record straight by saying, in words which may veil some matter that has never been aired, "Federal government and vested interests killed this project. Australian dealers sold out to New Zealand for thirty pieces of silver."

The RAAF is taking thirty-seven Airtrainers, an aircraft in which it showed some interest when it was offered by Victa. The trainers are known as the CTA Airtrainer, but the basic design is Millicer's.

A more recent announcement shows that twenty per cent of the machines are being built by the Commonwealth Aircraft Corporation's plant in Australia. R. L. "Bill" Abbott, general manager of CAC, said that, "It will put a lot of work into our plant."

Victa Aviation made several gallant attempts to break into the light aircraft market. They built a four-seater, the Pellarini R-2 prototype (VH-MVR) powered by a Lycoming 0-360 180 horsepower engine. It was test-flown at Schofields airstrip, near Sydney, on 16 February 1961, but development was abandoned in favour of a Millicer-designed four-seater, the Victa Aircruiser. This performed excellently but never progressed beyond the prototype stage. It received its type approval in March 1967 before it was sold to New Zealand.

Victa also produced a Gyroplane (Project 67/a) which was designed by John Blacker, a bright and dedicated young aeronautical engineer. The prototype, VH-MVB, began flight and ground trials at Camden during March 1962 but development did not reach certification. It seems that Victa had not realised the cost of developing a rotary-wing aircraft and the project did not reach Certificate of Airworthiness stage even though it augured well. If all had proceeded as planned, the Gyroplane might have been another moneyspinner at home and abroad.

Australian manufacture of the CTA Airtrainer has been part of the major rebirth of Commonwealth Aircraft Corporation. The appointment of R. L. Abbott as general manager had much to do with this regeneration. His background includes a successful period as manager of British Leyland, the Sydney motor vehicle manufacturers, and a long interest in aviation evidenced whenever he visited Winterbottom Holdings, the Western Australian agent for British Leyland which also held the agency for Mooney Aircraft. "Bill" Abbott spent many hours flying round the State with pilot/manager Ronald Tutt, and apparently took a great interest in aircraft construction and Tutt's emphatic views on aeronautics.

The Victa Gyroplane designed by John Blackler. Randy Green, prepared for the first test flight, discusses final details with John Howten (*Victa Limited*)

After Abbott's appointment, CAC showed an enthusiasm and urgency reminiscent of the war years. Like his predecessor Wackett, Abbott did not believe in waiting for things to happen, and began his own ventures. One such CAC project has been the manufacture of galleys for domestic airliners. CAC expects to do some contract work on the Nomad 22, and has engaged in considerable consultative and back-up work under the leadership of Abbott and Air Chief Marshal Sir Frederick Scherger, the chairman of CAC. Sir Frederick and Abbott visited Indonesia in July 1972 to study the air force requirements of that nation, and may help to maintain its Sabre jet fighters and the virtually useless fleet of military aircraft of Soviet origin, abandoned by the Indonesians since Suharto's rise to power. CAC has taken over Rex Aviation, the Cessna dealer, and is moving its sales orientation towards South-east Asia.

In November 1971 Abbott told *Aircraft* magazine that, "Words have been spilled over the 'Fourth Arm' of Australia's defence for the past twenty-five years . . . approving words, congratulatory words, soothing words, encouraging, almost promising words. It hasn't made a great deal of difference yet to the Australian aircraft industry. It would gladly give away the words for a planned, reasonably continuing programme of work."

Sir Frederick Scherger, former chairman of the Defence Chiefs of Staff Committee, had already fired a number of rounds. "There should be more co-operation between military forces and industry,

The Hawker Siddeley HS 748 (*RAAF Museum*)

starting from the ground up," he told the National Aviation Press Club.

The oddest aspect of the situation is that both CAC and GAF have proved their competence, ingenuity, and flexibility over and over again, and yet they still have to exist on hope and a struggle for orders. Abbott said, "What is needed is much better long-range planning and earlier decision-making at the Canberra end." He urged continuity of workload and the need to give the taxpayer value for money.

CAC received a substantial order to licence-build Bell helicopters for the military and civil market, but Abbott pointed out that if the government had decided on this order a year earlier it would have prevented a slump in Australian production facilities. The CAC has built the Mirage III-O and the Italian Macchi MB.326H under licence, and yet while the RAAF attempts to make long overdue decisions on a replacement fighter there seems little sign that a wholly-Australian aircraft will even be considered.

While the Australian aircraft industry battles for survival, a country that is smaller than Australia and has fewer natural resources has designed and built one of the world's best fighter aircraft. Sweden, with only 9 million people, possesses the Saab Viggen 37 (Witch) Mach 2.0 attack fighter, with a spectacular STOL ability and immense hitting power. Seven hundred Viggen are being manufactured, and the model is only one of a range of Saab aircraft. There is no reason, except perhaps a lack of self-confidence in public service circles, why Australia should not emulate Sweden in the creation of fine aircraft.

The reluctance of government to become involved in the aircraft manufacturing industry and its various problems and battles may perhaps be explained by the constant changes in aircraft design between 1950 and 1970. Any man who attempted to predict the future would have been brave indeed and there was always a risk of "going for broke" on a design which would have been obsolete before the first production model rolled off the line.

The post-war generation saw the appearance of scores of new aircraft which were lauded for their performance and seemed exactly right for the job but most of them are now almost forgotten. de Havilland manufactured many of these and there is no doubt that they were fine aircraft, though the DH Comet 1, the world's first jet airliner, had a brief and tragic life. Nine of the new airliners were produced and for a while they made every other passenger transport in the world seem old-fashioned, but three of them broke up in the air in less than a year. An intense accident research programme discovered that the disasters were caused by metal fatigue. The Comet was withdrawn, in an era when Boeing, Douglas, Lockheed, and Hawker Siddeley, together with some European firms, were wrestling for the new turbojet market. de Havilland has produced many fine aircraft since the demise of Comet 1 but has not recaptured the lead which was lost to Boeing. By 1975 more than 1,400 Boeing 727 were in service.

Australia saw comparatively few of the vast range of new aircraft manufactured during the 1950s, though the types which appeared in this country included the famous Vickers Viscount ordered by TAA. The first sketch plans for this machine were drawn during the second World War, when Britain was riding high on the aviation expertise garnered the hard way during the war and expected to make good use of it in post-war civil aviation. The Vickers name, used for the twin piston-engine Wellington bomber, was well known. The four-engine turbo-prop was a big step forward. The prototype flew on 16 July 1948, and by the end of 1950 the Viscount 701 was in full production. 444 of the machines were built during the next decade and as turboprop airliners they were rivalled only by the Fokker Friendship which took to the air in 1955 and remains a popular aircraft in the 1970s, with more than 600 on various routes around the world.

The F27 Fokker Friendship won rapid acceptance in Australia, not only because it is a superb aircraft design but also because of the personality of "Pim" Moree, the Fokker representative based in Sydney. He was universally liked and respected, and the whole Australian aviation world mourned when they heard of his death in a Friendship crash in the Philippines.

TAA ordered Viscount 701s in the early 1950s and when they appeared on interstate routes they seemed to launch Australian aviation into a new era of air transport, with travel time vastly reduced and passengers pampered in a manner never dreamed of before then. The rapidly increasing size of domestic airliners called for a rapid rethink in aviation facilities, and plans were made for the new airports which have since taken shape in all the capital cities and many of the smaller centres.

Another passenger aircraft to appear in the Viscount era was the DH Heron, a fine four-engine machine ideal for such airlines as Butler Air Transport. Arthur Butler had been a planemaker himself and knew a good aircraft when he saw one, and his airline had been re-equipped with Herons when Queen Elizabeth II made her first visit to Australia in 1953. Butler was very proud of his sleek and luxurious new Herons, and the story goes that he offered a Heron to William McMahon, then Minister for Air, to fly the Queen on a visit to Bathurst. According to Butler the suggestion was rejected out of hand, and when he met McMahon at the opening of Taree airfield he aired his dismay at the government decision. Butler claims that McMahon told him the Butler Air Transport pilots were not good enough to fly the Queen, a statement which made Butler fume. He glanced around, saw some journalists near by, and called to them that he had an interesting story.

McMahon said quickly: "I'll look into this matter on my return to Canberra and let you know our consideration of the offer." But the Queen still flew to Canberra in an RAAF Dakota.

One of the McDonnell F4E Phantom jets used on "lend lease" from the USA by the RAAF (*RAAF Museum*)

17

Changing Patterns

The postwar boom caused a huge expansion in Australian airfreight and passenger services. Tourists took off on overseas jaunts while Air Beef carried carcasses from the outback

Readers of the 1 May 1945 issue of the magazine *Walkabout* saw a full-page advertisement headed "QANTAS re-opens regular Air Service to NEW GUINEA" and with copy that read, "Re-establishing yet another of Australia's overseas air links, Qantas, operating D.C.3 aircraft, now provides a civil airline service to New Guinea via Brisbane, Rockhampton, Townsville, Cairns and Port Moresby." The slogan at the bottom of the advertisement was, "QANTAS EMPIRE AIRWAYS. On His Majesty's Service . . . and Yours."

It was just over four years since Kevin Parer had died during the victorious Japanese assault on Papua New Guinea. All but five of the Guinea Airways aircraft, which had been predominant in the PNG area after the hard years of pioneering, were destroyed on the ground at Lae by Japanese bombers. Guinea Airways retreated to the mainland and was to operate for nearly twenty years more, though its routes were restricted to South Australia and the Adelaide–Darwin route which it pioneered just before the war. Qantas took over the PNG services.

Qantas, the "Front Line Airline," was ready, willing and eager to extend and expand its overseas services. A restricted schedule between Britain and Australia had been flown throughout the war years, and the DC3 service from Sydney via Queensland ports opened well before the end of the war with Japan. Seats were restricted to civilians with essential work in the Territory and servicemen on active duty.

In May 1946 a competitor appeared on the traditional Qantas flying-boat route from Sydney to Southampton. She was the Short Hythe *Henley*, of the British Overseas Airways Corporation into which Imperial Airways had been merged. She was among the last flying-boats to make the long flight round the world. Countless long overwater flights made by landbased aircraft during the war had swept away the old fears about allowing such machines to fly over the sea passage between Darwin and Singapore. Qantas used converted Liberators and Lancastrians and ordered four Lockheed Constellations.

But the day of the flying-boat was not yet over. Qantas flying-boats flew for several years between Australia and Fiji, New Caledonia, the New Hebrides, Norfolk Island and Lord Howe Island, and as part of the services being established in the Papua New Guinea area. Long after these aircraft were withdrawn, Sandringham flying-boats operated first by ANA and then by Airlines of New South Wales flew from Sydney to Lord Howe Island.

Qantas was still a vigorous private enterprise, but the governments of Britain, Australia and New Zealand were beginning to press a little harder into the field of commercial aviation. Despite the uproar when TAA was established, government involvement in aviation was not a novelty. Imperial Airways had been a state-aided organisation since 1924, and in 1939, when it combined with British Airways to become BOAC, it became completely government-controlled. In 1938 the governments of Britain, Australia and New Zealand had formed Tasman Empire Airways Ltd. TEAL flying-boats flew a monopoly service between Sydney and Fiji, Sydney and Wellington, and Melbourne and Christchurch well into the 1950s.

One post-war result of this government air-mindedness was British Commonwealth Pacific Airlines Ltd, established on 24 June 1946 by Britain and the two dominions. Canada was invited to join, but preferred to begin her own Vancouver–Sydney service which opened in July 1949.

To begin with, BCPA was an airline without aircraft. The first BCPA service, which left Sydney on 15 December 1946 for Fiji, Honolulu, Canton Island, San Francisco and Vancouver, was flown by a DC4 chartered from ANA. The new airline did not obtain its own aircraft for more than a year, but on 25 April 1948 the first BCPA DC4 entered service. A year later, BCPA introduced the bigger, pressurised DC6 four-engine monoplane onto the route, which covered some of the longest stretches of open water in the world. It was only a little more than ten years since flying circles had been perturbed by the idea of flying passenger aircraft over the Timor Sea.

BCPA flew a regular trans-Pacific schedule, with one service terminating at Auckland and the others in Sydney, for nearly eight years. Veteran air travellers say that the service was more interesting than modern Pacific flying, even though it took longer, because of the frequent refuelling stops en route.

But although BCPA had three parents it was still something of a stepchild. The British government became increasingly involved with the expansion of BOAC services and decided to withdraw from TEAL and BCPA. On 31 March 1947 the Australian government bought out all British-owned shares in Qantas, and on 30 June took over all private shareholdings. Qantas became indeed "Australia's own airline" and it became apparent that BCPA would eventually be merged with the larger company.

However the merger did not come until May 1954, seven months after BCPA suffered a disastrous crash at San Francisco in which nineteen people died. Qantas absorbed the trans-Pacific route and the Australian government began negotiating with the Americans for the right to push the service on to New York and later to go to London. Eventually that permission was granted, but at the cost of allowing a second American carrier onto the Pacific route. Pan-American had flown from San Francisco to Sydney since 1945. They opened their service with DC4s, followed by the luxurious Boeing Stratocruisers, and the Americans were content to allow the status quo to remain undisturbed until the 1970s. Then, Australia had to pay for her "round the world via New York" service by allowing American Airlines into Sydney and Melbourne, with consequent wails from Qantas as to over-competition on the Southern Cross route.

Qantas had pushed far ahead after establishing its Bird of Paradise Service to Papua New Guinea. Still known as Qantas Empire Airways, it pioneered flights to Japan in 1947, South Africa in 1948, and Hong Kong in 1949. These were regarded as "proving" flights to begin with, especially since the airline was still limited to wartime-built Lincolns and Lancastrians, but when the DC4s and then the Lockheed Super Constellations arrived it became obvious that the whole world was within easy reach of Australia. It remained only to determine the most profitable routes for regular services.

Qantas Flight Hostesses, dressed in uniforms designed by Emilio Pucci, are a far cry from the women aviators dressed in oil-stained overalls of only thirty years earlier (*Qantas Airways photograph*)

The magnificent Constellation, with its four 2,200 horsepower Wright engines, heralded a new era in airline services. The aircraft had been on Lockheed drawing boards as early as 1939 and the prototype flew in 1943, but for some reason it was not developed as fast as the Douglas series which became the mainstay of military air transport during the war and of commercial aviation immediately after the war. When Constellations became available to airlines early in 1946 they were snapped up as fast as Lockheed could build them. They had a cruising speed of better than 300 m.p.h., were pressurised to fly at 20,000 feet, and had a range of 2,260 miles. More than 500 Connies and their successors the Super Constellations and Starliners were built for airlines in many parts of the world, and hundreds more were built for military aviation.

Aviation technology has developed so rapidly that even when Constellation passengers thought they were enjoying the last word in modern speed and luxury the aircraft were already becoming obsolete. Manufacturers in Britain, France, Holland, Russia and America were working on the engines and aircraft which were to revolutionise commercial aviation.

Sir Frank Whittle had been working on the theory and practice of jet-propelled aircraft since he was a cadet at Cranwell RAF College in 1928. His company Power Jets Ltd tested the world's first turbojet engine on 12 April 1935. The world's first jet aircraft, the Gloster E.28/39, was test-flown on 12 April 1937, and in 1944 the first jet-aircraft battles were fought when RAF Meteor fighters took the air against German flying bombs.

For a few years Britain led the world in jet aircraft design and manufacture. She lost it when three De Havilland Comets disintegrated from materials fatigue, though many fine jet aircraft for commercial or military use have been produced by the United Kingdom. But by the mid-1950s many airlines had lodged orders for the first Boeing and Douglas jet airliners. Qantas was among them, and the first Boeing 707s were delivered to Australia in 1959. Now, "Australia's own airline" operates nineteen Boeing 707–338Cs, one Hawker Siddeley HS125, and five Boeing 747s with one on order. Qantas has reserved four delivery positions on the controversial BAC/Aerospatiale Concorde, and Australia may be among the first nations to use this aircraft commercially.

According to the old saw, "Happy is the nation that has no history." The same may be said for an airline. Qantas makes the proud boast that "We've never lost a passenger," and since the beginning of the jet age in commercial aviation the history of Qantas has had no specially dramatic moments apart from the day when a phone rang in the Sydney office and a voice said, "I've put a bomb on one of your flights. If the aircraft goes above a certain height the bomb will explode."

There was no proof that the threat was authentic but the Qantas officials could not take risks. The outward-bound flight was recalled and instructed to circuit at an even height, while Qantas consulted the police and negotiated with the telephone hijacker. In an atmosphere which seemed unreal because it was so much like a TV thriller, the ransom money was handed over and the aircraft landed safely. There was indeed a bomb, designed to be triggered by barometric pressure, in the cargo hold, but there seems to be some doubt as to whether it would have operated. Eventually the hijacker was caught and some of the ransom money recovered.

Otherwise the recent history of Qantas has been that of peaceful expansion and growing reputation, and the airline's kangaroo insignia has become well known on international airports. The rate of expansion has been fast. In 1971 Qantas airliners could offer 1,500 seats a week to London. By 1975 it was possible for 5,000 passengers to fly to and from Europe each week.

Part of this expansion has been caused by the tremendous boom in Australian business and holiday travel. Time was when most Australians looked upon a trip to Europe either as a prerogative of the wealthy or the reward for a lifetime's saving. Now, nearly a third of overseas travellers are under thirty-five and many are teenagers. This trend has been helped by special student discounts and such package deals as the Qantas trips which can be taken over a minimum of three weeks or a maximum of six months, with specially reduced return fares and a whole range of attractive alternatives. London is still the terminal port for Qantas but passengers may leave or rejoin the Australian flights at Athens, Belgrade, Rome, Vienna, Paris, Frankfurt, or Amsterdam.

Like every other international airline Qantas has had to battle with problems of cost inflation. Perhaps it has some advantages because of government ownership,

One of the Lockheed P2V Neptunes used by the RAAF for long-distance ocean patrols around Australia (*RAAF Museum*)

T-jets and Fokker Friendships of Trans-Australia Airlines on the TAA tarmac at Tullamarine International Airport, Melbourne. TAA has ten passenger lounges along the concourse jutting from the main terminal building (*Trans-Australia Airlines*)

although the Australian government is by no means paternalistic in its attitude towards Qantas and Trans Australia Airways. The Labor administration which had to deal with strikes of Qantas and TAA personnel proved to be as stiffnecked as any capitalist consortium and had no hesitation in employing "scab labour," in the form of RAAF aircrews and aircraft, to fly members of parliament to Canberra when a strike of domestic aircrews had cut air communication with the Federal capital early in 1975.

It is to some extent remarkable that ministers and backbenchers should become so hot under the collar when they consider aircrew wages. The Australian parliament has never shown any reluctance to hike its members' salaries by several thousand dollars in one hit, but it emits heartrending cries when domestic and international aircrews ask for rises. Yet most members of parliament never face a responsibility remotely resembling that of an aircraft captain.

The story of TAA, ever since the controversy which attended its establishment, has also been one of steady improvement and expansion. A large part of this expansion has been developed by the creation of links with feeder and commuter airlines inside and outside mainland Australia, and each of these linking services would have a story of its own.

TAA lists eighty-three principal ports, including those in Timor, West Irian, Papua New Guinea, and the Solomons and neighbouring islands, which are served by its airline network. In addition to these, passengers can book through to 300 outback centres, large and small, by transferring to the relevant feeder airlines.

Some of these have fascinating names. In Queensland you can travel the State including "121 towns, stations, and communities throughout north Queensland," by Burnett Airways, Bush Pilots Airways, Henebery Aviation, Nation Wide Air Services (which links Rockhampton with Blackwater), New England Airways, North Coast Airlines, Pagas Airlines, Stradbroke Gardens Air Charter, Tangalooma Air Service, Thorpes Air Charter, and Union Air.

These services are little known outside their own State but they play an essential part in linking all the scattered settlements of the Queensland coast and inland together in a web of rapid communication. When John Flynn was planning the Flying Doctor service, the residents of Queensland's Gulf Country were cut off from the rest of the world when the huge complex of swamps and waterways filled up during and after the "Wet." Nowadays they can look to Pagas Airlines to "get the mail through," together with passengers and essential supplies, under any except completely impossible flying conditions.

Pagas Airlines also operate in South Australia, central Australia, and the Northern Territory, areas in

Loading cargo into the hold of a DC9 T-jet (*Trans Australia Airlines*)

which TAA services are extended by co-operation with Central Australian Airways, Connair, Opal Air, and Amber Air. Whether you want to go to Kangaroo Island to photograph the seal harems on Seal Beach, or to try your luck at opal gouging in Coober Pedy or Andamooka, there's a scheduled air service to take you there.

Ansett has similar connections with the feeder and commuter airlines of Australia and it's fair to say that, by this time, most Australian communities of any size have been brought within a few hours of each other by the network of major and minor airlines. New South Wales, for example, is very well covered. Aeropelican Air Charter Services connects Sydney with Newcastle, Masling Air Commuter runs a shuttle service to townships in the Snowy Mountains region, New England Airways and North Coast Airlines cover the North Coast, and Skyway, Macknight, and Kendell airlines link Sydney with Wee Waa, Gunnedah, Finley, Deniliquin, and Griffith. If you have business in Canberra, then you can choose from four domestic airlines making regular flights to the capital.

Kendell and Pagas airlines operate into Victoria, where the latter flies south to Flinders Island and north to Swan Hill. Bizjets flies woolbuyers to the woolsales at the huge woolstore outside Portland, and calls at Warrnambool en route. Air Charter Flying Services connects Melbourne with Sale.

Western Australia, once the exclusive preserve of Norman Brearley, now boasts several scheduled air services apart from the "giants." For example, Civil Flying Services operates on the Rottnest Island route and serves a huge semi-circle of the State ranging down to Esperance, Norseman, and Albany. Trans-West goes as far east as Kalgoorlie and as far north as Port Hedland and nearby townships. Tropic Air Services operates between Carnarvon and Denham.

The island State of Tasmania has two internal airlines: Air Tasmania linking Hobart with the north of the State, and Executive Air Services from Launceston to King and Flinders islands.

Much of the island flying north of Australia, to Papua New Guinea and nearby territories, is performed by the Friendships of Air Niugini together with a variety of small freighter and charter services such as Aerial Tours, Crowley Airways, Macair and Talair. The full story of flying in Papua New Guinea has never been properly told though portions of it may be found in such books as *Flight Plan PNG*, by Eileen Steenson, in which she writes about the adventures and misadventures of flying over PNG mountains and jungles.

All the above flying operations represent only a part of Australia's involvement in, and dependence on, commercial aviation, which includes the developing story of what used to be Australian National Airways.

Earlier chapters have outlined the gusto with which ANA, under the leadership of Ivan Holyman, pursued its aim of dominating Australian commercial aviation. From the day when ANA was formed, by co-operation between various shipping companies and existing airlines, this domination was sought in traditional private enterprise style. On the one hand, ANA tried to give such splendid service to the public that the airline would be preferred to any competitors. On the other hand, the board of ANA sought an interest in the shareholding of any substantial competitor, and at the same time looked for methods by which to pioneer new services.

There was no problem about the first part of this three-pronged advance. Despite the occasional disaster in the early years, ANA remained a well-respected airline. The second aim was pursued with a prudent eye to likely development. ANA was a silent partner and substantial shareholder in Butler Airways Pty Ltd, for many years the leading airline in New South Wales. Airlines of Australia Limited, the grandiose name under which a group of graziers and businessmen formed a company to link Lismore with Sydney, became a wholly-owned subsidiary of ANA. Australian National Airways also had its eye on other airlines, including Guinea Airways which stole a march on ANA by opening a Sydney–New Guinea service while ANA were waiting for aircraft to use on the route.

Attempts to expand ANA into the international arena did not work out. The company chartered aircraft to British Commonwealth Pacific Airways and Cathay Pacific Airways, and made various unscheduled overseas flights, but these moves did not develop into regular services. The directorate decided to concentrate on domestic competition and to give TAA and Ansett Airways Pty Ltd a run for their money while looking for new fields to conquer.

One of these pioneering efforts was the introduction of three front-loading Bristol Freighters into Australia.

Roger Meadmore's veteran Tiger Moth "Lovely" photographed in the air over South Australia. (*photo: Ron Langman*)

Top Left and Right: "A Spinner in a Sea of Gold" and "Sunburst." Two evocative aircraft photos by Joyce Battye. Bottom: A Hang-glider launches himself from a hillside. (*photo: Ron Langman*)

Wartime experience, following so soon after Guinea Airways' growth into the world's greatest carrier of air freight, had shown that aircraft could carry far larger cargoes than anyone dreamed of twenty years earlier. The Bristol Freighters introduced this huge cargo carrying ability to Australian aviation.

The next forward step was the brainchild of Ian Grabowsky, the ANA manager for planning and development. He was searching for new cargo ideas, and hit upon the remarkable notion of using air transport to increase beef production in the central Kimberley area of north-west Western Australia.

Traditionally, this part of Australia had followed a "wild west" operation in which great mobs of cattle were overlanded from the stations to the coast. At first they were taken to Derby and loaded aboard ship "on the hoof" for transport down to Perth. Later, after a meatworks was built at Wyndham, they were driven to that port and their carcasses were loaded for export.

The problem was that the cattle usually arrived in poor condition after their long overland walk. Also, the droving operation was complex and expensive. Skilled stockmen had to be employed, and even they could not prevent a fairly high percentage of losses while the cattle were on the track.

So Ian Grabowsky asked: Why not transport the beef instead of the livestock? Better still, why not transport it by air? The method would be perfectly straightforward. Kill the cattle in the area where they were raised, and speed the valuable meat to market without delay and in top condition.

Such a revolutionary idea, like all revolutionary ideas, was scoffed at to begin with. Luckily it attracted Gordon Blythe, whose grandfather took up Mount House and Glenroy cattle stations in the 1890s. He convinced a group of Kimberley pastoralists that the idea wasn't so strange after all, and they helped to finance a new company called Air Beef Ltd. The other shareholders were ANA and MacRobertson Miller Airlines.

The first Air Beef airlift was flown in May 1949 from Glenroy, where a killing works had been established, to Wyndham. The beef was carried in MMA aircraft and the scheme was proven viable from that moment on. Every winter from 1949 to 1962, MMA pilots and aircraft ran a ferry service from the Kimberley area to Wyndham with loads of beef carcasses and hides outward bound, and station stores and passengers on the return flight. The operation proved beneficial in more ways than one. It gave the isolated Kimberley area a rapid transport service, which enabled station owners to bring in materials much more easily than by the old style of carting them for hundreds of miles over outback roads.

Between 1949 and 1962 the MMA aircraft carried astronomical quantities of beef products. On a total of 2,842 return flights from Glenroy to Wyndham or Derby, covering 6,913 hours flying time, they airlifted more than 32 million pounds of meat, 2 million pounds of hides, and 8 million pounds of return cargo to Glenroy and other stations.

A side effect of the operation was the increase in local

Forerunner of the Air Beef Scheme. Beef slaughtered overnight at Mount House Station was flown to Perth in an MMA Lockheed 10A Electra to establish the feasibility of the scheme (*MacRobertson Miller Airways*)

Top: Backloading cargo to Kimberley cattle stations during the Air Beef season. Note the vehicle loaded into the DC3 (*MacRobertson Miller Airways*)

Bottom: Two old-timers that are still flying. The DH60 Gypsy Moth belonging to George Aylesmore of Gnowangerup, Western Australia, has her wings folded back while a mechanic services her old in-line engine. On right, the nose of a DH84 owned by Neville Hyder of Busselton, W.A., juts into the picture (*The West Australian*)

cattle production. Station owners saw that they could send their meat to market without difficulty and improved their herds considerably. A total of 55,529 cattle were slaughtered at the Glenroy abattoirs.

Air Beef was a financial and developmental success and a remarkable example of the way in which air transport could solve a problem. Unfortunately it became increasingly costly to operate and the Australian government showed its usual inability to comprehend the advantages of aviation. An attempt to gain a subsidy to help flying costs was unsuccessful and the State and Federal governments preferred to finance the establishment of "beef roads" so that cattle could be taken to market by motorised road trains. After 1962, huge refrigerated trucks carried the carcasses to Wyndham, and later these trucks were replaced by articulated road trains which now carry cattle on the hoof.

Long before Air Beef had made its last flight, the whole structure of ANA had undergone a complete change.

As has been stated, Sir Ivan Holyman had in 1953 decided against re-equipping the airline with Viscounts, and to choose instead a new version of the Douglas airliners which had served ANA so well. This venture, together with other developments, imposed a drain on capital. The airline faced a familiar business situation: that of having a viable operation capable of returning good dividends, but requiring a substantial

injection of capital to modernise its equipment. Sometimes this is called a "cash bind."

By the mid-1950s, ANA shareholders were looking glum. The time was ripe for a takeover, and Reginald Ansett made his bid.

When the news broke it was greeted with predictable reactions from both sides. The old guard, which had served ANA since it was Holyman's Airways Pty Ltd, deplored the gobbling-up of a thoroughly Australian company which still had a big future. On the other side there were those who said that ANA management had become inward-looking and hidebound, and that the company was certain to benefit by the kind of vigorous modern administration that Reg Ansett could provide. Probably the truth lay somewhere in between.

One of the old guard, though he was certainly not old-fashioned in his outlook, was Arthur Butler. He had founded Butler Air Transport Pty Ltd in 1934 to fly the Cootamundra–Charleville section of the England–Australia airmail service, and with the aid of ANA capital his company had grown to be the largest airline operating within the confines of a single State. Since 1948 New England Airways had been a subsidiary of Butler Air Transport.

Butler began operations with two seven-seater DH84s. By 1936 business looked flourishing enough to warrant removal of his headquarters from Cootamundra to Sydney, and he opened new services from the capital to Dubbo, Coonamble, Bega and Cessnock.

The second World War cut deeply into Butler's aerial services just as it did into all the scheduled flights of Australian airlines, but there was plenty of work for the aircraft engineering shops he had established in Sydney to service his own machines. At one stage he employed 250 people on Department of Aircraft Production contracts.

In December 1945 Arthur Butler was one of numerous airline operators who bid for war-surplus C47s, the military version of the DC3. He bought three and used his own workshops to convert them to civilian use.

The demand for air transport boomed during the post-war years and Butler Air Transport grew rapidly. He added two Avro Ansons to his fleet, and in 1953 replaced them with two fourteen-passenger DH114 Herons. His aircraft were kept hard at work and the mileage flown increased each year, but competition was increasing and he decided to phase out the DC3s in favour of two Viscount 747 series machines. This was a big step for an intrastate airline. In 1955, when he took delivery of the Viscounts, a number of the country airfields in New South Wales could not handle aircraft of this type. He had to look further afield and in December 1955 the first Butler aircraft flew into Melbourne. From then on they flew a regular service from Sydney via Wagga and Cooma on alternate trips. Later, they flew also to Adelaide via Broken Hill.

TOP: This Vickers Viscount has RAAF markings, but many of her sisters wore civilian dress on Australia's domestic airways before jet aircraft took over (*RAAF Museum*)

In 1957 he replaced the DC3s still flying on routes not serviced by the Viscounts with three Airspeed Ambassadors or "Elizabethans" purchased from British European Airways. These were among the last and best of the piston-engined airliners produced in Great Britain: beautiful high-wing twin-engined monoplanes capable of carrying up to fifty-five passengers.

When Ansett Airways took over ANA, the Butler shares held by ANA passed into their hands. Slowly at first, and then with increasing speed, Butler saw control of his airline slipping from his grasp. There were inevitable differences of opinion between Butler, Ansett nominees, and the airline personnel. Butler resigned and left the airline he had established to be run by other men.

He was an Australian aviator of the old school, always prepared to "look any man in the eye and tell him to go to hell." He was intensely proud of his airline and he must have found it hard to relinquish control.

The new management reorganised airline operations and little by little its flight schedules were withdrawn into New South Wales, a logical move to avoid competition with other Ansett operations. The fleet was altered to one more suitable to the revised schedules. Convair Metropolitans were introduced to replace the Viscounts and Ambassadors, and the Convairs were later replaced by Fokker Friendships. Two of the DC3s, were retained until the 1970s.

In December 1959 the name was changed to Airlines of New South Wales Pty Ltd, which later became Ansett Airlines of New South Wales. This company also controlled Ansett Flying Boat Services Pty Ltd, which operated the two Sandringham flying-boats that maintained the air services between Sydney and Lord Howe Island and also were used for charter flights to South Pacific islands.

When ANA was formed, its directors had followed a deliberate policy of expansion which included the takeover of other airlines. When ANA came under the Ansett wing this policy was continued by the new management, but perhaps with greater vigour and expertise. For some time the combined airline was known as Ansett-ANA, but eventually the suffix was dropped. By that time, Ansett had also taken over "Micky Mouse Airlines" of Western Australia.

MacRobertson Miller Airlines, whose longwinded cognomen was invariably abbreviated to MMA, has a solid place in Australian aviation history. It was founded in 1934 by Horrie Miller, some of whose exploits have been described earlier in this book, with the help of finance provided by Sir MacPherson Robertson the "Chocolate King" whose interest in aviation dated back many years although he took little active part in flying.

MMA first made its mark when Horrie Miller tendered for the government airmail contract from Western Australia to Daly Waters in the Northern Territory, to link up with the first Australia–England airmail service. He won it from Norman Brearley, who had pioneered the airmail routes in Western Australia, in a coup which aroused considerable comment in aviation circles.

But MMA was off the ground and during the next few years the new company flew faithfully and well. To begin with it used DH84 Dragons, carrying only eight passengers and cruising at 105 m.p.h. In 1938 Horrie Miller bought two DH86s from Qantas and ordered two Lockheed Electras from the USA. Another new airline was flying Western Australian skies at that time: Airlines (WA) Ltd, founded in 1935. The air coverage of the State was better than ever before as MMA concentrated on the Kimberley and other northern areas and the other new airline flew the south-west and some inland routes south of the Kimberley.

Nevertheless there were still some large gaps and flying conditions were fairly primitive. Radio navigation aids were almost non-existent and pilots flew with the aid of local knowledge of various landmarks or by following the "iron compass:" one or another of Western Australia's few railway lines. Horrie Miller scored high in this kind of flying because he knew the State well.

The northern areas of Western Australia suffer from summer weather conditions which so far as an aircraft pilot is concerned can only be described as frightening. Massive electrical storms were a real hazard to early flyers who soon learnt to avoid the cumulo-nimbus clouds with associated up draughts and down draughts that tossed an aircraft around like a leaf in a gale.

Airfields were poorly laid out and had minimal facilities, and the pilots had to be able to service their own aircraft because there were very few skilled aircraft mechanics outside Perth.

Despite such difficulties the two new airlines kept flying regularly and they were welcomed by outback folk to whom they were a wonderful breakthrough in the barrier of isolation. MMA became involved in numerous lifesaving and rescue operations apart from its scheduled flights, especially when MMA aircraft were chartered by the Flying Doctor services out of Broome and Port Hedland.

The Japanese entry into the war brought new difficulties to MMA operations in the north-west, especially after most of their aircraft were commandeered by the RAAF. When ANA's Douglas DC3s were taken over for a period in 1943, to help in the great airlift of troops and material into New Guinea that was to turn the tide of the Japanese advance, MMA Lockheeds were re-routed to fly from Perth to Sydney and even to Tasmania.

As soon as the war ended, Horrie Miller and his men buckled down to the task of rebuilding the airline. The company bought six Ansons and several Douglas C47s. They converted these into passenger aircraft and MMA launched out into better and faster services, with

TOP: Five Fokker F28 Fellowships of the MMA fleet lined up at Perth International Airport ready for the day's flying around Western Australia (*MacRobertson Miller Airways*)

BOTTOM: The last of a romantic era: a Sandringham flying-boat takes off from Sydney Harbour for Lord Howe Island (*Ansett Airlines of Australia*)

TOP: MMA DC3 high altitude photographic aircraft operating from Giles Weather Station in central Australia (*MacRobertson Miller Airways*)

BOTTOM: Two of Australia's aviation pioneers meet at Broome. Sir Reginald Ansett (left) talks to Horrie Miller (*The West Australian*)

the help of airfields improved by the RAAF and USAAF and radio navigational aids installed during the war and operated thereafter by DCA. But even the new aircraft were not altogether proof against north-western flying conditions and Robin Miller, in her book *Flying Nurse*, described a fearsome flight made aboard a DC3 when she was a girl.

MMA became even more widely known when company aircraft were used for aerial survey projects, seeding and fertilising from the air, and of course the Air Beef scheme. The faithful DC3s performed magnificently but at last the time came near for these sturdy old workhorses of the air to be retired, and MMA bought their first Fokker F27 Friendship. At about the same time the company merged with Airlines (WA) Ltd and the routes previously served by that airline saw the MMA insignia for the first time. The future looked good. Western Australia was the boom State of the late 1950s and the 1960s, with the huge mineral discoveries of Mount Tom Price and the Pilbara region being exploited and such projects as the Ord River reservoir on the drawing boards. The company's aircraft, still including six DC3s, were kept at full stretch, and MMA created a world record by achieving a daily utilisation rate of 10.66 hours for their F27 Friendship.

MMA had just ordered their second Friendship when MacRobertson Pty Ltd, the firm which owned the MMA shares originally taken up by Sir MacPherson Robertson, decided to sell the shares. MMA was receiving an annual development subsidy from the Australian government, and was obliged to make the first offer of these shares to the government, which declined. Instead they were purchased by Ansett Transport Industries in April 1963. Ansett also bought a large number of the shares held by Horrie Miller and his family and thus gained control of the company.

The takeover was of great significance both to Ansett and to MMA. It enabled Ansett to extend their operations considerably, and it brought new capital and new ideas to MMA just at the moment when the Western Australian boom period was gathering momentum. The backing of Ansett Transport Industries was of supreme importance in 1969, when it became apparent that Western Australian air services were ripe for a major re-organisation. The greater part of the government subsidy paid to MMA had been consumed by minor feeder routes, which it would have been impossible to operate except at a heavy loss unless a subsidy had been paid. Various light aircraft operators were anxious to become involved in these minor operations, for which they were ideally suited, and DCA saw this as a logical step. A co-ordinated plan was evolved, and MMA handed over the minor routes progressively while they concentrated on developing unsubsidised mainline routes.

The huge distances of Western Australia and the Northern Territory mean that jet aircraft are the natural choice for such mainline operations, including the Perth–Darwin route. Sir Reginald Ansett instigated the purchase of pure jet aircraft for the mainline operations and Ansett Transport Industries injected more than $20 million of capital to achieve the changeover. The first Fokker F28 Fellowship jet flew on MMA routes in September 1969. It was a leased aircraft but was followed by five production aircraft built to Ansett-MMA specifications, and by August 1971 MMA could advertise in *Aircraft*: "A first for intrastate airlines in Australia. Exciting and rapid development throughout W.A.'s MILLION square miles needs sophisticated modern AIR transport—and MMA provide it. 100% of all our MAINLINE SERVICES throughout WESTERN AUSTRALIA ARE BY PUREJET aircraft flying at over 500 m.p.h."

A downturn in the mineral market made one of the five aircraft redundant during 1972, but all five have been continuously in service since early 1973. A sixth was ordered for delivery late in 1972, but the order was cancelled. It has since been lodged again, for a jet with a special large cargo door to handle heavy equipment.

The Fellowships revolutionised air travel in Western Australia and the Northern Territory, where they fly as far as Gove and Groote Eylandt in Arnhem Land. The flight times of F27 Friendships, of which MMA operated six, have been cut in half and the Fellowships can fly non-stop between Perth and most of the large mining towns. Tom Price, and a few small towns where the airfields are not suitable for jet operation, still rely on short feeder services by smaller aircraft but these all link with the mainline jet services. One great advantage of the jets is that on many occasions they can fly above the weather which used to give passengers such a rough ride.

Western Australia north of the 26th parallel of longitude is one of the most sparsely populated areas in the world: a vast region inhabited only by about 60,000 people. Yet this tiny population is extremely air conscious, largely because aviation has solved the problems of distance and isolation. A recent letter from Ken Cohen, ex-Operations Manager and now Public Relations Executive of MMA, states, "One interesting point is that in our first full year of operation, 1935, we carried a total of only 893 passengers. It was 1946, some eleven years later, that the yearly carriage passed for the first time, 10,000 passengers. It then took another nineteen years, up to 1965, for the annual number of passengers to increase to 100,000, even though in the early 1960's we had experienced the onset of the mineral boom in this State, particularly iron ore and nickel."

MMA has kept pace with the development of Western Australia and has played a considerable part in this development. The fact that an experienced airline was ready to carry executives and personnel made the establishment of new mining towns that much easier to begin with. As soon as they began production, MMA served them in countless ways. Everything they need, from foodstuffs to heavy machinery, has to be brought in from outside and a great deal is carried by air. There are many occasions when speed is of vital importance. Apart from medical emergencies, the complex machinery of mineral extraction and processing is dependent upon outside facilities. When it breaks down, an outside expert or an essential spare part has to be produced in a hurry, or a company could lose thousands of dollars an hour. Fast-flying MMA aircraft help to keep the machinery running. They deliver the mail, transport families in and out of the mining towns, and supply a good many of the luxuries and necessities of daily life.

MMA supports its aircraft operations with seventeen fully-equipped branch offices and its own specially-trained staff. Each office is connected with telex networks and the Ansett Airlines computerised reservations system for travel throughout Australia. The airline has come a very long way since Horrie Miller took off in its first DH84.

In 1976 Horrie Miller still lived in Perth, not a great distance from that other great name from the early days of Australian aviation: Sir Norman Brearley. Sir Norman's aviation achievements were recognised by the award of the Oswald Watt Medal in 1974. The 1975 award was made to Sir Lawrence Wackett.

It is more difficult for a man to make his name in aviation nowadays. The adventurous days when a flyer could take off into the wild blue yonder without much regard for rules and regulations are more than half-a-century in the past. Now, a flyer is apt to be known only within the world of aviation itself. It is rarely indeed that a flyer enters the public consciousness to anything like the extent that Bert Hinkler did in the 1930s. Flying is now a fact of life and a pilot has to perform a particularly glamorous exploit to break into the headlines. The task of recording the continuing story of Australian aviation is left to such magazines as *Aircraft*, which performs it supremely well.

Aircraft are now "tools of trade" in many parts of Australia and are used for a remarkable variety of purposes. Aerial topdressing, for example. This began about twenty-five years ago, after its efficiency had been proven in the United States. Between 1955 and 1972 a number of firms were founded in each State and aircraft were used to distribute superphosphate and other fertilisers. They were particularly valuable in areas where such fertilising had previously been impossible or uneconomical, such as the steep pasturelands of the Australian Alps. The pilots flew over the pine plantations of South Australia and Victoria, the rice fields of New South Wales, and grain or grass country in many parts of Australia. Some countryfolk who saw topdressing aircraft in action for the first time made panic calls to their local police stations and yelled, "There's a 'plane on fire out our way! Bloody great cloud of white smoke streaming out behind it!"

The specially equipped aircraft were also used for

TOP: A daily event at Paraburdoo Airport, Western Australia. MMA Fellowship jets exchange loads with DH Twin Otters of the same company, which provides the airlink with the iron ore mining town of Mount Tom Price (*MacRobertson Miller Airways*)

BOTTOM LEFT: Sir Norman Brearley, C.B.E., D.S.O., M.C., A.F.C., F.R.Ae.S., photographed in August 1969 when he was seventy-eight years old (*The Sunday Times*)

BOTTOM RIGHT: Jimmy Woods, veteran Western Australian pilot and onetime owner of Woods Airways, known as the "smallest airline in the world." He also started one of the first helicopter companies in Australia. This photograph shows him with his Avro Anson *Willem de Vlamingh* (*The West Australian*)

Above: Pilot George Palmer loads a Cessna with supplies for the township of Gunpowder, north of Mount Isa, during the 1974 floods (*R. V. Eussen*)

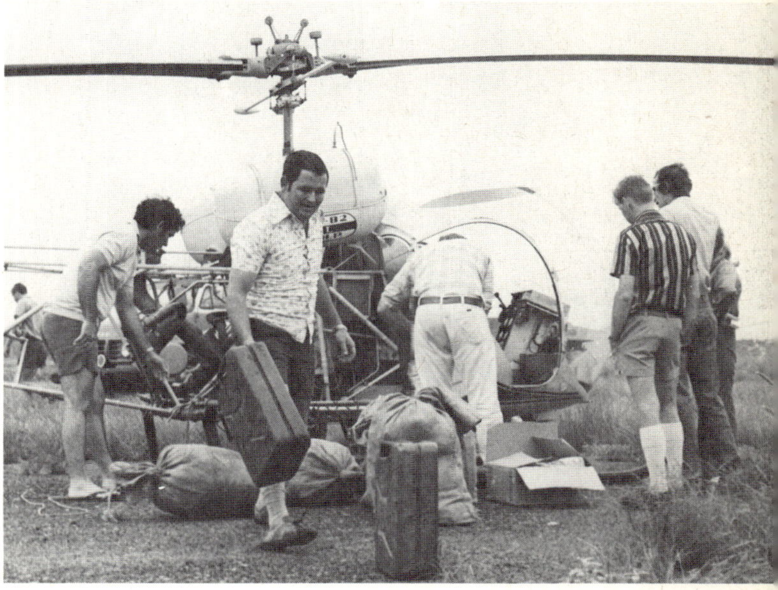

Helicopter rescue missions during the January 1974 floods in the Channel Country of north Queensland. At top right, Mike Gray takes off in his Bell helicopter. The photograph at right shows helpers loading urgent food supplies for Linda Downs Station. Below, the manager of Iffley Station, Peter Foster, carries a day-old calf to safety after a helicopter rescue from the banks of a dam surrounded by floodwaters. Mike Gray (right) flew many lifesaving missions and lifted people from the rooftops of their homes, cars, and caravans (*H. V. Eussen photographs*)

FROM TOP: A DH Drover used by the Royal Flying Doctor Service; an RAAF Dassault Mystère Falcon (*RAAF Museum*); RAAF Convair 440 Metropolitan (*RAAF Museum*); A Fairey Firefly of the RAN Fleet Air Arm (*RAAF Museum*)

pest control in some areas and for spraying such valuable crops as tobacco. It was hairy work for the pilots and not particularly well paid, though a number were badly injured or even killed. In rugged country they had to fly through valleys and bank steeply along hillsides. Over flatlands and plantations they had to fly at not much more than zero feet, and in closely cultivated areas they followed a procedure similar to that of pilots landing on an aircraft carrier, with a man on the ground using a couple of flags or paddles to guide them. Weather conditions had to be judged to a nicety because a wind in the wrong direction could blow the chemical spray or fertiliser onto someone else's paddock.

A good deal of this work came to an abrupt end in 1973, when the Labor government abolished the superphosphate bounty paid to farmers to help them fertilise their properties. This move contributed to the difficulties of a period in which a number of trained Australian pilots have found it hard to obtain full employment.

Farmers in the settled areas and graziers in the outback are becoming increasingly airminded and a good many of them own personal aircraft. This is a logical move for people who may live a couple of hundred miles from the nearest town, with no other way of getting there except by car over tracks ground out by bullock wagons. They use their aircraft for many purposes besides a quick flip into the township. The man who used to spend weeks in mustering a mob of cattle can now take off for a couple of hours flight around his property, to spot the areas favoured as camping grounds by isolated groups. When he returns to the homestead he can show a marked map to his stockmen, who set off by horse or motor-bike to bring the cattle in.

The work of the Royal Flying Doctor Service is well known, and the RFDS handles so many emergencies that they have become almost a daily routine, but somewhat less familiar is the work of the Lions Save Sight Project. This has sent two medical teams to the Northern Territory to check the vision of children and test adults for glaucoma, a disease which causes blindness. The first team to be financed by the Lions Clubs went north in 1972, and the second, consisting of four doctors and four orthoptists, visited Broken Hill, Wilcannia, eleven centres in the Northern Territory, and five settlements in Arnhem Land in 1975. Led by Professor F. Holland of the University of New South Wales, one of Australia's leading authorities on eye diseases, the team tested something like 10,000 adults for glaucoma and about 3,500 children for visual defects. The expedition was a prime example of the value of aircraft under Australian conditions. With the help of local organisers it took only two weeks to cover the route and perform all the examinations, a task which would have been next to impossible but for air transport.

Australian flyers use their machines for fire spotting, for fish spotting to help commercial fishermen find the deep sea shoals, and even for shark spotting to warn swimmers on city beaches. Some commercial travellers fly around country centres and thus spend their time selling instead of travelling. The flexibility of aviation enables it to be used in hundreds of ways to solve the problems of distance and isolation which bedevilled Australians from 1788 to the second World War.

For a population of about 13 million Australia has a substantial number of trained pilots and civil aircraft. The latest available figures show that 34,806 Australians are licensed to fly. 28,413 of these are certificated only for private flying: the remainder hold various types of commercial certificates. There is a total of 4,269 civil aircraft in Australia and of these 2,123 are for private use only.

A large number of the pilots are women—and women pilots are not necessarily weekend flyers only. They have made their mark in commercial aviation too, despite determined male opposition over the years. Robin Miller wrote of her attempts to join a commercial airline when she had obtained the requisite certification, and of the increasingly embarrassed objections raised by its male executives who finally fell back on the protest that she would not be physically strong enough to do all the work. In fact, women pilots have long since shown that they possess the flying flair and instinct to a high degree, together with a courage and determination which enables many of them to become flyers in face of great difficulties. Eileen Steenson wrote of the financial problems she had to overcome to obtain her pilot's licence, and many other women flyers have shown equal determination.

In an article in the *Australian* of 27 January 1976, Judy Olding related the 700 members of the Australian Women Pilots Association to the early days of "female flying" in Australia, when Millicent Bryant and Evelyn Follet were refused lessons by the Aero Club of New South Wales. It took Evelyn Follet twelve months to break down the barrier, and she later became co-founder of Adastra which is mentioned earlier in this book. Judy Olding mentions a number of women who made their name in Australian aviation, both in the air and on the ground, including Nancy Ellis Leebold who was one of our most talented aviation technicians. Mrs Margaret Kentley, who with Marie Richardson was the outright winner of the 1973 New South Wales Air Race, told Judy Olding, "A person is a pilot first—a male or female next."

Apart from the dedicated flyers and would-be flyers who spend everything that they can or can't afford on flying lessons, weekend flying, and light aircraft charter, and dream of the day when they'll own an aircraft of their very own, there are many devotees of powerless flight. One area of Australia, around Waikerie on the Murray, has proven particularly suitable for gliding because the huge expanse of flat country which surrounds it creates steady thermals.

And there has been a strange reversion to the earliest days of aviation. A few balloonists are sailing above the wide brown land, held up by envelopes filled with air heated by Portagas burners, but this is an expensive and complicated sport compared with the newer one of hang gliding. Hang gliding clubs are springing up all over the country, and on most weekends one may see devotees assembling the colourful plastic wings and simple aluminium frames of their gliders on hillsides and clifftops. They are even gliding from the very spot where Lawrence Hargrave experimented with kites and gliders so long ago, and perhaps his spirit stirs when it senses the silent breathtaking swoop of a hang glider overhead with its pilot's legs dangling like those of Otto Lilienthal's in sketches drawn in the nineteenth century. Lilienthal, one of the pioneers of flight, died in his attempts to build a one-man glider and hang gliding has already claimed its first victim. But the chance of death or injury has never deterred those who wish to "take unto themselves wings like a bird." The fascination of flight is eternal.

Into the new age. Led by Group Captain J. Newham, a flight of nine F111Cs featured in the opening ceremony of the Sydney Opera House (*Department of Air*)

18

Through Korea and Vietnam

Twenty-five troubled years of world conflict have demanded the utmost in RAAF efficiency. As usual, Australians in the air have responded magnificently

In 1946, RAAF veterans thought the situation had a depressing similarity to 1919. Another "war to end war" had been won and once more the accumulated expertise of wartime was being dissipated as fast as clerks could issue demobilisation papers. The force of 183,822 men and women and 5,845 aircraft built up during the war very soon faded away. Most aircrew and groundstaff found other jobs: many aircraft were sold or dismantled. A number of permanent RAAF personnel went to Japan as part of the British Commonwealth Occupation Force (BCOF) and these were augmented by short service recruits signed on for only two years. The whole mighty machine came almost to a halt.

But it was only a temporary standstill and was to some extent intentional, to give the top brass and the public servants a chance to decide what form the "peacetime" RAAF should take. Their deliberations soon took on a note of urgency. The world powers had had time to draw breath and make a new shift in alliances. Russia, the great wartime ally, was becoming the Cold War enemy, and was inspiring like-minded countries to challenge the power of capitalism.

The first moves came in June 1948, when Chinese guerrillas started off the twelve-year "Malayan Emergency" by slaughtering three rubber planters. In the same month, the Russians closed the ground routes into West Berlin. These events had immediate repercussions on the RAAF.

The government sent ten aircrews, with RAAF Dakotas, to help in the great airlift from West Germany that kept the beleaguered city supplied until mid-1949. The Dakotas carried food, fuel, medicines, and clothing into West Berlin and brought passengers on their return trips. Among other cargoes they carried a total of 7,000 tons of coal. It was unglamorous but dangerous work, flying literally at rooftop level and often in the worst possible weather conditions.

The government was not quite so quick to react to the needs of the British fighting the Chinese terrorists in what was then called Malaya, but it is fair to say that the British were still working out their strategy. This developed into a system of hounding the guerrillas through the jungle and spotting and attacking their bases from the air. Aircraft were used also to supply the ground forces so that they could spend prolonged periods in the jungle.

1 Squadron RAAF, flying Lincoln bombers, became operational in Malaya in July 1950 and served continuously until July 1958. 38 Transport Squadron, flying Dakotas, ran the airdrop service which parachuted rations, ammunition, and other equipment to the mobile columns. The RAAF units began the lengthy Australian involvement with Malaysian defence which was not phased out until the comparatively recent past. Air operations were very similar to those against the Japanese except that there was no active opposition. The bomber and transport pilots had to fly through tropical storms above the mountains and jungles of the Malayan peninsula and pick out their targets with pinpoint accuracy, no easy task when one jungle clearing looks much the same as another. 1 Squadron, based on Tengah, flew about 4,000 operational sorties during its time in Malaya.

The new form of the RAAF had taken shape before the squadron dropped its first bomb. In June 1947, the government approved an air force consisting of 144 aircraft divided between sixteen squadrons. Seven of these were to be used on home defence and nine as a striking force.

The striking force had ninety aircraft divided into three squadrons of heavy bombers; two long-range fighter squadrons; two transport squadrons; one tactical reconnaissance squadron; and one survey squadron.

Home defence had fifty-four aircraft employed as four interceptor squadrons; one bomber squadron; one maritime strike squadron; and one "general purpose" squadron equipped for air-sea rescue, reconnaissance and photographic work, and aerial bombardment.

The government recognised the need for a steady supply of skilled personnel at every level, and approved the establishment of the RAAF College at Point Cook and the Apprentice School in New South Wales.

These moves towards a "new order" were encouraging because they showed a far more emphatic recognition of Australia's military aviation needs than had been the case between 1919–39. They came only just in time. From 1950 onwards the RAAF has been one of the few air forces of the world to go through long periods of active operations. The principal problem, encountered almost immediately after this new planning was approved, was obsolescence of RAAF aircraft. The Mustangs, Lincolns, and machines of similar vintage with which the RAAF ended the war were rapidly being pushed into the background by new designs. It became obvious that re-equipment of Australia's air force would cost fantastic amounts. As stated in previous chapters, the government has never seemed able to find a truly Australian solution to this problem and still relies largely upon overseas developments.

Another forward move made during the post-war planning period was the decision to establish an independent Fleet Air Arm attached to the Royal Australian Navy, and this might be described as the second attempt to create such a force.

Spotter aircraft flown by AFC pilots were carried by RAN cruisers in the first World War, but Australian seaborne aviation went into the doldrums until 1926 when a naval aviation course opened at Point Cook. This coincided with plans for a seaplane carrier, HMAS *Albatross*, and when she was commissioned in 1929 the first graduates from the course were ready to take up their duties. They flew six Supermarine Seagull III seaplanes.

This embryo Fleet Air Arm was knocked on the head by the Depression, and in any case the use of seaplanes for fleet operations was already outmoded. The comparatively slow seaplanes were sitting ducks for rapid-fire ack-ack weapons or carrier-based aircraft. The *Albatross* was laid up and later traded-in for the cruiser *Hobart*, bought from the British.

Apart from seaplane and flying-boat operations,

A Canadian de Havilland Beaver purchased by the RAAF, still fitted with her snow skis (*RAAF Museum*)

Australian pilots gained comparatively little experience in naval aviation during the second World War unless they were serving as RN carrier pilots. When the RAN founded its own Fleet Air Arm it had to rely quite heavily upon the expertise gained by the Fleet Air Arm of the Royal Navy, which was a direct descendant from the Royal Naval Air Service established just before the first World War.

The establishment of an Australian Fleet Air Arm, in 1947, was part of an overall plan for future defensive and offensive strategy. The planners believed that, among other things, it would lift some of the burden of coastal defence from the RAAF.

The FAA became operational in May 1949 when HMAS *Sydney*, built by the Royal Navy as the *Terrible* in 1943–45, anchored in her home port with twelve Fairey Fireflies and twelve Hawker Sea Furies arrayed on her flight deck. She was Australia's first aircraft carrier but had a comparatively short life in that capacity. She was converted to a troop carrier to run between Australia and Vietnam, then rusted in Sydney Harbour while the government sought a purchaser who would guarantee not to use her for warlike purposes.

The *Sydney* and her two squadrons comprised the 20th Carrier Group. Another wartime carrier built by the RN but never commissioned, the *Majestic*, became HMAS *Melbourne*. Carrier designs change rapidly and she was almost rebuilt, to incorporate the new British invention of an angled flight deck which vastly increased the fly-off area, before she was commissioned into the RAN in 1955. The planned total of three carrier groups with a total of seventy-two aircraft was achieved when the *Melbourne* joined the fleet.

The *Sydney* arrived just in time to acquit herself nobly in the Korean War, but the *Melbourne* seemed fated to be an unlucky ship. She sank the RAN destroyer *Voyager* by collision on 10 February 1964 and the USN destroyer *Frank E. Evans* on 3 June 1969. 156 men died in the two disasters.

For Australian pilots, the Korean War began when 77 Squadron RAAF was alerted in late June 1950. The squadron, part of BCOF, was stationed at Iwakuni in Japan and was just coming to the end of its overseas tour when the North Koreans invaded South Korea. It was the last of eight RAAF units which had been stationed in Japan, and its personnel already were giving farewell parties in anticipation of their return to Australia.

But by 2 July 1950, 77 Squadron Mustangs were snarling over Korea on three missions in support of US Air Force bombers and transports. In the first month of combat operations they flew 172 missions against communist forces who were pressing hard on the US and Republic of Korea units. With rockets and cannon fire they blasted troop formations, transport columns, armoured vehicles, and artillery, and destroyed a number of the bridges which are essential links in the roads which wind around the stark Korean mountainsides.

These strikes were carried out from 77 Squadron's base at Iwakuni, which meant long flights for the Mustangs and their pilots and the use of auxiliary fuel tanks. This procedure was eased somewhat when the squadron began to use Taegu airfield, in South Korea, as a staging post for refuelling and arming. Extremely long flights were a feature of Korean operations and at one time the pilots were flying as much as 1,000 miles on a single sortie.

The tide of battle turned when the confused and disorganised US and ROK troops rallied against the invasion which had probed deep into South Korea. By October the North Koreans were retreating fast towards the Yalu River, hard-pressed by the Americans. 77 Squadron had taken a vigorous part in the repulse of the communists and had carried out a number of low-level attacks on enemy units actually assaulting Allied positions.

It seemed the war was almost over and that the North Koreans would retire to lick their wounds, but it hotted up again when the Chinese poured thousands of "volunteers" into Korea to support their communist allies. Together with troops, the People's Republic sent Russian-built armour and aircraft, especially MiG15 jet fighters. They brought a new dimension to the aerial war. Up to then the USAF and RAAF pilots had encountered little opposition except for some tentative sparring by North Korean pilots flying second World War piston-engined machines.

Australia's 20 Carrier Group did not take part in the first stage of the war but the "buzz" soon began to circulate around HMAS *Sydney* to the effect that she was ordered to Korea. As usual in wartime, rumours had the ship in action a dozen times before she was actually engaged. Attached to the US 7th Fleet, and with 805 and 808 Squadrons of Sea Furies and 817 Squadron of Fireflies aboard, she went into action on 11 October 1951.

On that day, the Sea Furies and Fireflies flew eighty-nine sorties against troop positions in the bleak hills around Kojo on the east coast, and for the rest of the month the *Sydney* was like a wasp's nest. All day and every day, her aircraft were taking off and landing as they executed hundreds of attacks along the coast in her patrol area. They flew in support of the Commonwealth Division of British, Australian, Canadian,

Soldiers of 5 Battalion, Royal Australian Regiment, boarding a helicopter of 9 Squadron RAAF in Vietnam (*Department of Air*)

New Zealand and Indian troops. The RAN pilots hit coastal shipping, bridges, road transport, enemy troop and artillery positions, and railways. By the time they returned to Japan at the end of October the Australians had flown nearly 900 sorties, for the loss of three aircraft and one pilot.

By that time the RAAF units in Korea had increased by one: 30 Transport Unit (later 36 Transport Squadron). They flew Dakotas in support of the Commonwealth Division. In less than eighteen months they covered more than a million miles and carried 100,000 military passengers and nearly 5,000 tons of freight and mail.

77 Squadron said farewell to their Mustangs in April 1951. The jet fighters introduced by the Chinese into Korea could fly rings around the mighty Mustang, one of the last and greatest of the piston-engined fighters, and it was time for her to retire. The final great effort of the Mustangs was in support of 3 Battalion Royal Australian Regiment at the battle of Pakchon in November 1950, when they blasted communist troops with rockets as the Diggers attacked on the ground.

The pilots of 77 Squadron returned to Iwakuni to retrain in Gloster Meteor VIII twin-jet fighters, which had combat speeds up to 500 m.p.h. and ceilings of 40,000 feet. They were a long jump ahead of Mustangs but still not the best available. It was thought that the USAF F86 Sabres were the only jets capable of standing up to MiG15s, but the Americans had none to spare for the Australians. Opinions changed somewhat when 77 Squadron flew their Meteors over Korea. The first month was uneventful, but on 29 August 1951 a force of thirty MiGs jumped eight Meteors near Chongju. At odds of more than three to one the Australians gave a good account of themselves. One Meteor plummeted, and another was so badly mangled by cannon fire that it could only just limp back to base, but after that the MiGs called it a day. There was no doubt that the MiGs had the edge on the Meteors, especially at high altitudes, but 77 Squadron fought back very briskly and during subsequent battles they downed three of the enemy.

The Meteors proved excellent for ground-attack operations, and for the rest of the Korean war the squadron was kept busy on these with occasional dogfights when the MiGs decided to intervene. In three years, 77 Squadron pilots flew 18,872 sorties in Korea, about 15,000 of them in Meteors, in a total of 4,836 missions.

Meantime the *Sydney* had been solidly engaged during a hectic period of operations which lasted from November 1951 until late January 1952, much of it in the abominable winter weather of those northern waters. As part of task groups comprising British, American, and Australian warships, of which she was sometimes in command, she patrolled the eastern and western coasts with her aircraft flying sorties whenever the weather allowed. When she returned to Australia,

Warrant Officer R. D. Guthrie, in front of his Meteor, shows flight mechanic Leading Aircraftsman S. L. Julius the map of a flight over Korea (*Department of Air*)

her pilots had flown 2,366 sorties for the loss of eleven aircraft.

The RAAF had learnt many useful lessons in Korea and these were applied to a further reorganisation. It began when Air Marshal Sir Donald Hardman was appointed Chief of Air Staff in 1952. It was his task to streamline the RAAF in a manner appropriate to the jet age which had now arrived in no uncertain terms.

The 1947 reorganisation had divided Australia into five RAAF command areas, continuing a system begun during the war. The backbones of these command areas were the Citizen Air Force squadrons: 21 CAF Squadron, Melbourne; 22 CAF Squadron, Sydney; 23 CAF Squadron, Brisbane; 24 CAF Squadron, Adelaide; and 25 CAF Squadron, Perth. Each of these squadrons bore the name of its own city, for example 21 'City of Melbourne' Squadron.

Citizen Air Force squadrons had a long and honourable history. Originally formed in 1936, they were absorbed by the wartime RAAF and reformed in 1948. In the early 1950s they were still equipped with such aircraft as the Mustang, Wirraway, and Tiger

Moth but they were regarded as an important part of Australian air defence. University students, young businessmen, and other flying enthusiasts were dedicated members of the squadrons and gave up many evenings and weekends to ground and air training. But they were far more than "weekend pilots." They achieved a high level of proficiency and were given plenty of opportunities to learn their trade in simulated combat conditions. They flew in such exercises as camera-gun dogfights, long distance formation flying, rocket attacks on ground targets, firing at drogues, and bombing.

The proportion of CAF men to permanent RAAF personnel varied somewhat from one squadron to another, but on average it was in the proportion of about sixty/forty. The RAAF personnel comprised ground staff, administrative staff, and instructors, many of whom had combat experience which helped them to hone the CAF into a sharp-edged weapon. At its peak, the strength of the CAF was close to 17,000 men.

TOP: LAC Alf Wade connects the firing circuit of a high-explosive rocket under the wing of a 77 Squadron Meteor in Korea (*Department of Air*)

BOTTOM: RAAF Meteors skim over a village in North Korea as they take off on an air strike (*Department of Air*)

TOP AND BOTTOM: Hang-gliders in Australia. Hang-gliding is the newest aerial sport in Australia and is popular because it enables aviation enthusiasts to enjoy the thrill of flight with comparative ease and economy. The gliders are made of light alloy tubing with plastic fabric "wings." Approximately 4,000 Australians now participate in this sport. (*photos: Ron Langman*)

Free ballooning is now a weekend hobby with some Australians: a swing back to the very early days when balloons were the only way of getting off the ground. The balloons are lifted by hot air supplied by Portagas burners. (*photo: Ron Langman*)

The Hardman reforms altered the organisational structure of the RAAF, and the five area commands were combined into one Home Command to be used as a strategic rather than a tactical weapon, but the CAF was still regarded as a vital home defence unit. Training and recruitment continued vigorously through the 1950s, but early in the decade it was obvious that most CAF aircraft were almost obsolete. Re-equipment with Vampire jet fighters began in 1954 and the citizen pilots re-qualified on the new aircraft—though some unregenerates delighted in showing that Mustangs could almost match the Vampires in a dogfight.

The CAF became such an integral part of the RAAF structure that the squadrons flew on manoeuvres with permanent RAAF units. The RAAF bombers carried out simulated high and low level attacks on Australian cities and the CAF had to fight them off. These exercises were carried out with such energy and enthusiasm that it needed only the crackling roar of cannon fire to make them the real thing.

TOP: RAF Vampires of the type flown by the Citizen Air Force squadrons of the RAAF (*Hawker Aircraft Limited*)

BOTTOM: "Pilot error." Airfield collision between a Gypsy Moth and a Vampire (*Hawker Siddeley photograph*)

The RAAF in Korea. Squadron Leader R. C. Cresswell, Commanding Officer of 77 Squadron, giving last minute instructions to Meteor pilots before the squadron took off to escort USAF Shooting Star jets on a photographic mission. Enemy jets were sighted and flak encountered during the mission (*Department of Air*)

But the Vampires eventually became as obsolete as the Mustangs. If the CAF squadrons were to continue as a fighting arm of the RAAF, ready to take the air within an hour or two of any overt threat to Australia, they had to be re-equipped with far more sophisticated fighters. The Air Staff looked wryly at the cost of such machines together with the necessary ground facilities to keep them in the air, to say nothing of the time and money needed to retrain the citizen pilots. They were not even certain whether "part time pilots" could learn and maintain the skills necessary for flying the incredibly complicated new machines.

So, reluctantly, they decided to strip the wings from the Citizen Air Force. In 1960, the CAF squadrons became ground training units.

The Vampire was Australia's front line fighter throughout the 1950s. This was the decade in which Britain and America were rallying the western democracies against the communist bloc: the days when the North Atlantic Treaty Organisation (NATO) became a solid political force. Australia's prime minister Robert Gordon Menzies, who according to his own description is "British to his bootheels," decided that Australia should play some part in this grand alliance and so, for the first time in history, RAAF squadrons went to the European area as operational defence units in peacetime. 75 and 76 Squadrons, then in the process of switching over from Mustangs to Vampires, were ordered to Malta as 78 Fighter Wing. They co-operated with the RAF and RN Fleet Air Arm units as a link in the chain of defences erected against communist ambitions.

The Air Staff encountered some brain-stretching problems when they began to shop around for new aircraft in the early 1960s. They had to think about a whole range of replacement machines, from transports to fighters. This was not only because the existing equipment was entering the all too familiar period of obsolescence, but also because air planning was turning towards the creation of an air force in which tactical mobility was to reign supreme. The world was entering the era of "brush fire wars" and it was necessary to have an air force which could operate quickly and efficiently under almost any conditions. By 1962 the Australian government had made its first commitment to Vietnam by sending a group of army instructors, and some thought had to be given to the possibility of RAAF involvement.

The first step into the new era was made by ordering eighteen Caribou transports from de Havilland Aircraft of Canada. The Caribou was to supplement the Dakota aircraft which has served so well for so long, and the new machine had considerable advantages. It could use unpaved airstrips and shorter runways for take off and landing, and load or unload much faster through the big ramp and doors in the stern of its cargo area.

The next problem was that of fighters and tactical bombers, and the solution of this one led the RAAF into its lengthy involvement with the F111C. Probably no other aircraft has been subject to such prolonged speculation and discussion as the "Eff One-Eleven." Everything about it, including the price, is so far beyond the average man's comprehension that he usually treated it as a joke.

The name was heard first in 1963, when newspapers reported that the RAAF had ordered twenty-four of the new aircraft at an all-up price of $112 million.

This was a bold gamble because the F111 was an untried aircraft and the RAAF put a great deal of trust in their interpretations of the manufacturer's designs and specifications.

Despite the difficulties associated with this aircraft it was worthwhile. If it had come off the RAAF would have had an aircraft years ahead of its time.

Unfortunately the manufacturers struck an apparently endless series of design problems. The greatest was that of the swing wing. One of the objects of the swing wing was to alter the characteristics of the aircraft while in flight so that she became supersonic, but the gremlins were so busy that the wing would not swing as it should have done. Barnes Wallis, the English aviation designer who has worked on countless projects since airship days, said that the Americans were on the wrong track. But they persisted, and at long last the RAAF could send pilots to the USA for F111 familiarisation and training. That was in 1968. In June 1973, just under ten years from the time when the RAAF lodged its order, the Australian pilots ferried the first six F111Cs across the Pacific to 6 Squadron airfield at Amberley, Queensland. The order has now been completed and the twenty-four aircraft have cost more than $300 million.

The F111C is an immensely complex piece of technology. Pilots and observers have to spend seven weeks in ground school, learning as much about her as they need to know, before they fit themselves side by side into the air-conditioned cockpit. After that they spend ten weeks in flight training.

A single F111C is capable of striking a blow immensely greater than all the bombs dropped by the RAAF in the second World War, but that's only if she's used for the nuclear capability for which she's equipped. Australia has not yet become a nuclear

One of the de Havilland (Canada) Caribou transports purchased by the RAAF in the 1960s (*RAAF Museum*)

The General Dynamics F111C (*RAAF Museum*)

power and so F111C operations are predicated on the use of conventional bombs, which still makes her a terrifying weapon.

Included in the range of her equipment of which the whole list sounds like something out of science fiction is the Terrain Following Radar, or TFR. It is a combination of radar and computer. The radar keeps a constant check on ground obstacles well ahead of the aircraft and feeds this information into the computer. The computer digests it, translates it into instructions much faster than any human mind or body could ever react, and passes a constant stream of instructions to the controls. These operate automatically and keep the F111C flying at the height pre-set by the pilot. She follows the contours of the ground with roller-coaster accuracy. If she's flying above a plain, she keeps straight and level. As soon as the ground rises into a hump or hillock or a range of hills, or whenever it's broken by a city or a forest or an oil refinery, the computer whirs and the servo-mechanisms obey and the aircraft rises the exact amount that will maintain her steady height above obstructions.

This means that the F111C can fly as low as 200 feet at supersonic speed, about 1,600 m.p.h. At such a speed, a human pilot could never move fast enough to keep a regular height above every obstacle. The advantage of this characteristic is that the F111C can fly below the radar beam of any detector developed so far; fly too low for anti-aircraft projectiles or missiles to be effective; and escape intercepting aircraft because they could not dive on her without plunging into the ground. In simple terms, she is a jet-propelled weapon that can strike and disappear almost before anyone knows what's happening.

TFR is only one component of the almost incredible box of electronic and mechanical tricks that is the F111C. She is one of the most complex and versatile aircraft ever built and her swing wing allows her pilot to make use of other flight characteristics as well as supersonic ability. The RAAF's twenty-four F111Cs have certainly put Australia well out ahead in the international aviation race, but there are some hefty problems. One is the huge cost of the aircraft. Another is the cost and complexity of the ground facilities required, and the amount of training necessary for aircrew and groundcrew. And, greatest of all, perhaps, is the fact that Australia is dependent upon the USA for replacements. If the nation should once more find itself

TOP: One of RAAF 10 Squadron's Neptunes, photographed during Operation Longex 72 which was a joint exercise of Australian, New Zealand, Canadian, and United States navies and air forces (*Department of Air*)

BOTTOM: RAAF Mirage fighter (*Hawker Siddeley photograph*)

Top: The first English Electric Canberra twin-jet bombers were ordered by the RAAF in May 1952 and this type of aircraft, with successive modifications, was used for a number of years including the Vietnam War period (*Hawker Siddeley photograph*)

Bottom: The Lockheed C130 Hercules. This type of aircraft is now used by the RAAF as their "heavyweight" transport and freighter (*RAAF Museum*)

out on a limb during a global war, then we'll be as dependent on America for our front-line aircraft as, in 1914 and 1940, we were on Great Britain.

While the F111 designers were still scratching their heads over the swing wing problem, Australia bought one of the American aircraft models in which their technology had triumphed: the Lockheed Neptune. This was a flying electronic power-house designed for one specific purpose: hunting and killing maritime intruders. With 12,000 miles of coastline to watch, the RAAF had a mighty problem. The Neptune, developed by the Americans to protect their own huge stretches of coastline, was the answer.

The Neptune's sub-hunting equipment evolved from that developed late in the second World War, including the airborne magnetometer and the sonic buoy. In simple terms, a magnetometer is an instrument which measures variations in the earth's magnetic field. A submerged U-boat, being a mass of ferrous metal, had its own magnetic field and thus created a local fluctuation in the general north-south pattern of magnetic lines of strain.

An aircraft which picked up such a local variation could "test its hunch" by dropping a sonic buoy. This contains a sonar instrument of the type used by warships to transmit underwater sound waves. They bounce back from a submarine's hull and are picked up by the warship's receiver. When the sonic buoy's sonar gear transmitted and received a similar signal, the result was automatically radioed to the circling aircraft and it promptly plastered the area with a pattern of depth bombs.

A submarine's greatest asset, invisibility, was pierced by these inventions. Aircraft which carry their modern equivalents can keep an electronic eye upon vast areas of ocean in a manner which is enormously valuable to an island continent like Australia. The Neptune patrols, supplemented and now being superseded by the Orion patrols, keep a regular and reliable check upon our ocean approaches and fly almost to the edge of the Antarctic ice.

The continuous re-equipping of the RAAF which has been necessary ever since the first order was lodged for F-111s has included the purchase of Mirage III-O fighters for the air defence and ground attack squadrons; Canberra and Phantom F4-Es for the bomber squadrons; Hercules C130A and E series, Caribous, Mystere 20s, HS748s, and BAC111s for the transport squadrons; Neptune SP2Hs and Orion P3-Bs for the maritime squadrons; Winjeels and Macchis for training; and a new departure: Iroquois and Chinook helicopters for the newly formed helicopter squadrons.

The most noticeable feature in this array of aircraft is, of course, the lack of Australian-origin machines for operational purposes. Some of the foreign aircraft have been built under licence for the RAAF in Australia, but the vital element which would have inspired the creation of Australian machines to rival those from overseas seems to have been lacking.

Nevertheless the RAAF has chosen well from the great variety of military aircraft available, and for a small nation in terms of population we have a flexible and well-balanced air force. The Hercules C130As and Es, for example, are excellent transport aircraft and have given enduring service on operations as diverse as carrying troops and supplies to Vietnam and evacuating children during the final collapse. Built by Lockheed, the Hercules series of transports has been in production since 1954.

The restructuring of the RAAF continued over about twenty years from 1952 onwards. In 1952, three Field Commands were formed and these have since been reduced to two: Operational Command and Support Command, under the control of the Air Board. The present structure is founded on a series of bases, where units having similar functions or requiring similar facilities are grouped together. The flying squadrons are divided into strike reconnaissance; air defence; ground attack; medium and short range transport; helicopter; and maritime. As in the other branches of the armed forces, the "sharp end" of the RAAF is backed up by a complex support organisation including hospitals, stores depots, communications units, repair and overhaul depots, schools and training

FROM TOP: The CAC27 Winjeel, built by the Commonwealth Aircraft Corporation as one of the RAAF basic training aircraft; the CA3D Macchi trainer; and the Macchi MB326H (*RAAF Museum photographs*)

Top left: Preparing to load a Red Top air-to-air missile on a Gloster Javelin (*Hawker Siddeley photograph*). Top right: An RAAF Chinook helicopter picks up a patrol group. Centre left: RAAF Bell Iroquois helicopter. Bottom: Cessna 180 in RAAF service (*RAAF Museum photographs*)

units and organisations including the Women's Royal Australian Air Force, the RAAF Academy, Staff College, Diploma Squadron, and Apprentice School, the Airfield Construction Squadron which builds and maintains airfield facilities, and the special conversion units which re-train pilots to transfer from one type of aircraft to another or to qualify in newly-acquired aircraft.

The efficiency of the RAAF has stayed at a constantly high level because it has been a "working" air force for more than a quarter of a century, ever since the brief slump in operational purpose immediately after the second World War. It has gained vital experience in modern military operations in south-east Asia, and is frequently called on to help in civil rescue operations within Australia during such incidents as the aftermath of Cyclone Tracy in 1975 and the inland floods of 1976. Such experience, when everyone from pilots to ground crew is flying "for real" and not simply as part of an exercise, has helped the RAAF to achieve an enviable pitch of professionalism.

The most recent operational involvement, of course, was the "forgotten war" of Vietnam, that so many people did not even want to think about while it was going on and gladly wiped from their memories as soon as it was over. The RAAF was there from August 1964 until 1972, when it began a controlled withdrawal. According to the *Official Year Book of the Commonwealth of Australia*, No. 58, 2 Canberra Bomber Squadron operated in South Vietnam from April 1967 to June 1971 and flew over 13,000 sorties. 9 Helicopter Squadron was there from June 1966 to December 1971 and flew over 237,000 sorties in support of Australian troops. It carried over 414,000 military personnel and nearly 27 million pounds of freight and made 4,000 medical evacuations. RAAF Transport Flight Vietnam, later 35 Caribou Squadron, flew in South Vietnam from August 1964 to February 1972 and flew over 81,000 sorties. The Caribous carried over 677,000 passengers, 91 million pounds of freight, 12 million pounds of mail, and 106 medical evacuations.

Like all wartime statistics, these cold figures do not convey the immense effort and extreme risks of one of history's most unpleasant wars. A total of 4,443 RAAF personnel participated in the Vietnam war and earned 120 decorations, including sixty awards of the Distinguished Flying Cross and three Bars to the D.F.C., and 138 were mentioned in desptaches.

One officer, Flight-Lieutenant Garry Gordon Cooper of Adelaide who is now the captain of a Cathay Pacific airliner, became one of the most highly-decorated of all living Australian servicemen because of his exploits in Vietnam. But he is permitted to wear only the D.F.C., which is one of the British awards approved for Australians, and his campaign medals. The other decorations were awarded by the Americans and he is not officially allowed to wear them.

The RAAF rescues an outback stockman with a foot injury during the inland floods of 1975 (*Department of Air*)

This strange situation was reported in Australian newspapers in March 1975, after Garry Cooper had left the RAAF. The newspaper articles claimed that, during his eight-month tour of Vietnam duty, he was recommended for about thirty US awards for gallantry. One of the recommendations was for the Congressional Medal of Honour, the American equivalent of the Victoria Cross and equally valued by fighting men. The recommendation was made by Major-General Julian J. Ewell of the US Army, who said that Cooper had displayed great coolness and heroism against overwhelming odds and "saved one of my most valuable brigade commanders."

Cooper was flying with the brigade commander, as air liaison officer in a helicopter piloted by an American. The helicopter was struck by enemy fire, the pilot was killed and the other two men wounded. Cooper, dazed and covered with blood, managed to reach across the dead pilot and grasp the aircraft controls. He brought the helicopter down to a crash landing, dragged the brigade commander out of the wreckage, and then fought off a Viet Cong patrol. He killed ten of the enemy before American troops broke up the fight.

On other occasions, he flew what the Americans called "bird dogging" operations. The "bird dogs" flew tiny unarmed Cessnas over the jungle ahead of advancing troops, in an attempt to spot enemy ambushes. If they spotted suspicious movements they dropped a smoke marker, but very often the first sign of the enemy was machine-gun bullets ripping through the fuselage.

It is always slightly invidious to single out the exploits of specific men, because it means that those of so many others must go unrecognised. And, as always in wartime, it is the overall effort of an entire force that counts the most.

The whole story of the RAAF involvement in Vietnam has been very well told by George Odgers in *Mission Vietnam*, published by the Australian Government Publishing Service for the Department of Defence (Air Office). The book does not only cover the operations of the units previously mentioned but also the activities of the Base Support Flight, No. 1 Operational Support Unit, and the RAAF Contingent at Vung Tau.

Odgers traces the story from 8 August 1964, the day on which the RAAF Transport Flight, Vietnam, flew their Caribou transports from the RAAF base at Butterworth in Malaysia to Vung Tau in South Vietnam. Vung Tau was an airfield about forty miles south-east of Saigon. The Caribou flight was welcomed by the local American commander and by South Vietnamese bearing a banner lettered "Vietnamese People Warmly Welcome the Australian Air Detachment."

Vung Tau had been a holiday resort but the RAAF crews had little time to relax. The Transport Flight was integrated operationally into the US Air Force airlift system and it was controlled by the 2nd Air Division's 315th Tactical Airlift Wing. The Caribou crews were taken on briefing flights by the Americans, to enable them to compare area maps with the Vietnamese terrain, and on 14 August 1964 the RAAF men flew their first operational missions.

The red kangaroo insignia of the RAAF soon became well known throughout South Vietnam and the Transport Flight was nicknamed "Wallaby Airlines" by the Americans. As the first months went by the aircrews became accustomed to handling an amazing variety of cargoes. They carried troops, civilian passengers, ammunition, mail, and food. Some of the food was "on the hoof" in the form of live ducks, pigs, and cows. "A knowledge of cattle handling, while not a requirement, certainly would have been an asset to many an RAAF loadmaster in Vietnam," Odgers says. The cows were loaded in wooden crates to be parachuted to their destinations, but sometimes they broke loose in the cabins and caused chaos.

The South Vietnamese troops often took their families along when they were being shifted from one base to another, so the loadmasters had to look after passengers ranging from infants to grandparents and find space for all their household goods.

Many of the flights were fairly uneventful but those which took the Caribous into the northern provinces

Helicopters in RAAF service during the 1970s. FROM TOP: Bell Sioux; Bell Kiowa; Bristol Sycamore; Sud Alouette (*RAAF Museum photographs*)

could become very tricky at times. Viet Cong guerrillas were active on the ground and the transport aircraft were tempting targets. They had fighter escorts into the danger areas but these could not prevent a well-concealed guerilla from loosing off a cartridge belt at a Caribou, and all the aircraft drew fire during their northern flights.

But the first Caribou to be written off was lost through "operational error" rather than enemy action. Piloted by Flying Officer Brian Hammond, she was on a supply mission to a US Special Forces outpost at A Ro, in the mountains less than ten miles from the Laotian border. The parachute supply drop system failed but the outpost commander told Hammond that they were very short of supplies and suggested that he could land. He told him that the airstrip was suitable for the Caribou. Hammond landed successfully but had to use sustained reverse thrust to brake the aircraft, and this smothered the windscreen with mud spray. He could not see a soft patch on the edge of the strip and the starboard wheel sank into it. The Caribou was badly damaged and a salvage party had to fly in to strip her of components that could be used again.

Early in 1965 the Viet Cong assault was stepped up continuously and the Caribous were in the air for as many as ten sorties a day. The aircrews were on an eight months tour of duty and some of them flew more than 1,000 sorties during that period. US ground troops began to pour into Vietnam and in June the 1st Battalion Royal Australian Regiment arrived and went into action north-west of Saigon. The Caribou crews often landed their aircraft in areas under attack by the Viet Cong. One aircraft landed short at Ha Tien, a fortified settlement near the Cambodian border, and suffered damage to its undercarriage. A work party of RAAF and US Army technicians flown in to repair it had to take cover from Viet Cong attacks, and when they had the Caribou almost ready to fly out again an American supply aircraft dropped a case of ammunition through her wing. A new wing with American markings was flown to Ha Tien by helicopter and fitted to the RAAF aircraft.

On 12 March 1966 the Viet Cong had fought their way so close to Saigon that they were able to mount a guerilla attack on Vung Tau airfield. About sixty mortar rounds exploded among the parked aircraft and two of the Caribous were hit, but the base continued operations as usual and the other aircraft took off on

A 9 Squadron Iroquois helicopter and a 35 Squadron Caribou transport flying over Vung Tau, Vietnam (*Department of Air*)

TOP: Men of the Royal Australian Regiment in South Vietnam wait to be carried into an operational area by the helicopters landing behind them (*Australian Army*)

CENTRE: Helicopters of 9 Squadron RAAF (*Department of Air*)

BOTTOM: Start of a mission. Checking a 9 Squadron Iroquois as groundcrew load material for transport to a forward unit in the jungles of South Vietnam (*RAAF Public Relations*)

RAAF gunships in Vietnam. TOP: Loading high-explosive rockets before taking off on a mission. BOTTOM: The door gunner of a 9 Squadron Iroquois en route to the Task Force headquarters at Ba Ria from Vung Tau (*RAAF Public Relations*)

their missions while ground crews repaired the damage.

In mid-1965, Australian involvement in Vietnam was increased by the arrival of a combined Army-RAAF Task Force which included 9 (Helicopter) Squadron. Formed at RAAF Base Williamtown in 1962, 9 Squadron was the first RAAF helicopter unit although the service had been experimenting with helicopters since 1947. The decision to buy Bell Iroquois helicopters was made in 1961 and RAAF pilots went to America to gain flying experience on the new machines.

The role of the helicopter was seen as that of air support for the army, including troop movement, casualty evacuation, reconnaissance, supply, and search and rescue operations. The Americans had begun to use helicopters for such operations during the Korean War and in the Vietnam conflict they converted many helicopters into "gunships:" heavily armed aircraft, with armour protection for some crew members, which could hover above a battlefield and pour fire down on the enemy.

After forming 9 Squadron, the RAAF formed another helicopter unit, 5 Squadron. This unit went to the Australian base at Butterworth in Malaysia and its crews gained valuable experience when they flew on counter-insurgency operations along the Thai-Malaysia border.

The helicopter war. FROM TOP: Gunship returns fire from an enemy jungle patrol; 9 Squadron Iroquois lifts off after landing 2 RAR troops on a cordon-and-search mission near Long Hai village; Iroquois on tactical manoeuvres with Centurion tanks of 1 Armoured Regiment in Vietnam (*Department of Air photographs*)

The 9 Squadron helicopters went to Vietnam aboard HMAS *Sydney* and most of the squadron personnel flew in by chartered aircraft. On 12 June 1965 they arrived at Vung Tau airfield, which had been greatly expanded by the work of RAAF 5 Airfield Construction Squadron.

9 Squadron aircrews and groundcrews had hardly had time to draw their first breaths of the humid Vietnam air before they were working frantically to send two machines on their first mission. Intelligence had advised that the Viet Cong was planning an attack on the 5th Battalion RAR at Nui Dat, and the Diggers were short of ammunition. On 13 June, the first Iroquois helicopter took off with part of a load of 9,000 pounds of ammunition.

In July, the helicopters became involved in the "cloak and dagger" operations of the Australian Special Air Service. The SAS embarked on daring long-range patrols into the jungle, often behind the enemy, and the patrols had to be inserted and extracted. Two helicopters went to pick up a six-man patrol in Phuoc Tuy province and as soon as they settled onto a muddy landing strip they were fired on by a Viet Cong unit. The helicopter gunners returned fire as the SAS men scrambled aboard and the choppers escaped unharmed.

The duties of the helicopter crews covered just about every aspect of modern warfare including that of propaganda, when Flight Lieutenant Max Hayes was sent out to drop leaflets urging the Viet Cong to surrender; to entertainment of the troops when the choppers carried the Col Joye concert party in and out of Nui Dat. In between these extremes came all the grimmer business of operations against a skilled and savage enemy fighting in terrain that he knew like the palm of his hand. The value of the choppers was demonstrated during Operation Hobart, when all the 9 Squadron helicopters were used to airlift 6th Battalion infantry forward in support of their comrades. Each chopper carried four soldiers at a time at treetop height over a distance of about 3,000 yards, landed them with the main force and returned for more. The whole airlift was completed in twenty-five minutes.

The evacuation of wounded, known as "dustoff" operations because the choppers were hardly long enough on the ground to blow the dust off the landing pad, was a regular part of chopper work although it was usually carried out by American machines attached to 9 Squadron. The Australians joined in whenever opportunity served and made some daring landings in the middle of actions.

In extreme contrast to these mercy missions were the search and destroy operations in which the choppers worked with the ground troops, and those in which they supported the Diggers by keeping them supplied with ammunition. This may sound fairly unglamorous but in fact it was extremely important. The fact that ammunition and other supplies could be delivered by helicopters enabled the infantry to press forward without having to carry the burden of supplies, and with the knowledge that replenishment would be available whenever it was required. During and after the battle of Long Tan, when D Company of the 5th Battalion RAR fought off the Viet Cong Regiment 274 and killed 245 of the enemy in the process, the helicopters played a double role. Despite a violent tropical storm they located the besieged company and supplied them with ammunition, and later on they returned to evacuate the wounded.

High above the Vietnam jungle an RAAF medical corpsman treats a wounded Viet Cong prisoner (*RAAF Public Relations*)

George Odgers' book tells many stories of 9 Squadron operations and he says that. "In 18 months operations to 31 December 1968 the squadron had become thoroughly proficient in the art of helicopter warfare. The rotary winged aircraft had come a long way since the Korean war, dramatically changing the character of infantry operations by giving the ground troops an undreamed of mobility." One remarkable point is that during the whole period only one helicopter had been lost.

While the helicopters were building up thousands of hours of experience in this unusual style of warfare the RAAF had introduced somewhat more conventional strike aircraft to Vietnam. They were the Canberra bombers of 2 Squadron, the second oldest in the RAAF with a fighting history dating from 1916.

The RAAF ordered its first Canberra in 1951, the year when this new English Electric jet made flying history by cutting the England-Australia flight time down to twenty-one hours. The RAAF ordered the bombers as replacements for the Lincoln, and the Government Aircraft Factory built the Australian version under licence from English Electric. By April 1967, the month in which 2 Squadron flew its Canberras into Vietnam, the bomber was a little long in the tooth by world standards but was still a highly efficient aircraft. The eight bombers of 2 Squadron had been attached to the Australian component of Far East Air Forces (Malaysia) and stationed at Butterworth until they were transferred to Vietnam. They were stationed at Phan Rang, one of the airfields built by the Americans to accommodate jet combat aircraft, and became part of 35th Tactical Fighter Wing of the US 7th Air Force.

The task of 35th TFW was to provide comprehensive air support to assist ground operations anywhere within South Vietnam. The Canberras became involved in "Combat Sky Spot" strikes, which were night attacks directed by ground radar. Typical targets were Viet Cong supply dumps, base camps, tunnel installations, and bunkers. It was a systematic process in which each of the eight Canberras made a bombing attack once in each twenty-four hours, seven days a week, and it was a somewhat mechanical type of warfare because the Australian aircrews were rarely able to obtain any information about the success, or otherwise, of their attacks. They were pleased by a changeover to day bombing in September 1967. The Americans had been somewhat doubtful of the efficiency of the Canberra as a tactical day bomber, which required the Australians to work under the control of Forward Air Controllers. The FACs flew in light observation aircraft and fired smoke rockets to mark targets. When the Australians carried out their first attack in co-operation with an FAC the bomb damage assessment was 100% on target and the Americans' doubts disappeared.

Before long, 2 Squadron built up a reputation for unparalleled accuracy. The aircrews were accustomed to the precision flying demanded by night operations, and when they switched to daylight bombing this stood them in good stead. Whenever a bombing error in excess of sixty feet was reported, the aircrew was debriefed in detail to discover the cause of error. If it did not seem to be of human origin then the aircraft's bombing system was checked over. The Canberra aircrews used precision bomb sights of second World War vintage and were the only squadron in Vietnam to employ a level bombing technique which incorporated this bombsight.

The war rose to a sudden climax during the night of 30–31 January 1968 when the North Vietnamese launched what became known as the Tet offensive. ("Tet" is the Chinese new year holiday). The offensive aimed at ending the war with a single massive blow and North Vietnamese strategy included the simultaneous invasion of about 105 South Vietnam towns and cities and attacks on a large number of military bases and airfields. Guerrilla troops had infiltrated as far as Saigon and the offensive erupted when they attacked targets in and around the city including Tan Son Nhut airfield which was used by Australian aircraft. Two rockets landed close to the RAAF Air Movements Office at the airfield and when dawn broke the Australians saw that the barbed wire perimeter was littered with the corpses of Viet Cong killed by South Vietnamese defenders. Street fighting raged close to RAAF Headquarters in Saigon and the staff remained in the office for twenty-four hours a day, armed with sub-machine guns and hand grenades in readiness to fight off attacks.

All RAAF units, especially 9 Helicopter Squadron, were kept busier than ever until the Tet offensive faded away into the usual chaos of ruin and refugees which was characteristic of that unhappy war. 9 Squadron was just receiving UH-IH helicopters, a more powerful model of the Iroquois. It became known as the H or "Hotel" model as it replaced the older "Bravo" machines. The new helicopters were in action before the end of February in a massed formation airlift of Australian troops, and in March–April they took part in Operation Pinaroo mounted by the Australian Army against the Viet Cong defensive complex in the rugged Long Hai Hills.

A senior American officer said that "The Vietnam war hangs on the rotor blade of a helicopter." This was a pardonable exaggeration but there is no doubt that, during the Vietnam operations, the helicopter came into its own as a weapon. It was the ideal aircraft to use in co-operation with troops moving through a comparatively undeveloped country and as the war progressed it became an offensive weapon in the form of "gunships." These were developed by the Americans and the first Australian gunship was authorised in 1968. It was a "Hotel" Iroquois equipped with an American gunship kit consisting of two seven-tube rocket pods, two 7.62 mm guns, and two sets of twin-mounted side firing M60 machine guns. After flight and firing tests the RAAF gunships, called "Bushrangers" by 9 Squadron, were ready for work. The first three Bushrangers took off on 11 April 1969 and after that they were steadily involved in operations which

Caribou transports of 35 Squadron RAAF over South Vietnam (*Department of Air*)

included a "hot" extraction of an SAS patrol which had been surrounded by Viet Cong in the jungle and could not reach a place suitable for a helicopter landing. The only solution was to winch the Australian Special Air Service men out of the jungle one by one. The patrol marked their position with a smoke grenade and one helicopter hovered above them while three Bushrangers moved in to smother the Viet Cong with fire. The enemy responded vigorously but they became so busy in fighting off the Bushrangers that the SAS men were winched up in safety, and the choppers flew away in a storm of fire which continued until they were more than 2,000 yards from the pick up point.

This was one of many episodes in which 9 Squadron crews distinguished themselves for courage and flying skill. Leading Aircraftsman Ronald Wilson, medical orderly on a Dustoff evacuation helicopter, braved the danger of unexploded mines in order to carry wounded men out of a minefield to the chopper. LAC Kerry Shipp was winched down twice from his helicopter into hostile terrain in an attempt to rescue the crew of an American aircraft shot down in the jungle. Flying Officer Treloar took his machine down into the middle of a fierce action and hovered only twenty-five feet above the ground as his crew hauled wounded Australian infantrymen up into the aircraft. The enemy poured fire at them from only thirty yards away and one of the wounded men was hit again as he lay on the floor of the chopper. The crew heard him exclaim, "The miserable little bastards, they're even shooting at me in an aircraft." Flight Lieutenant Woolf's helicopter was brought down by machine gun fire while he was engaged on a similar rescue operation, but fortunately he and his crew escaped with minor injuries although the chopper was wrecked.

In the meantime the men of "Wallaby Airlines" flew on equally important and sometimes equally hazardous missions. The huge airlift of weapons, men, and materials was a vital part of the Allied campaign in Vietnam and 35 Transport Squadron RAAF achieved far more than could be expected of a single squadron because of their keenness and because they were willing to push flying time and cargo loads to the limit. By mid-1970, the squadron had carried more than 500,000 passengers and 78 million pounds of freight. Often they operated from primitive airstrips in constant danger from ground fire. Many of the Caribou transports were hit, several were destroyed by enemy fire, and a number of aircrew were wounded. The RAAF men flew to the limits of their own and their aircraft's endurance although the machines were maintained in top condition by the ground crews. On one memorable day a Caribou crew flew twenty-eight

LEFT, FROM TOP: A 2 Squadron Canberra drops its load of six 750-pound bombs over a target in South Vietnam; Hunting jet Provost; RAAF BAC111 transport; de Havilland (Canada) Otter in RAAF service (*Department of Air and RAAF Museum photographs*)

RIGHT, FROM TOP: Helicopter salvage operation. A Chinook lifts a wrecked Iroquois. The lower photograph is a Swiss-built Pilatus Porter Army co-operation aircraft of the RAAF (*Department of Air and RAAF Museum photographs*)

Parts of Australia are ideal for gliding, especially the flat country of the inland where heated air rising into cloudless skies creates thermals that have enabled gliders to make some record-breaking flights. This photograph was taken near Waikerie, South Australia, on the Murray River, in 1976 (*photo courtesy the Melbourne Herald*)

sorties during eleven-and-a-half hours and stopped the engines only for refuelling.

An incident on 19 January 1969 was typical of the risks run by Caribou crews. Flight Lieutenant Thompson was in command of an aircraft on a supply mission to Katum, a small outpost near the Cambodian border. Thompson had landed the aircraft and was taxi-ing to a parking slot when a mortar bomb exploded in front of the port wing and broke the windscreen. A Viet Cong patrol concealed in the jungle kept up a continuous mortar fire as Thompson kept the Caribou rolling forward. Bombs exploded all round the aircraft, a chunk of shrapnel pierced Thompson's leg, and his co-pilot, Flying Officer Rod McGregor, was cut by flying glass from the windscreen. Bomb splinters burst two of the tyres and cut the hydraulic lines to the undercarriage and flaps.

Thompson made a quick decision to unload the aircraft and get her in the air again. The two loadmasters worked at furious speed to unstrap and discharge the cargo and Thompson coaxed the Caribou off the ground. The undercarriage would not retract and had to be pumped up by hand, and Thompson used the emergency system to make a safe landing at Bien Hoa.

The RAAF component in Vietnam included Airfield Defence Guards, a Photo Interpreting group, Medical staff of doctors, nurses, and orderlies, Air Movement Control sections, Airfield Construction units, and in fact a representative selection of RAAF personnel of almost every category. They were a closely-knit team of high morale, and they devoted a great deal of their spare time and money to relieving the plight of civilian refugees. At Vung Tau the RAAF contingent raised $20,000 among themselves to build an orphanage consisting of ten home units and this was only one example of the financial and practical help given by Australian servicemen. RAAF Chaplains organised many projects to help South Vietnamese health care, education, agricultural production, social welfare, and water supply to the scattered little villages.

The Vietnam wars which lasted from 1945 to 1975 have now faded into the sad history of the human race and very few people need to be reminded of their traumatic effects upon western democracy. In a purely military sense, however, the Vietnam experience was an important one for the RAAF. It was a gruelling test of men and machines and they all responded magnificently.

There cannot be a definite end to the story of Australians in the air because it is one that is added to every day. As in the past, Australians continually pit themselves against the challenges of flight. Sometimes the acceptance of these challenges has a tragic outcome, as in the case of twenty-three-year-old Russell Williams of Woodville, South Australia. He had been a devotee of flying since his boyhood and qualified as a commercial pilot and instructor by the time he was nineteen. He had logged over 3,000 flying hours, and when he took off in a Cessna 180 on the sunny afternoon of Saturday 27 March 1976 he was fairly certain of acceptance for passenger jet training by Ansett Airlines.

The suburbs of Adelaide resounded to the sound of aircraft that afternoon, because it was the fiftieth anniversary year of the Royal Aero Club of South Australia and local pilots were celebrating with an aerial pageant watched by about 50,000 people at Parafield airfield. Russ Williams' flight was part of the celebrations. He was making an attempt on a light aircraft altitude record and hoped to make it a feat worthy of the anniversary.

Ground observers watched the Cessna climb steadily into the sky over the Adelaide Hills until Williams reached an estimated 11,500 feet. Then they saw a puff of white smoke under the fuselage and a flicker of orange flame. The Cessna exploded a few seconds later. The cause of the explosion will not be known until hundreds of aircraft fragments have been examined but it is believed that an oxygen leak was responsible.

Every human activity has its triumphs and disasters. Russ Williams strove for triumph and met disaster, and joined the brave company of Australian aviators who died in attempts to fly a little higher, a little further, a little better. Their deaths have never been wasted because they have always taught something to those who follow after.

For a small population, Australia has had a remarkable number of flyers of international repute and an even greater number whose names are little known but who have contributed greatly to the advancement of flying. The efforts of all such men and women have helped to make Australians into one of the most air-minded people in the world and it is strange that the continent does not possess a truly national aviation manufacturing industry. Possibly that will come in time and allow full scope for the energy and ingenuity that has always been shown by Australians in the air.

TOP: Troops board an RAAF Chinook during manoeuvres near Rockhampton. Queensland (*Defence Public Relations*)

BOTTOM: The "new style" uniform adopted by the RAAF in 1974 (*Department of Air*)

The wheel turns full circle. The first flights over Australia were in lighter-than-air balloons, and free ballooning has now returned as a sport. This balloon is owned by Roger Meadmore, of Adelaide, who is also the owner of the veteran Tiger Moth seen in the photograph (*Ron Langman*)

Selected Sources

Advertiser, The Adelaide (newspaper) various dates.
Aero (magazine) various dates.
Aeroplane, The (magazine) various dates.
Affleck, Arthur *The Wandering Years* (1964).
Aircraft (magazine) various dates.
Anderson, Peter N. *Fighting Mustangs RAAF & RNZAF* (1975); *Mustang Squadrons Australia and New Zealand* (1975).
Australian Aviation Yearbook (1968, 1969, 1970, 1971).
Australian Encyclopaedia (1958) relevant articles.
Australian Government: Department of Transport (Air Transport Group) publications *Statistics of Civil Aviation Industry* (1974, 1975); *Statistics of Air Transport Industry* (1974, 1975); *Domestic Air Transport Statistics* (1974, 1975); *Domestic Airlines Traffic on Scheduled Operations* (1975). *Year Books of the Commonwealth of Australia* (1950 *et seq.*) relevant articles.
Australian War Memorial (publishers) *These Eagles* (1942); *RAAF Saga* (1944).
Bilton, John, *et al. The Royal Flying Doctor Service of Australia* (1961).
Brearley, Sir Norman *Australian Aviator* (1971).
Brisbane Courier, The (newspaper) various dates.
Brodie, Bernard, article on Tet Offensive in *Decisive Battles of the Twentieth Century* (1976).
Charles, David *The Story of Aircraft* (1974).
Clarke, Ronald *The Battle of Britain* (1965).
Collier, Basil *The Battle of Britain* (1962).
Cooper, Bryan *The Story of the Bomber 1914–1945* (1974).
Copley, Gregory *Australian Aviation Yearbook* (1970, 1971).
Copley, Gregory (ed.) *Aero Aviation Magazine* (various dates).
Cutlack, F. M. *The Australian Flying Corps in the Western and Eastern Theatres of War* (1928).
Eustis, Nelson *The Greatest Air Race* (1969).
Encyclopaedia Britannica (1955) relevant articles.
Fenton, Clyde *Flying Doctor* (1952).
Firkins, Peter *Australians in Nine Wars* (1971); *Of Nautilus and Eagles: History of the Royal Australian Navy* (1975).
Flight (magazine) various dates.
Fysh, Sir Hudson *Qantas Rising* (1965), *Qantas at War* (1963), *Wings to the World* (1970)
Gibson, Ron J. *Australia and Australians in Civil Aviation: an Index to Events. Vol. I, 1823–1920* (1971).
Gillison, Donald *Royal Australian Air Force 1939–1945* (1964).
Godwin, John *Battling Parer* (1968).
Harrison, Kenneth *The Brave Japanese* (1966).
Herington, John *Air War against Germany and Italy* (1954).
Historian Publishers *Historical British Aircraft* (1973).
Hocking, D. M., and Haddon Cave, C. P. *Air Transport in Australia* (1951).
Hudson, Harry *Flynn's Flying Doctors* (1956).
Idriess, Ion *Flynn of the Inland* (1932).
Illustrated Sydney News (magazine) various dates.
Inlander, The (magazine) various dates.
Isaacs, Keith *Military Aircraft of Australia 1909–1918.*
Joy, William *The Aviators* (1965).
Lockwood, Douglas *Australia's Pearl Harbor* (1966).
Melbourne Age, The (newspaper) various dates.
Melbourne Herald, The (newspaper) various dates.

McGrath, David *Chronology of Australian Aviation History* (in *Australian Aviation Yearbook*, various editions).
MacKenzie, R. D. *Solo: the Bert Hinkler Story* (1962).
McPheat, W. S. *John Flynn: Apostle to the Inland* (1954).
Miller, H. C. *Early Birds* (1968).
Miller, Robin *Flying Nurse* (1971).
Monday, David *Aircraft* (1973).
Nelson, Robert, *et al. A Pictorial History of Australians at War* (1970).
New American Encyclopaedia (1965) relevant articles.
Odgers, George *Air War against Japan 1943–1945* (1957); *Mission Vietnam* (1974).
Perlmutter, Tom *War Machines Air* (1975).
Reed, A. H. & A. W. (publishers) *Australia's Airmen at War*.
Robertson, Bruce *Battle of Britain* (1970).
Robins, James G. *The Wooden Wonder: A Short History of the Wooden Aeroplane* (1975).
Scott, C. W. A. *Scott's Book* (1934).
Smith, Sir Charles Kingsford *My Flying Life* (1937), and with Charles T. P. Ulm *The Story of the Southern Cross Trans-Pacific Flight 1928* (1928).
Smith, Sir Ross *14,000 Miles through the Air* (1922).
Smith's Weekly (journal) various dates.
Shennan, Anthony *The P.38 Lightning*.
Stannage, J. *Smithy* (1951).
Sydney Morning Herald, The (newspaper) various dates.
Taylor, John W. R., and Munson, Kenneth *History of Aviation* (1973).
Wackett, Sir Lawrence *Aviation Pioneer* (1972).
West Australian, The (newspaper) various dates.
Wykeham, Peter *Fighter Command* (1960).
Younger, R. M. *Australia and the Australians: A New Concise History* (1970).

Index